Information Security and Cryptography

For further volumes:
http://www.springer.com/series/4752

Marc Joye · Michael Tunstall
Editors

Fault Analysis in Cryptography

Springer

Editors
Marc Joye
Technicolor
1 avenue de Belle Fontaine
Cesson-Sévigné Cedex
35576
France

Michael Tunstall
Department of Computer Science
University of Bristol
Woodland Road
Bristol
BS8 1UB
UK

ISSN 1619-7100
ISBN 978-3-642-29655-0 ISBN 978-3-642-29656-7 (eBook)
DOI 10.1007/978-3-642-29656-7
Springer Heidelberg New York Dordrecht London

Library of Congress Control Number: 2012939112

ACM Computing Classification: E.3, B.1

Printed on acid-free paper

Springer is part of Springer Science+Business Media (www.springer.com)

Juliette Joye

Foreword

Fault attacks is an active area of research in cryptography, currently explored in hundreds of research papers and dedicated conferences. This book is the first comprehensive treatment of the subject covering both the theory and practice of these attacks as well as defense techniques.

Fault attacks exploit the fact that computers sometimes make mistakes. These mistakes can result from a programming error, as in the case of the infamous Intel floating-point bug. Or they can result from direct interference by an attacker, say by running the computer in a hostile environment. This book explores what happens to cryptographic algorithms when the computer implementing the algorithm makes a calculation error. Very often these errors, called faults, can have disastrous consequences, rendering the system completely insecure. As an extreme example, a single mistake during the calculation of an RSA digital signature can completely expose the signer's secret key to anyone who obtains the faulty signature. Over the years it has been shown that a wide range of cryptographic algorithms succumb to fault attacks. This book does a beautiful job of presenting powerful fault attacks against a wide range of systems.

Preventing fault attacks without sacrificing performance is nontrivial. Over the years a number of innovative ideas have been proposed for efficiently verifying cryptographic computations. Many defense strategies are described in the book, some of which are already deployed in real-world cryptographic libraries. Nevertheless, many implementations remain vulnerable. I was thrilled to see the material covered in the book and hope that it will make fault defense the standard practice in the minds of developers.

Dan Boneh
Stanford University

Preface

One of the first examples of fault injection in microprocessors was unintentional. May and Woods noticed that radioactive particles produced by elements present in the packaging materials used to protect microprocessors were energetic enough to cause faults [277]. Specifically, it was observed that α particles were released by uranium-235, uranium-238, and thorium-230 residues present in the packaging, decaying to lead-206. These particles were able to create a large enough charge that bits in sensitive areas of a chip could be made to flip. These elements were only present in two or three parts per million, but this concentration was sufficient to change the behavior of a microprocessor.

Further research into the physical effects that could affect the behavior of microprocessors included studying and simulating the effects of cosmic rays on semiconductors [435]. While the effect of cosmic rays are very weak at ground level because of the Earth's atmosphere, their effect becomes more pronounced in the upper atmosphere and outer space. This is important as faults in airborne electronic systems have potentially catastrophic consequences. This provoked research by organizations such as NASA and Boeing to "harden" electronic devices so that they are able to operate in harsh environments.

Since then other physical means of inducing errors have been discovered but all of these have had somewhat similar effect. In 1992, for example, Habing determined that a laser beam could be used to imitate the effect of charge particles on microprocessors [173]. The different faults that can be produced have been characterized to enable the design of suitable protection mechanisms.

The first academic publication that discussed using such a fault to intentionally break a cryptographic algorithm was described by Boneh, DeMillo, and Lipton in 1997 [56]. It was observed, among other things, that an implementation of RSA that uses the Chinese Remainder Theorem to compute a modular exponentiation is very sensitive to fault attacks (see Sect. 8.2). A similar publication followed this that described a fault to intentionally break a secret key cryptographic algorithm [49]. More specifically, this attack applied techniques from differential cryptanalysis that would allow an attacker to exploit a fault to break an implementation of DES (see Sect. 3.3).

Aumüller, Bier, Fischer, Hofreiter, and Seifert published the first academic paper detailing an implementation of one of these attacks [18]. They describe an implementation of the attack by Boneh et al. breaking an implementation of RSA computed using the Chinese Remainder Theorem.

Since then numerous attacks and countermeasures have been proposed and implemented. This book presents a summary of the state of the art in the theoretical and practical aspects of fault analysis and countermeasures. *Happy reading*!

Rennes (France), Bristol (UK), April 2011

Marc Joye
Michael Tunstall

Contents

Contributors

Kahraman D. Akdemir Worcester Polytechnic Institute, USA

Abdulaziz Alkhoraidly University of Waterloo, Waterloo, Canada

Alessandro Barenghi Politecnico di Milano, Italy

Alexandre Berzati CEA Leti, France

Guido M. Bertoni STMicroelectronics, Italy

Kaouthar Bousselam Université de Montpellier II, France

Luca Breveglieri Politecnico di Milano, Milan, Italy

Cécile Canovas-Dumas CEA Leti, France

Christophe Clavier XLIM & Université de Limoges, France

Jean-Luc Danger Telecom ParisTech, France

Giorgio Di Natale LIRMM / CNRS, France

Agustín Domínguez-Oviedo Tecnológico de Monterrey, Mexico

Nadia El Mrabet Université de Caen, France

Marie-Lise Flottes LIRMM / CNRS, France

Toshinori Fukunaga NTT Information Sharing Platform Laboratories, Japan

Christophe Giraud Oberthur Technologies, France

Shigeto Gomisawa The University of Electro-Communications, Japan

Louis Goubin Université de Versailles Saint-Quentin-en-Yvelines, France

Sylvain Guilley Telecom ParisTech, France

Arash Hariri The University of Western Ontario, Canada

M. Anwar Hasan University of Waterloo, Waterloo, Canada

Paolo Ienne École Polytechnique Fédérale de Lausanne, Switzerland

Mark Karpovsky Boston University, USA

Chong Hee Kim Université Catholique de Louvain, Belgium

Israel Koren University of Massachusetts, USA

Yang Li The University of Electro-Communications, Japan

Marcel Medwed Université Catholique de Louvain, Belgium; Graz University of Technology, Austria

Phong Q. Nguyen École Normale Supérieure, France

Kazuo Ohta The University of Electro-Communications, Japan

Elisabeth Oswald University of Bristol, UK

Dan Page University of Bristol, UK

Mauro Pellicciolі Politecnico di Milano, Italy

Gerardo Pelosi Politecnico di Milano, Italy

Jean-Jacques Quisquater Université Catholique de Louvain, Belgium

Francesco Regazzoni Université Catholique de Louvain, Belgium; University of Lugano, Switzerland

Arash Reyhani-Masoleh The University of Western Ontario, Canada

Matthieu Rivain CryptoExperts, France

Bruno Rouzeyre Université de Montpellier II, France

Kazuo Sakiyama The University of Electro-Communications, Japan

Jörn-Marc Schmidt Graz University of Technology, Austria

François-Xavier Standaert Université Catholique de Louvain, Belgium

Berk Sunar Worcester Polytechnic Institute, USA

Junko Takahashi NTT Information Sharing Platform Laboratories, The University of Electro-Communications, Japan

Mehdi Tibouchi École Normale Supérieure, France

Elena Trichina STMicroelectronics, Italy

Frederik Vercauteren Katholieke Universiteit Leuven, Belgium

Zhen Wang Boston University, USA

Part I
Introductory Material

Chapter 1
Side-Channel Analysis and Its Relevance to Fault Attacks

Elisabeth Oswald and François-Xavier Standaert

Abstract Side-channel attacks are a class of attacks where an attacker deduces the internal state of a device by observing information that leaks during the normal functioning of the device. In this chapter we describe side-channel analysis and its relevance to fault attacks. Side-channel analysis is typically used to extract information about cryptographic keys. However, we will be concentrating on how it can be used as a means to identify target operations and as a trigger mechanism for fault attacks.

1.1 Introduction

Side-channel attacks are a class of attacks where an attacker will attempt to deduce what is occurring inside a device by observing information that leaks during the normal functioning of the device. The first publication that mentions a side-channel attack is [422]. In 1956, MI5 mounted an operation to decipher communications between Egyptian embassies. The communications were enciphered using Hagelin machines [207]. Enciphering occurred by routing electronic signals from a keyboard through seven rotating wheels to generate a ciphertext. The "key" was the initial setting of these seven wheels. The machine was reset every morning by the clerk

---

F.-X. Standaert is an associate researcher of the Belgian Fund for Scientific Research (FNRS-F.R.S.).

E. Oswald (✉)
Department of Computer Science, University of Bristol,
Bristol, UK
e-mail: Elisabeth.Oswald@bristol.ac.uk

F.-X. Standaert
Crypto Group, Université Catholique de Louvain,
Louvain-la-Neuve, Belgium

M. Joye and M. Tunstall (eds.), *Fault Analysis in Cryptography*,
Information Security and Cryptography, DOI: 10.1007/978-3-642-29656-7_1,

who would be sending messages. MI5 managed to plant a microphone in close proximity to one of these machines. This allowed the initial settings to be determined by listening to them being made every morning. This would have allowed MI5 to decipher intercepted communications with another Hagelin machine set to the same key. In practice MI5 was only able to determine a certain number of wheel settings because of the difficulty of distinguishing the noise of the wheels being set from background noise. This made deciphering more complex, but not impossible, as the number of possible keys was significantly reduced by the partial information.

In this chapter we describe side-channel analysis and its relevance to fault attacks. Side-channel analysis is typically used to try to reveal information on cryptographic keys. However, we will be concentrating on identifying target operations and on how to trigger a mechanism to inject a fault.

The first academic publication of a side-channel attack described an attack based on observing the time required to compute a given operation [239]. However, the overall time required to compute an operation is not relevant to the discussion of side-channel analysis with regard to fault attacks.

Subsequent publications involved analyzing acquisitions of the instantaneous power consumption [240] or electromagnetic emanations [153, 331]. In each case there are two types of attack, which we will discuss in more detail in this chapter.

- The analysis of one acquisition to determine information on the operations being computed; we elaborate on this in Sect. 1.3.
- The statistical analysis of multiple traces to reveal information, described in Sect. 1.4.

1.2 Background

In this section we describe the equipment that one would require to take acquisitions that could be used to conduct side-channel analysis. We focus on the aforementioned analysis of the power consumption and the electromagnetic field around a microprocessor.

In Fig. 1.1 we show an example of the equipment required to acquire power consumption traces while executing an arbitrary command. A smart card was inserted into an extension card that was then inserted into a standard smart card. An oscilloscope was used to acquire a power consumption trace, visible as a red trace on the laptop computer and as a cyan trace trace on the oscilloscope. This trace represents the voltage drop across a resistor in series with the ground pin of the smart card, typically acquired with a differential probe. A yellow trace is also visible on the oscilloscope that shows the I/O pin of the smart card that is used to trigger the oscilloscope.

It was observed in [153, 331] that the same information is present in the change in the electromagnetic field surrounding a microprocessor. Traces showing the change in the electromagnetic field can be acquired by using a suitable probe. This typically

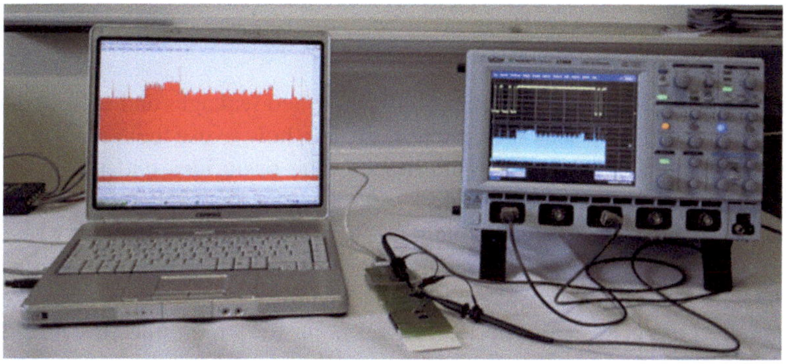

Fig. 1.1 An example of the equipment one could use to take acquisitions of the instantaneous power consumption of a smart card

Fig. 1.2 An example of a probe used for acquiring traces of the electromagnetic emanations above an ARM7 microprocessor

consists of a tube formed from a single wire that is then placed perpendicular to the surface of a microprocessor, an example of which is shown in Fig. 1.2.

In this chapter we discuss side-channel analysis using the power consumption as a running example. The same analysis can be made using acquisitions from an electromagnetic probe. However, these acquisitions are typically more noisy and are therefore less suitable for presenting clear examples.

1.3 Simple Power Analysis

Having introduced the basics of commonly exploited leakages and sketched how to capture and record these leakages, we now explain how to use them in simple power analysis (SPA) scenarios to aid fault attacks. The basic idea underlying all these attacks is to identify operations using a side-channel, and to then inject faults into the identified functions. We first explain this idea in more detail by briefly reviewing some examples of what is possible using SPA. We then discuss how someone wishing to implement a fault attack could use these techniques in practice. An attacker may be required to identify an appropriate trigger point for a fault injection mechanism, i.e. to automatically examine a power trace to trigger a fault inducing mechanism.

1.3.1 Case Study: RSA Private Operation

The principal operation of the RSA [349] signature scheme is a modular exponentiation in $(\mathbb{Z}/N\mathbb{Z})^*$. That is, a signature s is generated from a message m by computing $s = \mu(m)^d \bmod N$, where d is the private key, N is the product of two large primes, and μ is an appropriate padding function. This signature can be verified by checking whether $\mu(m)$ is equal to $s^e \bmod n$. We define $d \equiv e^{-1} \pmod{\phi(N)}$ where ϕ is Euler's totient function.

One of the most widely known algorithms for implementing an exponentiation is the square-and-multiply algorithm, where an exponent e is read from left to right bit by bit. Starting with an accumulator set to 1, a squaring operation is performed if a bit is equal to 0, and a squaring operation followed by a multiplication (with the value being raised to e) is performed if a bit is equal to 1. This algorithm is detailed in Algorithm 1.1.

Algorithm 1.1: Binary left-to-right exponentiation

Input: $m, x < m, d \geq 1, \ell$ the binary length of d (i.e. $2^{\ell-1} \leq d < 2^{\ell}$)
Output: $A = x^d \bmod m$

$A \leftarrow x$; $R \leftarrow x$
for $i = \ell' - 2$ **Down to** 0 **do**
  $A \leftarrow A^2$
  if $(\mathrm{bit}(n, i) \neq 0)$ **then** $A \leftarrow A \cdot R$
end
return A

In Fig. 1.3 we show a power consumption trace captured while a microprocessor was computing a modular exponentiation using Algorithm 1.1. A series of operations can be seen in the power consumption that are separated by downward peaks. From the ratio of operations, dictated by Algorithm 1.1, we can determine that the

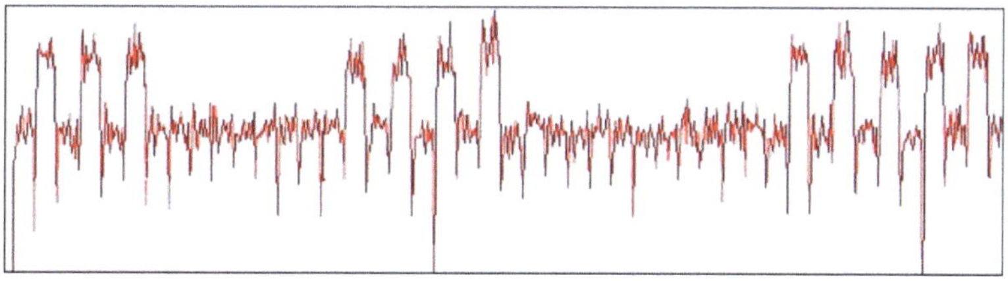

Fig. 1.3 The power consumption of a microprocessor while computing a modular exponentiation

lower power consumption corresponds to a squaring operation and the higher power consumption corresponds to a multiplication. If we denote a squaring operation as `S` and a multiplication as `M`, then the series of operations in Fig. 1.3 is

`SMSMSMSSSSSSSSSMSMSMSMSSSSSSSSSSSSSMSMSMSMSM.`

These directly correspond to the first 41 bits of the exponent used to generate the power consumption trace in Fig. 1.3, which, from the sequence of operations, can be identified as

$$1111000000001111000000000000 11111\,.$$

1.3.2 Case Study: AES Encryption

Implementations of block ciphers in small embedded devices are another example of where SPA can be straightforwardly applied in order to identify individual operations or trigger a fault insertion apparatus. Differential fault analysis attacks typically target a specific round of a block cipher. Hence, solutions that allow determining the exact time at which these operations are computed will be useful in this context. In this section we use the Advanced Encryption Standard (AES) as an example. In this section multiplications are considered to be polynomial multiplications over $\mathbb{F}_{2^8}$ modulo the irreducible polynomial $x^8 + x^4 + x^3 + x + 1$. It should be clear from the context when a mathematical expression contains integer multiplication.

The structure of the Advanced Encryption Standard (AES) [142], as used to perform encryption, is illustrated in Algorithm 1.2. Note that we restrict ourselves to considering AES-128 and that the description omits a permutation typically used to convert the plaintext $P = (p_0, p_1, \ldots, p_{15})_{(256)}$ and key $K = (k_0, k_1, \ldots, k_{15})_{(256)}$ into a 4×4 array of bytes, known as the state matrix. For example, the 128-bit plaintext input block P and 128-bit ciphertext C are arranged as

$$\begin{pmatrix} p_0 & p_4 & p_8 & p_{12} \\ p_1 & p_5 & p_9 & p_{13} \\ p_2 & p_6 & p_{10} & p_{14} \\ p_3 & p_7 & p_{11} & p_{15} \end{pmatrix} \text{ and } \begin{pmatrix} c_0 & c_4 & c_8 & c_{12} \\ c_1 & c_5 & c_9 & c_{13} \\ c_2 & c_6 & c_{10} & c_{14} \\ c_3 & c_7 & c_{11} & c_{15} \end{pmatrix}.$$

Algorithm 1.2: The AES-128 encryption function

Input: The 128-bit plaintext block P and key K
Output: The 128-bit ciphertext block C

```
X ← AddRoundKey(P, K)
for i ← 1 to 10 do
    X ← SubBytes(X)
    X ← ShiftRows(X)
    if i ≠ 10 then
        X ← MixColumns(X)
    end
    K ← KeySchedule(K)
    X ← AddRoundKey(X, K)
end
C ← X
return C
```

The encryption itself is conducted by the repeated use of a number of round functions:

- The `SubBytes` function is the only nonlinear step of the block cipher. It is a bricklayer permutation consisting of an S-box applied to the bytes of the state. Each byte of the state matrix is replaced by its multiplicative inverse, followed by an affine mapping. Thus the input byte x is related to the output y of the S-Box by the relation, $y = A\,x^{-1} + B$, where A and B are constant matrices. In the remainder of this paper we will refer to the function S as the SubBytes function and to S^{-1} as the inverse of the SubBytes function.
- The `ShiftRows` function is a byte-wise permutation of the state.
- The `KeySchedule` function generates the next round key from the previous one. The first round key is the input key with no changes; subsequent round keys are generated using the `SubBytes` function and XOR operations. Algorithm 1.3 shows how the rth round key is computed from the $(r-1)$th round key. The value h_r is a constant defined for the rth round, and $\ll$ is used to denote a bit-wise left shift.
- The `MixColumns` function is a bricklayer permutation operating on the state column by column. Each column of the state matrix is considered as a vector where each element belongs to $\mathbb{F}(2^8)$. A 4×4 matrix

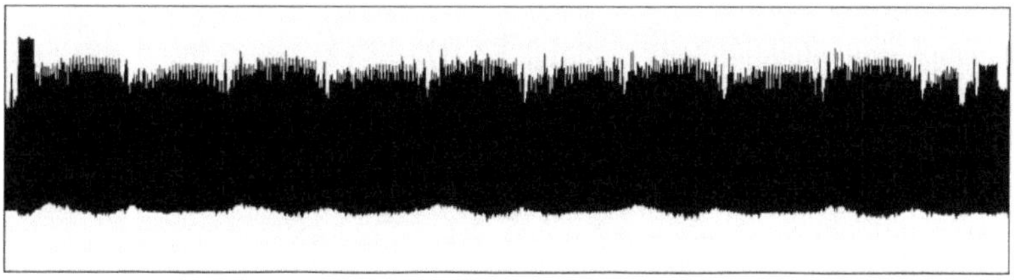

Fig. 1.4 Power consumption trace of a single AES encryption performed by a smart card

$$M = \begin{pmatrix} 2 & 3 & 1 & 1 \\ 1 & 2 & 3 & 1 \\ 1 & 1 & 2 & 3 \\ 3 & 1 & 1 & 2 \end{pmatrix}$$

whose elements are also in $\mathbb{F}(2^8)$ (here expressed as a decimal digit) is used to map this column into a new vector.

- `AddRoundKey`: Each byte of the array is XORed with a byte from a corresponding array of round subkeys.

Algorithm 1.3: The AES-128 `KeySchedule` function

Input: $(r-1)^{th}$ round key ($X = x_i$ for $i \in \{1, \ldots, 16\}$).
Output: r^{th} round key X.

for $i \leftarrow 0$ **to** 3 **do**
  $x_{(i \ll 2)+1} \leftarrow x_{(i \ll 2)+1} \oplus S(x_{(((i+1) \wedge 3) \ll 2)+4})$
end
$x_1 \leftarrow x_1 \oplus h_r$
for $i \leftarrow 1$ **to** 16 **do**
  if $(i-1) \mod 4 \neq 0$ **then**
    $x_i \leftarrow x_i \oplus x_{i-1}$
  end
end
return X

Figure 1.4 shows a power consumption trace taken while a device was performing an AES encryption. It shows a pattern that is repeated ten times, corresponding to the ten AES rounds as described in Algorithm 1.2. The last pattern is slightly different because the MixColumns function is not computed in the last round.

More information can be observed if the power consumption of one round is analyzed in more detail. Figure 1.5 shows the power consumption of an ARM7 microprocessor while it is computing the first round of the AES. Two patterns of 16 peaks can be seen on the left-hand side that correspond to the plaintext and secret key being permuted to enable efficient computation given the matrix representation

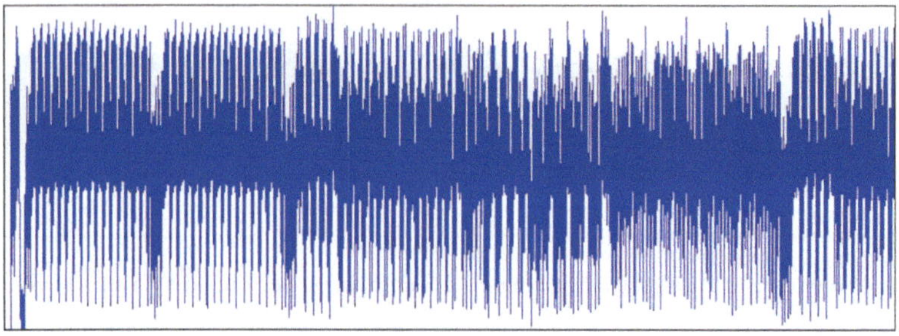

Fig. 1.5 Power consumption trace of a single round of an AES encryption performed by an ARM7 microprocessor

in the specification of the algorithm. This is followed by a pattern of four peaks that correspond to the XOR with the first key byte (the ARM7 has a 32-bit architecture). Following this there is a pattern of 16 peaks that correspond to the SubBytes function. This is followed by two patterns of four peaks that corresponds to the generation of the next subkey. The ShiftRow function occurs between these functions but is not visible in the power consumption. The XOR with this subkey can be seen on the right-hand side of Fig. 1.5, which means that the remaining area between this XOR and the generation of the subkey is where the MixColumns function is computed. However, no obvious pattern can be seen without plotting this portion of the trace with a higher granularity.

In general, such a visual inspection of the leakage traces can be used as a preliminary step before a more powerful attack (such as DPA or DFA), and allows an attacker to determine the parts of the traces that are relevant in a straightforward manner.

The drawback of SPA is its limited granularity, as demonstrated by our example above. It is typically straightforward to identify the AES rounds, as in Fig. 1.4, but identifying to round operations (i.e. AddRoundKey, SubBytes, MixColumns, and ShiftRows) may be more difficult. For example, in Fig. 1.5 the MixColumns is identified by a process of elimination rather than by observing a pattern in the power consumption. Furthermore, devices with larger data buses and ASIC (or FPGA) implementations that process 128 bits in parallel may also make visual inspection more difficult.

1.3.3 Case Study: File Access

As demonstrated in the previous sections, an attacker can potentially recover meaningful information by analyzing a single acquisition of a side-channel. This can be further extended to include branches in the commands being executed by a device. One example of this is the ten rounds of the AES that are visible in Fig. 1.4. If this

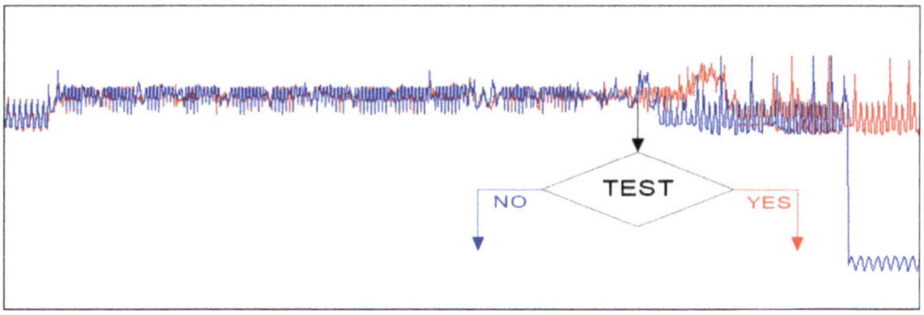

Fig. 1.6 The power consumption of a microprocessor while evaluating the file access rights on a GSM SIM. The red trace corresponds to a successful file access and the blue trace corresponds to an unsuccessful file access

is implemented as a loop there will be a test at the end of each round to determine if ten rounds have been computed. Another example would be file access rights. Some devices, such as GSM SIMs, consist of a file structure with associated file access conditions. In Fig. 1.6 the power consumption of a SIM that allows access to an arbitrary file is plotted in red, and the power consumption of the same SIM that denies access to an arbitrary file is plotted in blue. Superposing the traces, as in Fig. 1.6, would allow an attacker to determine the moment that file access is granted or denied by the microprocessor.

1.4 Differential Power Analysis

Differential power analysis (DPA) is a natural extension of SPA, in which the adversary conducts intensive data acquisition (e.g. thousands of traces) in order to recover a cryptographic key. This process usually starts by selecting the subkeys s that are to be recovered by the attack. In the context of block ciphers (our running example), these subkeys typically correspond to small parts (e.g. bytes) of the master key. Then, the attack can be described as a combination of the following three steps [269]:

1. For different plaintexts x_i and subkey candidates s^*, the adversary predicts some intermediate values in the target implementation. For example, one could predict S-box outputs z_i and get values $v_i^{s^*} = \mathsf{S}(x_i \oplus s^*)$.
2. For each of these predicted values, the adversary models the side-channel leakage. For example, if the target block cipher is executed on a CMOS-based microprocessor, the model is typically the Hamming weight (HW) of the predicted values. One then obtains the modelled leakage $m_i^{s^*} = HW(v_i^{s^*})$.
3. For each subkey candidate s^*, the adversary finally compares the modelled leakages with the acquisitions produced with the same plaintexts x_i and a subkey, using a statistical distinguisher of his choice. One commonly chosen distinguisher is Pearson's correlation coefficient [72], i.e. an attacker computes the correlation between the modelled leakage and each point in the acquired traces.

In first-order DPA attacks against unprotected devices, each $m_i^{s^*}$ is compared with a single point in the traces. Hence, the comparison step is independent of all other points in the acquired trace. In practical attacks, this comparison is applied to many points in the leakage traces and the subkey candidate that performs best is selected by the adversary.

Note also that many distinguishers can be used to detect data dependencies in the leakages. Classical solutions include Kocher's difference of means test [240], Pearson's correlation coefficient [72] and template attacks [82].

1.4.1 Profiled Fault Triggering with DPA

If the effectiveness of SPA in triggering a fault insertion against certain classes of devices becomes limited, it is alternatively possible to exploit DPA for this purpose. However, it is important to note that using DPA implies a modification of the attack scenario. That is, while SPA-based triggering techniques could be used "on the fly", DPA usually makes sense as a preparation stage, to be applied before the application of a fault attack. In fact, most successful DPA attacks come both with the identification of the points of interest in a trace, and with secret key information.

An example is shown in Fig. 1.7, where the lower trace is a single power consumption trace taken during the computation of an AES on an ARM7 microprocessor (as previously considered in Sect. 1.3). The upper trace shows the correlation between the Hamming weight of the output of one byte of the SubBytes function and a series of power consumption traces. The peaks in the correlation trace show at which points in the power consumption trace the predicted byte is manipulated by the microprocessor. This complements the information one can observe using SPA. The first peak corresponds to where the first byte is produced in the SubBytes function and indicates which of the 16 peaks corresponds to that byte being produced. The subsequent peaks in the correlation trace indicate the points in time where the same byte is manipulated in the MixColumns function.

Note that if the key is unknown, only the intermediate values after the first (and before the last) round(s) can initially be predicted by the adversary, i.e. before the diffusion in the cipher makes them dependent on too many key bits. But once a single successful DPA has been applied (i.e. once a master key is known), it is possible to predict any intermediate value in the cipher and consequently to detect its precise execution time for a subsequent fault attack.

1.5 Advanced Scenarios

The previous sections discuss case studies in which power analysis can be used to complement fault injection, essentially for enhancing the triggering aspects. However, most embedded devices for security applications are now protected with various

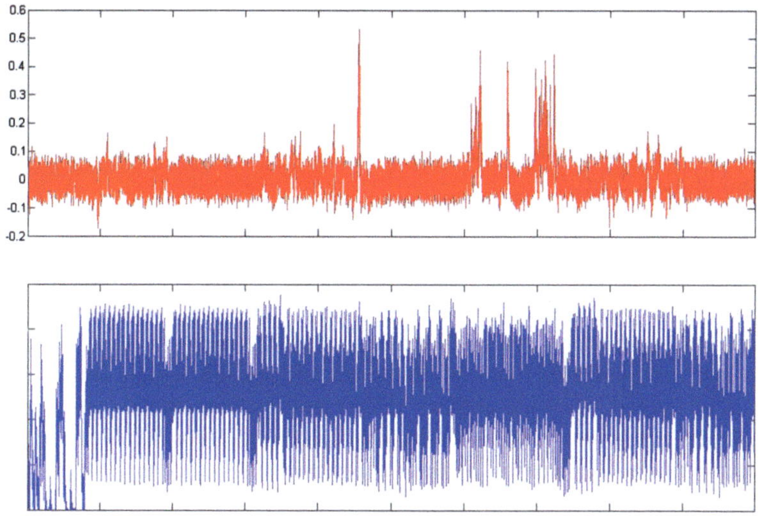

Fig. 1.7 An example of Differential Power Analysis applied to AES. The lower trace is an example power consumption trace, and the upper trace is the correlation between the Hamming weight of the output of one byte of the SubBytes function and a series of power consumption traces

countermeasures against physical attacks. The precise description of these countermeasures and the advanced attacks that can be applied to circumvent them is out of the scope of this chapter. But for completeness, we list certain types of countermeasures, with pointers to further readings.

Regular Algorithms. The information available to an attacker analyzing one trace needs to be limited to the identification of operations, without revealing any information on cryptographic keys. The simplest countermeasure is to use regular algorithms that will compute the same sequence of operations for all possible key values. When computing an RSA signature an alternative is to ensure that the same function is used to compute both operations [85]. This was proposed to allow the continued use of Algorithm 1.1, which is preferable on devices with limited resources. The rest of the countermeasures described in this section apply to DPA.

Data Randomization. The most investigated countermeasure to side-channel analysis is masking (see, e.g. [168, 347]). The intuitive idea behind masking is to split the sensitive data into a number of shares and to perform the encryption operation on the different shares separately. As a consequence, any successful DPA will have to combine the information from several shares (and, if not computed in parallel, time samples) in order to observe key dependencies in the leakages. Attacks on implementations using masking are referred to as 'higher-order' attacks, and Chari et al. showed in [81] that the number of traces required to recover a key from the leakages of a masked implementation increases exponentially in the number of shares, given that a sufficient amount of noise is limiting the quality of the measurements [385]. Masking does not directly impact fault attacks, besides the fact that it increases the difficulty of finding the time samples corresponding to certain operations with DPA. A fault inserted on a byte of either the masked data or the mask will propagate through the cipher as a fault inserted into the unprotected data [60]. Clearly, under specific fault models, e.g. inducing faults which set mask bytes to 0, masking becomes ineffective. Another idea which uses faults to counter masking is described in Sect. 2.2.6.

Random Ordered Processing. A countermeasure that complements data randomization is to process operations, as much as possible, in a random order. The operations required to compute a block cipher will, by necessity, be computed in a deterministic order. However, the individual commands required to compute these operations can be computed in an arbitrary order. If, for example, we consider the SubBytes function, there are 16! possible orders in which the 16 bytes can be treated. The side-channel information is then spread over 16 different points in the trace, which reduces the amount of information available to an attacker. The most common application of this countermeasure is to block ciphers; it has also been shown to be applicable to group exponentiation, often used in public key cryptographic algorithms [401]. This does not prevent an attacker from targeting a particular operation, but will prevent a specific command within that operation with any degree of certainty.

Random Delays. Another countermeasure that affects the location of information in acquired side-channel traces is the use of functions that pause for a random amount of time before continuing. These functions can be inserted between operations to reduce the amount of information available to an attacker. When conducting a side-channel attack, this means that an attacker will have to synchronize the acquisitions before applying a DPA attack. When attempting to inject a fault at a given point in an algorithm, this becomes more of a problem, as the location of a target operation will be distributed around a given point. An attacker will typically seek to inject a fault at a given point and repeat the attack until the fault is injected in the desired operation. Some proposals have been made to try a maximize the size of this distribution [105, 402]. However, given that this countermeasure slows rather than prevents an attacker, it is debatable whether any complex operations are required.

Technology Scaling. We note finally that, for any type of physical attack, technology scaling is likely to have a significant impact [192]. While they cannot be considered as countermeasures in themselves, the resulting modifications of the semiconductor physics (e.g. the increasing importance of static leakages, or variability issues) imply modifications of the typical leakage models and the types of faults that can be assumed for the target devices.

Data Redundancy. Adding redundancy to the data in order to detect faults during the computation of a cryptographic operation is an appealing solution to preventing fault attacks. As detailed in the later chapters of this book, such redundancies can be effective solutions to preventing various types of errors, with high detection rates. But as data randomization only helps to prevent side-channel attacks, data redundancies only help to prevent fault attacks. In fact, if the type of redundancy present in an implementation is public, it can even be used by the adversary, in order to enhance DPA attacks (to reject certain guesses for the intermediate values).

1.6 Conclusion

In this chapter we have given several practical examples of how different types of side-channels can be used to enable fault attacks. In our examples, side-channel attacks are used to determine trigger points for inducing faults. This can be done using simple side-channel attacks, which are typically combined with on-the-fly fault insertion techniques. As an alternative, one can apply differential side-channel attacks to determine (the order of) operations and launch fault attacks with this information. Lastly, we have sketched some advanced scenarios, all of which are related to different types of countermeasures. We have described how each countermeasure may help or hinder fault attacks.

Part II
Fault Analysis in Secret Key Cryptography

Chapter 2
Attacking Block Ciphers

Christophe Clavier

Abstract Differential Fault Analysis (DFA) was one of the earliest techniques invented to attack block ciphers by provoking a computational error. In the basic DFA scenario the adversary obtains a pair of ciphertexts both of which encrypt the same plaintext. One of these ciphertexts is the correct result while the other is an erroneous one resulting from a faulty computation. Though applications of DFA to DES and AES have proven to be quite effective, other techniques have also been invented which can threaten block ciphers in different ways. This chapter presents some of these fault analysis methods, including Collision Fault Analysis (CFA) and its close variant Ineffective Fault Analysis (IFA). These methods depart from DFA by the fault model they rely on, by their ability to defeat classical countermeasures against DFA or DPA, or by their application to specific implementations.

2.1 Introduction

Differential Fault Analysis (DFA) was one of the earliest techniques invented to attack block ciphers by provoking a computational error. In the basic DFA scenario the adversary obtains a pair of ciphertexts that are the result of encrypting the same plaintext. One of these ciphertexts is the correct value while the other is an erroneous one resulting from a faulty computation. Since the two encryptions performed identically up to the point where the fault occurred, the two ciphertexts can be regarded as the outputs of a reduced-round iterated block cipher where the inputs are unknown but show a small (and possibly known) differential. By analyzing the propagation of this differential over a small number of rounds, the adversary seeks to obtain information about the key material involved in the last round.

C. Clavier (✉)
XLIM (UMR 6172), Université de Limoges, Limoges, France
e-mail: christophe.clavier@3il.fr

M. Joye and M. Tunstall (eds.), *Fault Analysis in Cryptography*,
Information Security and Cryptography, DOI: 10.1007/978-3-642-29656-7_2,

Though applications of DFA to DES [49] and AES [127, 160] as described in Sects. 3.3 and 4.2 have proven to be quite effective, other techniques have also been invented which can threaten block ciphers in different ways. This chapter presents some of these fault analysis methods, which depart from DFA by the fault model they rely on, by their ability to defeat classical countermeasures against DFA or DPA, or by their applicability to specific implementations.

An important class of such attack methods is referred to as Collision Fault Analysis (CFA) and has a close variant referred to as Ineffective Fault Analysis (IFA). The first section of this chapter is devoted to the presentation of these two techniques and their successive usages ranging from a first trivial CFA/IFA on unprotected implementations of AES-like algorithms to more elaborate attacks which can either attack implementations protected against both DFA and High-Order DPA, or even reveal the key of a so-called externally encoded DES whose precise specification is unknown to an attacker. Other fault attacks on block ciphers are presented in the second section, including some that explicitly aim to reduce the number of rounds, and others which exploit a perturbation of the initial randomization of substitution tables in DPA-resistant implementations.

2.2 Attacks on Block Ciphers by Exploitation of Identical Outputs

2.2.1 Three Resemblant but Different Fault Analysis Methods

2.2.1.1 Collision Fault Analysis

While DFA exploits a differential between a genuine and a faulty ciphertext where the fault occurred in one of the last few rounds of an encryption function E, CFA gains information from a collision event where the two ciphertexts C and $C^{\lightning}$ respectively obtained from a normal and a faulty encryption are equal. An attacker typically first obtains the faulty encryption $C^{\lightning} = E^{\lightning}(M_0)$ of an arbitrary plaintext M_0, and then searches for some particular M which gives the same ciphertext $E(M) = C = C^{\lightning}$ without any fault. Intuitively, it may appear difficult to produce a collision between two outputs of a cryptographic function designed to closely behave as a random function. This difficulty is circumvented by producing the fault very early in the encryption process in order to avoid the avalanche effect between the encryptions of M and M_0, and by relying on a specific fault model.

One usually adopts a fault model where the fault has a predictable effect on a small portion—a bit or a word—of the intermediate result. Such a classical fault model assumes that a fault occurring during a logical or an arithmetic operation, e.g. an XOR between two bytes, produces a result equal to some constant value (typically considered to be zero), whatever the input values.[1] When produced at the very beginning of the algorithm a fault will typically corrupt an intermediate value that only depends on one bit or one byte of the input. As the remainder of the intermediate value is not modified, the attacker just has to change that precise bit or byte until a collision with the corrupted value occurs. All other intermediate data being uncorrupted, this local collision propagates to the ciphertexts.

2.2.1.2 Ineffective Fault Analysis

As described above, a CFA attack consists of searching for a plaintext whose corresponding ciphertext collides with some corrupted ciphertext. IFA is slightly different, and applies where an attacker tries to find an input M with the property that when a fault is induced on some precise operation during the encryption process, the intermediate data targeted by the fault is not corrupted, resulting in an identical ciphertext. This kind of analysis gains information from faults which do not locally modify the intermediate result, so-called ineffective faults from which the analysis name is derived.

While usually relying on the same kind of fault model, CFA and IFA differ in many respects. While CFA recovers some piece of information about the key with only one fault, IFA needs to compare pairs of ciphertexts ($C = E(M), C^{\lightning} = E^{\lightning}(M)$) until $C = C^{\lightning}$, so many faults[2] are required before an ineffective one is obtained. On the other hand, the higher complexity of IFA in terms of the number of required fault injections is compensated for by the property that the operation targeted by IFA does not need to occur near the beginning of the algorithm. Since the attacker compares pairs of executions with the same input, any operation during the encryption process can be targeted to identify an ineffective fault. In light of the fact that a fault appears to be ineffective if and only if the natural[3] result of the targeted operation is zero, under the previously described fault model, it is clear that IFA can be considered as a kind of probing tool. Indeed, for any given plaintext it is possible to decide whether the result of any arbitrary targeted operation[4] is zero or not.

Another property specific to IFA is that an attacker does not actually require the value of any faulty ciphertext. The only information that is required is whether the fault had an effect or not. As a consequence, IFA is not thwarted by the classical countermeasure against DFA, which consists in checking the computation and withholding the output if a fault is detected. Indeed, whether a fault is detected or not

---

[1] An exception to this usually adopted fault model for CFA is given by the collision/differential fault attack from Hemme in the first rounds of DES, described in Sects. 2.2.3 and 3.5. This attack is applicable even if the fault produces a random modification of an S-box output.

[2] For instance, under a byte-oriented fault model, 128 (or 256) faults are required on average per ineffective fault event when the input of the targeted operation is chosen (or not chosen).

[3] I.e. without fault.

[4] Provided that this operation is susceptible to the considered faults.

provides an attacker with the same information as whether or not the fault corrupted the ciphertext.

Finally, note that IFA relies on the fact that the attacker can safely decide that a fault attempt has been ineffective because of the natural value of the targeted data rather than because the fault did not occur. When performing IFA, it is thus very important that the fault injection tool be considered highly reliable.

2.2.1.3 Safe-Error Analysis

A third fault analysis method which exploits the identity of outputs of cryptographic algorithms is the analysis of so-called safe-errors, which we refer to as safe-error Analysis (SEA). This method was first introduced to break a private exponentiation of the RSA cryptosystem [204, 427, 429]. The basic principle consists of modifying some internal data and inferring the value of a private exponent bit from whether this modification resulted in an identical or a different output. Two examples of such analysis on a binary square and multiply exponentiation are: (i) the perturbation of a multiplication in an *always multiply* version of the left-to-right binary exponentiation algorithm, and (ii) including a dummy register in a right-to-left version of the algorithm. In both cases, depending on the particular value of the current exponent bit, the modified data may or may not be used in the sequel of the computation. The latter case corresponds to a safe-error which results in no modification of the algorithm output.

At first, ineffective fault analysis and safe-error analysis look quite similar. Indeed, they both infer information about a secret from whether an induced fault affected the output or not. Actually, there is a conceptual difference between IFA and SEA. In safe-error analysis the data that is targeted by the fault is actually modified, and the output is not modified simply because the modified data is not used. In contrast, an ineffective fault targets data involved in the computation of the algorithm, but because of the fault model and the data value, this data is not modified. In summary, IFA probes a data value while SEA probes a data usage.

A consequence of the difference between these two techniques is that IFA is highly dependent on the effect a fault has on the data being processed while SEA is not. A safe-error will be safe whatever the resulting value of the targeted data, so SEA is applicable in the general random error model. In contrast, ineffective fault analysis usually requires a more restrictive *stuck-at* fault model.

In the remainder of this section, we describe a series of research papers which consider CFA and IFA techniques to recover secret keys of block ciphers. We present them in the chronological order, which roughly corresponds to a trend of increasing ability to defeat countermeasures.

2.2.2 Bit-Wise Collision/Ineffective Fault Analysis on AES

The first attack from the CFA/IFA family was proposed by Blömer et al. [55], where they apply this technique to the AES algorithm. They assume a fault model where an attacker is able to force to 0 any chosen individual bit of an intermediate result. While they advocate its practicality, this fault model can be considered, according to the authors, as a rather strong one.

The principle of their attack is quite simple. An attacker first encrypts an all-zeros plaintext and obtains the corresponding reference ciphertext. Then for each arbitrary bit of the output of the first `AddRoundKey` operation, a faulty execution is performed where the fault sets this particular bit to 0. If the resulting ciphertext and the reference one are the same then the corresponding key bit is 0. If they differ the corresponding key bit is 1. By scanning all the bits of the first `AddRoundKey` output, the whole AES key is recovered with 128 fault injections.

Due to the bit-oriented nature of the fault model, this attack can be considered either as a degenerate form of CFA[5] or as an IFA. The computation checking countermeasure does not prevent the attack since only whether or not the fault corrupted the reference ciphertext is informative, not the value of the eventually corrupted result.

Interestingly, one can notice that this attack applies not only to the AES cipher, but more generally to the whole class of block ciphers whose very first operation, and the only one involving the plaintext, is an XOR with the key. Note also that the attack does not require the knowledge of the transformation subsequent to the first XOR with the key. It is thus applicable to any (even unknown) algorithm belonging to this class.

2.2.3 Collision Fault Analysis on DES

Designing a collision fault analysis on a Feistel network presents a specific difficulty. This is because, when encrypting a modified version of the initial plaintext M_0 used for the faulty execution, the input differential is involved in both the left and the right paths of the function. It is thus impossible to provoke a modification *only* on the input of one targeted operation and obtain a collision of intermediate results at the very beginning of the function.

Hemme dealt with this difficulty [178] by using the concept of characteristic, borrowed from differential cryptanalysis [46].[6] A characteristic is a set of prescribed differentials that an execution may follow from one round to another. Hemme considers only characteristics whose differential after $k = 2$, 3 or 4 rounds is the same

[5] The probability that the faulty ciphertext is not corrupted is as much as $\frac{1}{2}$.

[6] Hemme's attack can be regarded as a CFA since the adversary eventually finds a plaintext which encrypts to the same ciphertext as the faulty one. On the other hand it can also be considered as a kind of DFA since the adversary must analyze which differential characteristic could lead to the colliding message. The reader will find a more detailed description of this attack in Sect. 3.5.

as the differential that may have been produced by a fault. Assuming an attacker can modify the output Y_k of the kth round to $Y_k \oplus \varepsilon$ where ε belongs to a prescribed set defined by the fault model,[7] a pool of characteristics is created which includes, for each possible ε, the most probable characteristic ending after k rounds with a null differential on the left side and a differential equal to ε on the right side. After obtaining a faulty ciphertext, and for each ε, the attacker tries to exhibit a normal execution which follows the k-round $(0, \varepsilon)$-characteristic belonging to the pool. Provided that ε is the actual differential effect of the fault, a normal execution following the $(0, \varepsilon)$-characteristic on k rounds starts the $(k + 1)$th round with the same intermediate data as the faulty one, and results in a colliding ciphertext. Each obtained collision gives some information about the first round key K_1. This information is gathered by counting how many times each key candidate is suggested by a colliding pair. An improved version of the attack which makes use of an extended pool of characteristics can retrieve the whole K_1 with about 400, 1×10^4 and 5×10^6 faulty executions[8] when k is respectively equal to 2, 3 and 4.

One benefit of this attack is the possibility to attack the DES by fault analysis even when the implementation is protected against classical DFA by recomputing the last few rounds of a block cipher with a verification of the result. Note that this is an advantage only when the attacker does not have access to a decryption function that uses the same key. In the case where the attacker can both encrypt and decrypt any data of his choice, the redundancy of the last few rounds only is not sufficient to prevent DFA. Indeed, an attacker could then decrypt any input C while inducing a fault at the beginning of the algorithm (preferably during the second round) and obtain the faulty output M'. Then he can encrypt M' without fault and get C'. With respect to M', C' is a genuine ciphertext, while C is a faulty one analogous to what would have resulted from a classical DFA on the penultimate round.

2.2.4 *Defeating a DPA-Resistant AES by Collision Fault Analysis*

Amiel et al. presented several fault attacks which apply to DPA-resistant implementations of AES and DES [12]. In this section, we describe two of these that are collision fault analyses applied to AES.

---

[7] The author chose the set of 32 one-bit errors at the end of the round function. While this choice is naturally suited for a bit-oriented fault model, the author noticed that it also happens to be the best one for the less restrictive byte-oriented random error fault model.

[8] The attack also necessitates 4×10^4, 1.11×10^6 and 8.10×10^8 normal executions respectively.

2.2.4.1 Attacking the First AddRoundKey

A simple CFA attack on AES consists of an adaptation of the attack described in Sect. 2.2.2 to a byte-oriented fault model. Given the faulty ciphertext $C^{\lightning}$ produced by encrypting M_0, and producing a fault which forces to zero the output of the ith XOR operation in the first `AddRoundKey`, the attacker exhaustively encrypts all 256 plaintexts $M = (m_0, \ldots, m_{15})$ which coincide with M_0, except for the byte m_i, until one of the ciphertexts collides with $C^{\lightning}$. The plaintext byte m_i leading to a collision verifies that $m_i \oplus k_i = 0$, so the attacker decides that k_i is equal to this particular m_i. By scanning all 16 XOR operations, this attack allows a key to be completely recovered with only 16 faulty executions.

While quite efficient, this attack does not apply when the implementation is protected against first-order DPA by a Boolean masking scheme. In such implementations, all intermediate bytes are masked by an XOR with a random value R which is different from one execution to another, but which is the same from one byte to another. During the first `AddRoundKey`, each byte m_i of the plaintext is thus XORed with a masked key byte $\tilde{k}_i = k_i \oplus R$. When one tries to apply the above CFA on the result of the ith iteration of the `AddRoundKey`, the physical zero value forced on the XOR output actually corresponds to a logical value equal to R. While exhausting all m_i, the collision occurs when the logical XOR output is also equal to the random mask R of the faulty execution. This occurs for m_i such that $m_i \oplus k_i = R$, so the attacker erroneously decides that the key byte is equal to $k_i \oplus R$ instead of k_i. As the value R is unknown to the attacker, and different from one faulty execution to another, the attack fails.

Amiel et al. [12] devised a variant of the CFA that can break such DPA-resistant implementations of AES. The method is based on the observation from [21, 300] that a fault may allow an attacker to prematurely terminate a **for** loop before it has gone through all its iterations. If the memory location where the result of the XOR between the plaintext and the key is stored has not been used previously, then there are good odds that it is in an uninitialized state and contains bits all set to 0.[9] Now, suppose that one produces a fault two iterations before the end of the loop of the `AddRoundKey`. Then all 14 first result bytes are correct, but the last two are both have a physical value equal to zero, meaning a logical value equal to some random mask R that is identical for these two bytes. Encrypting all 2^{16} plaintexts where m_{14} and m_{15} take all possibilities will produce a collision for (m_{14}, m_{15}), verifying that

$$m_{14} \oplus k_{14} = R \quad \text{and} \quad m_{15} \oplus k_{15} = R.$$

An attacker will not know R but can nevertheless infer that

$$k_{14} \oplus k_{15} = m_{14} \oplus m_{15}.$$

The size of the key space is thus reduced from 2^{128} to 2^{120}.

---

[9] Or all set to 1 depending on the logical representation of the physical state.

By repeating the attack with the three last result bytes unaffected by the loop, the colliding ciphertext is obtained when logical values $m_{13} \oplus k_{13}$, $m_{14} \oplus k_{14}$, and $m_{15} \oplus k_{15}$ are all equal to a same unknown value R[10]

$$m_{13} \oplus k_{13} = R, \quad m_{14} \oplus k_{14} = R \quad \text{and} \quad m_{15} \oplus k_{15} = R.$$

Combining the first two equations now also reveals the value of $k_{13} \oplus k_{14}$.

Note that the exhaustive search on (m_{13}, m_{14}, m_{15}) does not cost 2^{24} encryptions but only 2^{16} since one can restrict the search to tuples of the form $(m_{13}, m_{14}, m_{14} \oplus \delta_{14,15})$ where $\delta_{14,15} = k_{14} \oplus k_{15}$ is known.

The attack continues repeatedly in the same way, each time revealing the value of $k_i \oplus k_{i+1}$ for $i = 12, 11, \ldots, 0$, ending with the knowledge of the complete key up to an XOR with a constant of the form $(c, c, \ldots, c)$. The correct value is identified by exhaustively trying the 256 remaining candidates. Overall the attack necessitates 15 faulty encryptions and $15 \times 2^{16} \approx 2^{20}$ normal encryptions.

A common complementary countermeasure against DPA consists in computing as much as possible in a random order. Note that the above attack can also be adapted to the case where the first `AddRoundKey` loop is executed in a random order. One needs to precompute a dictionary of approximately 2^{23} ciphertexts, which may be somewhat time consuming. A trade-off can speed up this precomputation by generating only a fraction of the dictionary at the cost of an increased number of faulty executions. An interesting practical property of the attack on the random order implementation is that the attacker does not need to change the instant of the perturbation from one fault injection to another.

Both attacks have been practically performed on smart cards. These experiments are described in detail in [400].

2.2.4.2 Attacking the Key Transfer from NVM to RAM

Prior to its use in the computation of a block cipher, each key byte must be XORed with the same random mask value R that is used to randomize any substitution table. This key masking with the value $(R, R, \ldots, R)$ commonly occurs when the key is transferred from nonvolatile memory (NVM) to RAM. For security purposes the key stored in NVM is also typically masked, where the mask is full size, meaning that every mask byte will be different, but static and diversified from one device to another. Denoting this full-size long-term mask by $r = (r_0, r_1, \ldots, r_{15})$, the masked key in NVM is equal to

$$\tilde{K}_{NVM} = (k_0 \oplus r_0, \ldots, k_{15} \oplus r_{15})$$

while the masked key in RAM is equal to

[10] Note that the value of R does not necessarily remain the same from one faulty ciphertext to another.

Algorithm 2.1: Key masking

Input: $\tilde{K}_{NVM} = (k_0 \oplus r_0, \ldots, k_{15} \oplus r_{15})$
$r = (r_0, r_1, \ldots, r_{15}), R$
Output: $\tilde{K}_{RAM} = (k_0 \oplus R, \ldots, k_{15} \oplus R)$

for $i \leftarrow 0$ **to** 15 **do**
  $\tilde{K}_{RAM,i} \leftarrow \tilde{K}_{NVM,i} \oplus R$ $\tilde{K}_{RAM,i} \leftarrow \tilde{K}_{RAM,i} \oplus r_i$
end
return $\tilde{K}_{RAM}$

$$\tilde{K}_{RAM} = (k_0 \oplus R, \ldots, k_{15} \oplus R)\ .$$

The mask conversion performed during the transfer follows Algorithm 2.1 with the difference that key bytes are actually processed in a random order. By faulting the loop so that only the first iteration has executed, the uninitialized memory space storing $\tilde{K}_{RAM}$ physically contains 15 bytes equal to zero and another one at a random index i equal to $k_i \oplus R$. This corresponds to a logical value of the form $(R, \ldots, R, k_i, R, \ldots, R)$, which is the actual key used in the faulty encryption of M_0.

The attack consists in precomputing a dictionary of about 2^{20} ciphertexts produced by the encryption of the reference plaintext M_0 under all keys of the previous form where i, k_i and R take all possible values. A search of the faulty ciphertext in the dictionary immediately reveals k_i. Each new faulty execution gives the opportunity to learn the value of a key byte at a different index. An average of only 54 faults allows all 16 key bytes to be determined. Note that the dictionary does not depend on the actual value of the attacked key.

2.2.5 Ineffective Fault Analysis on Externally Encoded DES

Contrarily to Kerckhoffs' principle, many applications of modern cryptography still adopt the *security by obscurity* paradigm. A particular way of designing a proprietary algorithm consists in surrounding a well-known and widely used block cipher $\mathtt{E}$ with two secret external encoding permutations $\mathtt{P}_1$ and $\mathtt{P}_2$ (one-to-one mappings over the input and output spaces respectively), leading to the new, secret, obfuscated block cipher $\mathtt{E}' = P_2 \circ \mathtt{E} \circ \mathtt{P}_1$. By basing the construction on a well-known block cipher $\mathtt{E}$ we allow the design to inherit proven or empirical cryptographic strength. Also, the two secret encodings $\mathtt{P}_1$ and $\mathtt{P}_2$ ensure that inputs to and outputs from $\mathtt{E}$ cannot be known by an attacker, so physical attacks requiring this knowledge should not be feasible.

A particular case studied by Clavier [93] depicted in Fig. 2.1 considers $\mathtt{E}$ instantiated as the DES function. Despite the impossibility of applying classical DFA [49],

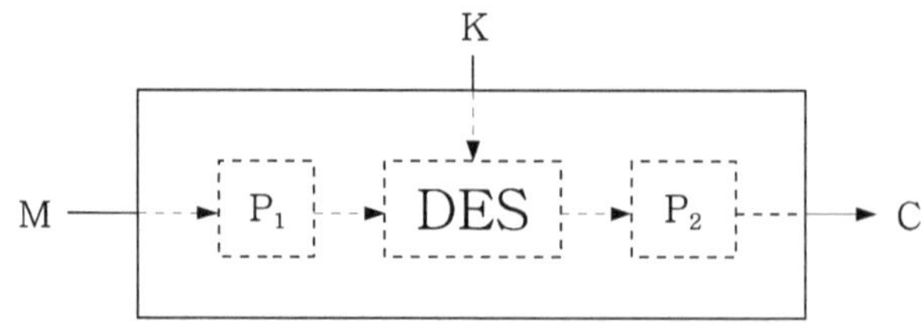

Fig. 2.1 A DES obfuscated by secret layers P_1 and P_2

which needs the output of the block cipher, and CFA [178], described in Sect. 2.2.3, which needs the control of the DES input, the author devised an ineffective fault analysis which recovers the secret key and applies to any member of the large class of unknown (to the attacker) cryptographic functions.

The attack assumes a classical software implementation of DES on an eight-bit architecture. Also, we assume an attacker is able to precisely control which instruction is executed when a fault is injected. As for other CFAs/IFAs the fault model assumes that a fault injected during the execution of an XOR between two eight-bit operands results in a zero[11] output whatever the input operand values were. Finally, the attacker is supposed to have control over the input given to the encryption function $\mathtt{E}'$ as well as knowledge of its output.[12]

The attack is somewhat complex and makes use of *pairs* of related ineffective faults. We now give a sketch of the principle of a basic version of the attack. As we assume an eight-bit architecture, there are 12 XOR operations per round: eight for the computation of the inputs of each S-box—denoted by `xor_key`[j] ($j = 1, \ldots, 8$)—and four others at the end of the round for combining its 32-bit output with the left part—denoted by `xor_left`[i] ($i = 1, \ldots, 4$). The attack intensively uses IFA to probe the output of these different eight-bit XOR operations that may appear at any round.

First, suppose that for some arbitrary plaintext M, a fault injected during some `xor_left`[i] at round $(h-1)$ turns out to be ineffective (i.e. the ciphertext obtained with M by faulting this XOR is identical to the one obtained with the same input without fault). This implies that the corresponding output byte is zero. Thus, eight of the 32 bits at the input of the next round h are known to be 0. The subsequent permutation expands them to 12 bits, which are involved in four adjacent S-Boxes at round h as in Fig. 2.2. Now, suppose that for another execution with the same plaintext M, a fault on `xor_key`[j] (for $j \in \{2i-1, 2i\}$) at round h also turns out to be ineffective. Then we know that the input x_j of this S-box is one of the four preimages of $S_j(0)$.[13] As x_j is the XOR between a key byte k_j and a six-bit value having five

---

[11] Note that the attack works equally well if the faulted XOR output is supposed to be any arbitrary known constant instead of zero.

[12] These assumptions may be relaxed, since an attacker only needs to be able to replay many different arbitrary inputs, and to detect whether two outputs are equal.

[13] As the S-box is a compressive function from six-bit inputs to four-bit outputs, any preimage of $S_j(0)$ behaves exactly as the input 0—which has been forced in the faulty execution—and thus produces the same ciphertext.

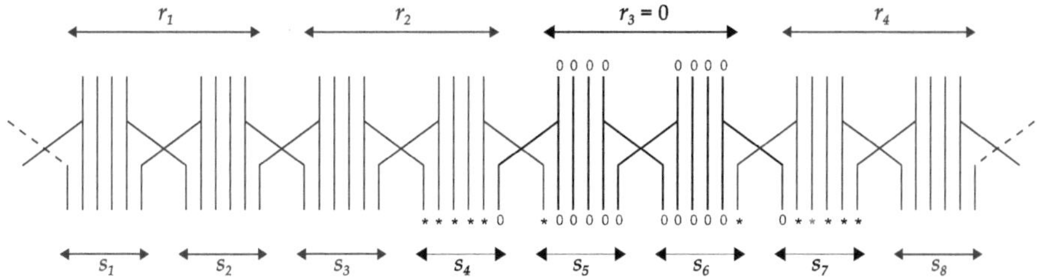

Fig. 2.2 A zero byte through the expansive permutation

bits known to be 0, we clearly see that k_j belongs to a set of only eight possible values, which reveals three bits of information about the key.

Overall, the attack repeatedly tries to detect a plaintext producing such a rare pair of related ineffective faults. As much information as possible is gathered by exploiting this kind of event on every possible loci—defined by a couple (`xor_left`[i], `xor_key`[j])—at every round. According to simulation results, the median residual entropy on the key is reduced from 56 to 26.49 bits (or 22.32 bits) after using 5×10^4 faults (or 1×10^6 faults). The author also described an improved version of this attack which gains information from more complex kinds of events. This improved attack achieves a median residual entropy of 13.95 bits (or 6.68 bits) after using 5×10^4 faults (or resp. 1×10^5 faults).

Compared to fault analysis on known cryptosystems, this attack requires a large number of fault injections. This can be considered as a fair price to pay for the 'magic' property of being able to retrieve the key of such unknown functions regardless of the two secret external encodings P_1 and P_2.

2.2.6 Passive and Active Combined Attacks on AES

The Boolean masking countermeasure that the two CFA attacks described above in Sect. 2.2.4 can bypass is only intended to protect against first-order DPA. State-of-the-art implementations must actually protect against high-order analyses, as introduced by Messerges [284]. The basic principle of such a countermeasure is to use a Boolean masking scheme based on one random byte that is different from one execution to another (as is the case for a first-order countermeasure), also from one S-box operation to another and from one round to another.

The recent attack presented by Clavier et al. in [94] actually succeeds in breaking even this kind of protection. The state-of-the-art implementation considered by the authors implements the S-box description from Oswald et al. [313] and is designed to resist HO-DPA attacks such as those presented in [11, 270]. The method used to break this implementation is a passive and active combined attack. The passive part of the attack consists in an adaptation of Correlation Power Analysis (CPA), originally described by Brier et al. [72]. As with the original analysis on unprotected implementations, an attacker predicts, under some guess about a key byte, the series—one value for each power curve—of intermediate data manipulated during the AES computation. The difference with this classical scheme is that the targeted intermediate data is not the output of a first round S-box, but rather the random value by which this S-box is masked. Provided the value of this random mask can be predicted, under a key guess, for each execution, a CPA can be performed which will show a correlation peak for the correct guess at each instant the mask is manipulated. The need to identify, up to the corresponding key byte, the series of values taken by some random mask is precisely the reason why the passive part of the attack is preceded by an active phase which identifies the random series by means of a collision fault analysis.

As often, the target of this CFA is an output byte of the first `AddRoundKey` which is forced to zero by a fault. In the present implementation the plaintext M and the key K have been respectively masked, before they are XORed together, by two 16-byte randoms rm and rk respectively. Iteration i of the `AddRoundKey` thus computes $(m_i \oplus rm_i) \oplus (k_i \oplus rk_i) = (m_i \oplus k_i) \oplus r_i$ with $r_i = rm_i \oplus rk_i$ being the random value used to mask the ith S-box in the first round. As forcing to zero this XOR output is equivalent to introducing a differential equal to $\delta = m_i \oplus k_i \oplus r_i$, the attacker is able to identify $k_i \oplus r_i$, which is precisely the value for the ith plaintext byte that produces a collision with the faulty ciphertext. Thus, for each faulty execution, the attacker identifies the value $k_i \oplus r_i$ by means of a CFA, and stores the corresponding power curve. When sufficiently many $k_i \oplus r_i$ values and corresponding curves have been obtained, it is possible to perform a CPA where the series of r_i is predicted by simply guessing k_i. On average, 126 faults are enough to collect traces with 100 different r_i values, while 1,568 are required to perform the CPA with all 256 occurrences. Note that this attack is applicable even if the 16 XOR operations are computed in a random order. In that case the attacker has to search for the colliding ciphertext in an extended set of 2^{12} plaintexts where the input byte to modify can be in any position. The plaintext producing the collision automatically informs the attacker about the index i of the XOR corrupted by the fault. The attack strategy is slightly different since the different sets of power curves for all possible indices must be collected at the same time rather than successively.

The attack described above does not apply in the case where the computation verification countermeasure is implemented since no corrupted ciphertext is then given to the attacker. In that case it is still possible to adapt the attack by preferring an IFA strategy. When the plaintext byte m_i is set to the value u, each injected fault is an attempt to obtain a power curve for which $k_i \oplus r_i = u$. The probability of success of each attempt is 2^{-8} (or 2^{-12} in the case of random order), so the expected number of faults required to collect n power curves with different masks for each attacked key byte amounts to $2^{12} \times n$ (or $2^{16} \times n$). This is certainly a large number of faults, so this attack is hardly practical, but a door is opened here to break strongly protected implementations which include at the same time: high-order Boolean masking, computation verification and random order execution.

Note that an interesting and unexpected property of the ineffective fault variant is that it is easier to perform when a computation verification countermeasure is implemented than when one is not. Indeed, when such a countermeasure is present the attack becomes a known plaintext attack instead of a chosen plaintext attack.

2.3 Other Fault Attacks on Block Ciphers

2.3.1 Reducing the Number of Rounds

Most block ciphers are *iterated cryptosystems*, which are families of cryptographically strong functions that iterate n times a weaker round function. As the security of the block cipher is an increasing function of the number of rounds, an obvious attack path could be to try to produce a perturbation in the normal sequencing which would reduce the number of rounds. This idea has originally been formulated in [15], where the authors suggest corrupting the appropriate loop variable or conditional jump by means of a glitch in either the clock or the power supply to the chip. In the following we give the descriptions of two concrete experiments which put this idea into practice.

2.3.1.1 Round Reduction Using Faults on AES

Choukri and Tunstall demonstrated in [89] that reducing the number of rounds of a block cipher is indeed possible. They used glitches on the power supply as the fault injection media, and conducted the attack on a Silvercard (`PIC16F877`) which does not contain any senors to protect against this sort of attack. The AES algorithm implementation used does not contain any countermeasure intended to prevent any sort of attack. The aim was to show that precise faults can be induced within a chip that can lead to the desired effect.

After a search among a wide range of combinations of relevant parameters such as the clock speed, the size of the glitch, the applied voltage and the time position in the computation of the AES,[14] various different configurations happened to be successful in inducing a fault that reduced AES computation to only one round. The exploitation of such a result is trivial and requires only two pairs of known plaintexts/ciphertexts. For each S-box an exhaustive search for the key byte value verifying

$$S(m_1 \oplus k) \oplus S(m_2 \oplus k) = \texttt{MixColumns}^{-1}(c_1 \oplus c_2),$$

leads to two expected hypotheses for each key byte, leading to an overall exhaustive search amongst 2^{16} keys.

[14] The identification of the correct time position has been helped by the capture of a power curve that clearly shows the different rounds.

2.3.1.2 A Case Study of Fault Attacks on Asynchronous DES Cryptoprocessors

Asynchronous circuits represent a class of circuits which are controlled not by a global clock but by the data. The class of circuit described by Monnet et al. [292] use so-called Quasi-Delay-Insensitive technology and multi-rail encoding, which are often thought to offer native resistance against faults. The authors designed two DES hardware implementations, a reference one and a hardened one, and studied their susceptibility to fault attacks using round reduction.

The round counter of the reference version is an asynchronous state machine that uses a 1-out-of-17 code: each round number is coded using one wire. Sixteen wires are used to encode the 16 rounds. The hardened version implements the same counter but protected with alarm cells. These cells are able to detect any wrong code generated in the counter module, i.e. any state using two or more wires is detected. Alarms inform the environment when a wrong code is detected.

Faults were induced using a laser beam, which offers the advantage of its directionality that allows the precise targeting of a small circuit area (e.g. $5\,\mu\text{m}^2$). In a time and space scan of the counter block which represented over 5^3 shots for each circuit, about 40 % of the shoots revealed errors. Some of them were identified as having modified the sequence of rounds.

From the properties of the DES key scheduling, the authors analyzed how a modification of the sequence of rounds alters the corresponding sequence of round keys actually used during the encryption. They give an example of the exploitation of a pair of faulty executions which led to close sequences of round keys. A detailed analysis of how to exploit these faults is given in [92], where how to recover the DES key is described for each possible case where the two sequences of round keys differ from each other by suffixes of length of at most 2. A significant number of the faulty pairs obtained fell into these exploitable cases. While the alarm cells of the hardened version actually detected most of the alterations of the counter block, a few executions with a modification of the round sequence remained undetected.

2.3.2 Corrupting the Randomization of a DES S-box

Before each execution of a DES protected by Boolean masking, all substitution tables are first randomized so that they comply with the original S-box tables in an implementation where all intermediate data are masked. This randomized S-box precomputation is typically performed as described in Algorithm 2.2. Assuming that an attacker is able to modify some S-box entry during this randomization phase, the subsequent DES execution could then be altered. Amiel et al. [12] precisely studied

Algorithm 2.2: S-box randomization

Input: The original S-box table $S = (s_0, \ldots, s_{63})_{16}$
The input and output masks $R \in (0, \ldots, 63)$ and $r \in (0, \ldots, 15)$
Output: The randomized S-box table $\tilde{S} = (\tilde{s}_0, \ldots, \tilde{s}_{63})_{16}$

```
for i ← 0 to 63 do
  s̃_{i⊕R} ← s_i ⊕ r
end
return S̃
```

how to exploit a modification of an S-box entry in both cases, where the index of the corrupted value is known and unknown to the attacker.[15]

2.3.2.1 Modifying Known S-box Values

When all eight S-Boxes are randomized as in Algorithm 2.2 an attacker can precisely know which S-box entry it is modifying. For some fixed arbitrary plaintext, it is possible to scan the randomization loop of some attacked S-box, induce a fault at each successive iteration, and observe whether the ciphertext has been corrupted or not. This results in the list of indices of all entries that are used for some S-box in one round or another of this execution. Such list contains an average of 14.255 indices which can all be taken as possible candidates for the S-box entries used in the first round. From the relevant bits of the plaintext, each index value can then be turned into a hypothesis on the S-box first round subkey. The different lists of candidates for each subkey can be further reduced by repeating the attack with one or more other plaintexts. The expected number of remaining candidates for the first round key K_1 is $2^{30.7}$ (or $2^{15.4}$) when the attack exploits one plaintext (or two plaintexts), which results in $2^{38.7}$ (or $2^{23.4}$) candidates for the whole key K.

In practice, embedded implementations of DES are unlikely to have the 512 four-bit S-box values written as separated bytes in memory. To avoid this waste of memory space, DES S-Boxes are usually compressed by storing the data into four 64-byte tables where the odd-numbered S-Boxes are stored in the high nibbles and the even-numbered S-Boxes are stored in the low nibbles.[16]

The number of key hypotheses generated by this attack against a DES using compressed S-Boxes is shown in Table 2.1 for different numbers of plaintexts. Since the number of faults required with two plaintexts when S-Boxes are compressed is the same as those required with only one plaintext when they are not, one can notice that the attack is more efficient on compressed S-Boxes. As an example, with 512

[15] Note that some figures given in [12] happen to be imprecise. The figures we give in the sequel to this section are borrowed from [92], where they have been more rigorously computed.

[16] While there are a few other ways in which the S-box data could be compressed, we do not consider this as something an attacker must previously know since he can conduct his attack under all possible compression methods until the correct one is found.

Table 2.1 Number of key hypotheses generated by attacking compressed S-Boxes

Plaintexts	Hypotheses per S-box pair	Hypotheses for the first round key	Whole key space
1	25.3	$2^{37.3}$	$2^{45.3}$
2	10.8	$2^{27.4}$	$2^{35.4}$
3	5.29	$2^{19.2}$	$2^{27.2}$
4	3.23	$2^{13.5}$	$2^{21.5}$

Algorithm 2.3: S-box randomization in random order

Input: The original S-box table $S = (s_0, \ldots, s_{63})_{16}$
The input and output masks $R \in (0, \ldots, 63)$ and $r \in (0, \ldots, 15)$
Output: The randomized S-box table $\tilde{S} = (\tilde{s}_0, \ldots, \tilde{s}_{63})_{16}$

for $i \leftarrow 0$ **to** 63 **do**
  $\tilde{s}_i \leftarrow s_{i \oplus R} \oplus r$
end
return $\tilde{S}$

faults, the whole key space is reduced to $2^{35.4}$ keys with compressed S-Boxes instead of $2^{38.7}$ with uncompressed S-Boxes.

Note also that the attack can be further slightly optimized by observing that, based on the faulty ciphertext, it is possible to identify cases where an S-box entry has been used only in one of the last two rounds. In these cases, the S-box entry should not be considered as a candidate for the first round subkey.

A countermeasure for this specific attack is to randomize the order in which the S-Boxes are randomized. This applies to the order in which the S-Boxes are treated, and to the order in which the S-box elements are masked. The data masking can be done as shown in Algorithm 2.3. The counter i is XORed with a random number before being used so the order in which the S-box elements are treated is unknown.

When only the order in which the S-Boxes are treated is randomized, one can still attack by searching for S-box indices that never change the ciphertext when modified. If the same S-box index is repeatedly changed, but the ciphertext never changes after numerous executions with the same plaintext, it can reasonably be assumed that this index value does not represent a key hypothesis for any part of the first round key. Faulting 22 times on the same index with no ciphertext modification ensures a probability of non-detection as small as 0.01. The expected number of indices that are used for at least one S-box in at least one round is 55.5. For each of them four faults are enough on average to show that this index is used. For the others the fault must be repeated 22 times. Thus an average of 409 fault injections are required for each plaintext. Table 2.2 presents the complexity of the resulting exhaustive search for different numbers of plaintexts.

Table 2.2 Number of key hypotheses generated by attacking compressed S-Boxes initialized in random order

Plaintexts	Hypotheses per S-box pair	Hypotheses for the first round key	Whole key space
1	55.5	$2^{46.4}$	$2^{54.4}$
2	48.2	$2^{44.7}$	$2^{52.7}$
3	42.1	$2^{43.2}$	$2^{51.2}$
5	32.5	$2^{40.2}$	$2^{48.2}$
10	18.5	$2^{33.7}$	$2^{41.7}$
15	12.4	$2^{29.1}$	$2^{37.1}$
20	9.71	$2^{26.4}$	$2^{34.4}$

2.3.2.2 Modifying Unknown S-box Values

If the attacker does not know which S-box index he has modified (for instance, if Algorithm 2.3 is actually used), then the previous attack does not work. In that case it is possible to derive an attack based on the exploitation of the event that the modified S-box value has been used only in the fifteenth round. This kind of event is easily identifiable from the normal and faulty ciphertexts by analyzing the differential at the end of the round.[17] Each time this event occurs it can be exploited by simply using the classical DFA method [49]. The probability of such event is 0.0123 when the S-Boxes are not compressed, and 0.0192 when they are. The number of faults needed for this attack is thus significantly larger than for a DFA where the fault directly targets the fifteenth round, but the advantage of the present attack is that it is not prevented by the computation verification countermeasure unless the S-box randomization is performed again between the two DES computations.

[17] It may happen that a corrupted S-box value used only in the fourteenth round is misinterpreted as being used only in the fifteenth round. We refer to [92] for a detailed analysis of these false positives which, as mentioned in [163], do not have much impact on the success of the attack.

Chapter 3
Differential Fault Analysis of DES

Matthieu Rivain

Abstract This chapter reviews the techniques an attacker could employ to conduct a Differential Fault Analysis of the Data Encryption Standard (DES). Biham and Shamir proposed the first such attack on a block cipher based on the differential cryptanalysis of DES. This attack was later extended to permit an attack based on faults in the early or middle rounds of DES.

3.1 Introduction

When Bellcore announced in September 1996 [32] that computational errors could be exploited to efficiently break some widespread public key cryptosystems, the sensitivity of symmetric ciphers to such fault analysis was left as an open issue. One month later, Biham and Shamir filled this gap by introducing *Differential Fault Analysis* (DFA) on the *Data Encryption Standard* (DES) [48]. This attack was subsequently presented in a paper published at Crypto '97 [49], and several papers followed improving DFA and extending it to other ciphers.

This chapter reviews the Differential Fault Analysis of DES. After describing the DES algorithm (Sect. 3.2), we present the original attack, which exploits computational errors occurring in the final rounds of the cipher (Sect. 3.3). We then present a generalization of DFA which can exploit faults occurring in middle rounds of DES (Sect. 3.4). Eventually, a DFA technique based on internal collisions is presented which targets the early rounds of the cipher (Sect. 3.5).

M. Rivain
CryptoExperts, Paris, France
e-mail: matthieu.rivain@gmail.com

M. Joye and M. Tunstall (eds.), *Fault Analysis in Cryptography*,
Information Security and Cryptography, DOI: 10.1007/978-3-642-29656-7_3,

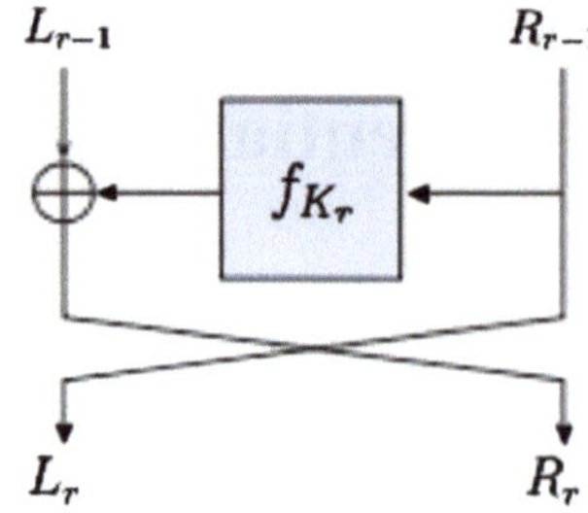

Fig. 3.1 Round transformation in the Feistel scheme

3.2 The Data Encryption Standard

The Data Encryption Standard (DES) is a block cipher that was designed during the 1970s. In May 1973, the US National Bureau of Standards called for proposals for an encryption algorithm that could be used by companies to secure their communications. At that time, no candidate was retained as an acceptable encryption standard. Following a second call in August 1974, IBM submitted a cipher called Lucifer designed by Feistel and his colleagues. After a few modifications of the design by the US National Security Agency, the cipher was selected as an official standard in 1976. The use of DES quickly became internationally widespread even though it was soon identified to be vulnerable to exhaustive key search due to its short key length. To tackle this issue, Triple-DES was proposed as a replacement for DES, which consists of three successive applications of DES with different keys. Although a new Advanced Encryption Standard (AES) was adopted in 2002, Triple-DES is still widely used, for instance, in the smart card protocol EMV.[1]

DES uses a 56-bit key (usually represented on 64 bits including 8 parity check bits) and it operates on 64-bit message blocks. It has an iterative structure applying the same round transformation F 16 times, which is preceded by an initial bit-permutation IP and followed by a final bit-wise permutation FP. Every round transformation is parameterized by a 48-bit round key K_r that is derived from the secret key K through a key schedule process. To summarize, a ciphertext C is computed from a plaintext P as

$$C = \mathsf{FP} \circ \left(\bigcirc_{r=1}^{16} F_{K_r} \right) \circ \mathsf{IP}(P).$$

The round transformation follows a Feistel scheme: the block is split into two 32-bit parts, L (the left half) and R (the right half), and F is defined as

$$F_{K_r}(L, R) = \left(R, L \oplus f_{K_r}(R)\right),$$

where f is a function mapping 32 bits to 32 bits and is parameterized with a 48-bit round key. This structure is illustrated in Fig. 3.1.

[1] Europay MasterCard Visa.

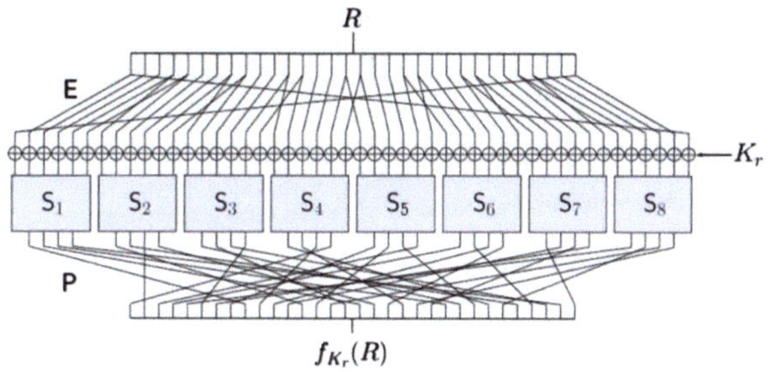

Fig. 3.2 DES f-function

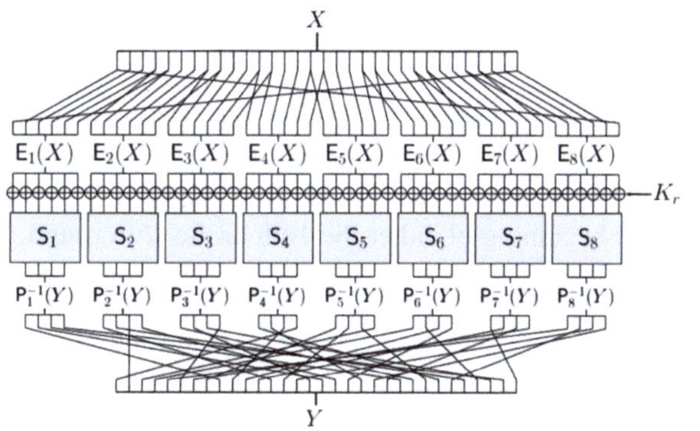

Fig. 3.3 Splitting of the f-function

The DES f-function (see Fig. 3.2) first applies an expansion layer E that expands the 32 input bits into 48 output bits by duplicating 16 of them. The round key is then introduced by bit-wise addition. Afterwards the block is split into eight six-bit blocks, each entering into a different substitution box (S-box) S_i producing a four-bit output. Finally, the 32 bits from the eight S-box outputs are permuted through a bit-wise permutation P which yields the 32-bit output block.

Notation: For the sake of simplicity and without loss of generality, we shall omit the initial and the final permutations in the rest of the chapter. Namely, we will consider $C = \bigcirc_{r=1}^{16} F_{K_r}(P)$. We shall also denote by (L_r, R_r) the value of the DES internal state at the end of the rth round. Specifically, we have

$$(L_r, R_r) = F_{K_r}(L_{r-1}, R_{r-1}),$$

with $(L_0, R_0) = P$ and $C = (L_{16}, R_{16})$.

We shall further denote by E_i, P_i^{-1} and $K_{r,i}$, the ith six-bit coordinate of the expansion function E, the ith four-bit coordinate of the inverse permutation P^{-1}, and ith six-bit coordinate of the round key K_r respectively. As illustrated on Fig. 3.3, these notations allow us to rewrite an equation $Y = f_{K_r}(X)$ with eight independent equations:

$$\mathsf{P}_i^{-1}(Y) = \mathsf{S}_i(\mathsf{E}_i(X) \oplus K_{r,i}),$$

for $i \in \{1, \ldots, 8\}$.

3.3 Basic Attack

Differential Fault Analysis exploits errors occurring during the encryption of several plaintexts under the same secret key to recover the latter. The attacker is assumed to observe several pairs of ciphertexts $(C, C^{\lightning})$, each corresponding to a plaintext P which is correctly encrypted (yielding C) and erroneously encrypted (yielding $C^{\lightning}$). DFA exploits the difference between C and $C^{\lightning}$ in order to infer information on the secret key.

The original attack described by Biham and Shamir in [49] assumes that one bit of the right half of the DES internal state is flipped at a random position during some round in the faulty encryption. We detail hereafter this attack when the fault occurs at the beginning of either the 16th or the 15th round.

Notation: In the following, $L_r^{\lightning}$ and $R_r^{\lightning}$ will respectively denote the corrupted value of the left part L_r and the right part R_r at the end of the rth round and $C^{\lightning} = (L_{16}^{\lightning}, R_{16}^{\lightning})$ will denote the faulty ciphertext. We shall further denote by $\varepsilon \in \{0, 1\}^{32}$ the induced error. For instance, if the error is induced in the right part of the DES internal state at the end of the rth round, we have $R_r^{\lightning} = R_r \oplus \varepsilon$. Eventually, $(\Delta L_r, \Delta R_r)$ shall denote the XOR-difference between the correct and faulty DES internal states at the end of the rth round and Δf_r shall denote the XOR-difference between the correct and faulty outputs of the f-function at round r.

3.3.1 *Attack on the 16th Round*

Let us assume that some bit of R_{15} is flipped at the beginning of the 16th round, which yields $R_{15}^{\lightning} = R_{15} \oplus \varepsilon$. The error propagation is represented in Fig. 3.4. Then, the XOR-difference between R_{16} and $R_{16}^{\lightning}$, satisfies

$$\Delta R_{16} = f_{K_{16}}(L_{16}) \oplus f_{K_{16}}(L_{16}^{\lightning}). \tag{3.1}$$

This equation is used by the attacker to find the value of the last round key K_{16}. With a few pairs $(C, C^{\lightning})$ can be uniquely determined.

Let us now explain how to solve such an equation. As illustrated in Fig. 3.3, the structure of the f-function implies that (3.1) holds for every S-box independently. More precisely, every six-bit coordinate $K_{16,i}$ of the round key enters in separate S-box and satisfies

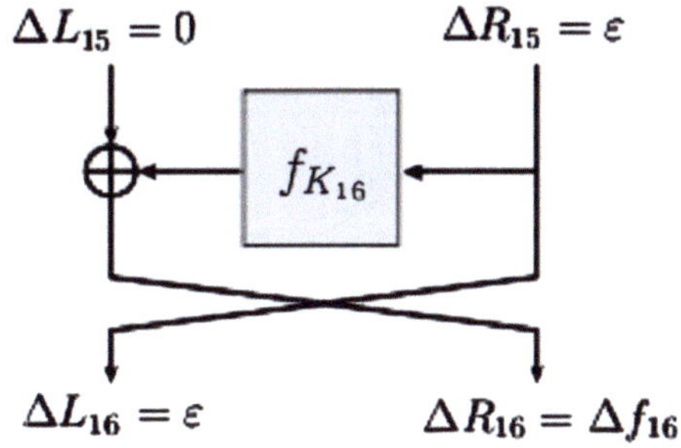

Fig. 3.4 Error propagation on round 16

$$\mathsf{P}_i^{-1}(\Delta R_{16}) = \mathsf{S}_i(\mathsf{E}_i(L_{16}) \oplus K_{16,i}) \oplus \mathsf{S}_i(\mathsf{E}_i(L_{16}^{\lightning}) \oplus K_{16,i}). \tag{3.2}$$

For every $i \in \{1, \ldots, 8\}$, the attacker then tests all the possible values in $\{0, 1\}^6$ for $K_{16,i}$ and discards those which are not consistent with respect to (3.2). Note that if we have

$$\mathsf{E}_i(L_{16}) = \mathsf{E}_i(L_{16}^{\lightning}) \quad \Leftrightarrow \quad \mathsf{E}_i(\varepsilon) = 0,$$

namely if the error vector ε does not affect the bits entering in the ith S-box, then the XOR-difference in (3.2) is 0. In that case, every value in $\{0, 1\}^6$ for $K_{16,i}$ is consistent with respect to (3.2) and no information is inferred about $K_{16,i}$. On the other hand, if the error affects the bits entering the ith S-box, we shall say that the ith S-box is *active*, only a few values for $K_{16,i}$ (four on average) are consistent with respect to (3.2). This way, the subkey $K_{16,i}$ is recovered with a few pairs $(C, C^{\lightning})$ for which the ith S-box is active.

Depending on the flipped bit, one or several S-boxes may be active. Since every single input bit of the DES f-function enters one or two S-boxes, a single-bit fault in the right half of the DES internal state activates one or two S-boxes in the following round (see illustration in Fig. 3.5). Note that this attack applies whatever the induced error ε, as long as it only affects R_{15}. In that case, the higher the number of flipped bits (i.e. the higher the Hamming weight of ε), the higher the number of active S-boxes (on average), and the lower the number of pairs $(C, C^{\lightning})$ required to recover the full round key.

3.3.2 *Attack on the 15th Round*

Let us now assume that the fault corrupts the value R_{14} at the beginning of the 15th round such that $R_{14}^{\lightning} = R_{14} \oplus \varepsilon$. The error propagation is represented in Fig. 3.6. In that case, we have

$$\Delta R_{16} = f_{K_{16}}(L_{16}) \oplus f_{K_{16}}(L_{16}^{\lightning}) \oplus \varepsilon. \tag{3.3}$$

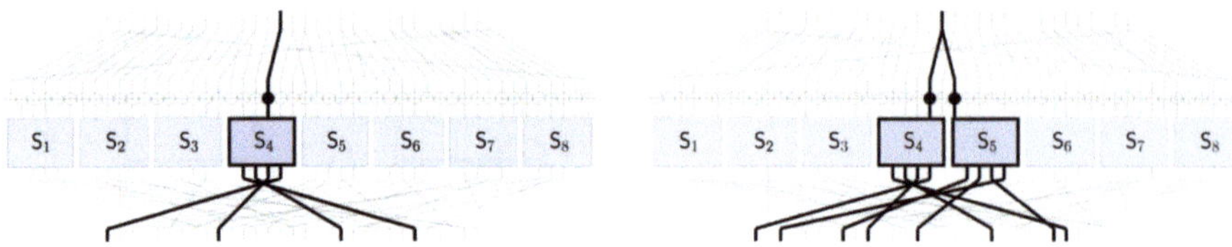

Fig. 3.5 Single-bit error propagation through the f-function with one active S-box (on the *left*) and two active S-boxes (on the *right*)

Fig. 3.6 Error propagation on round 15

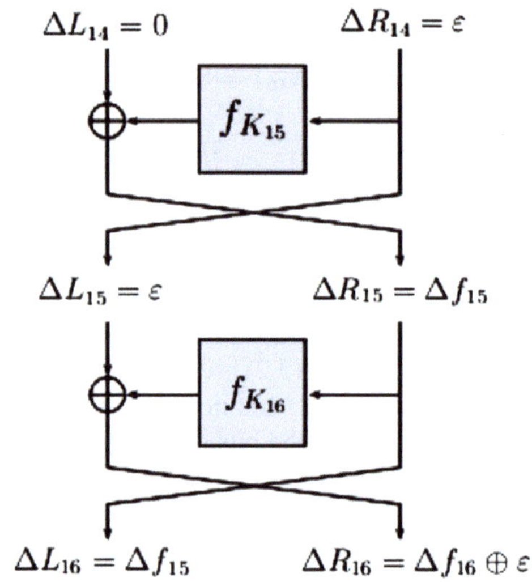

Here, the round key K_{16} is not the only unknown value of the equation: the attacker does not a priori know the error vector ε. So the attacker has first to determine ε or at least to isolate it from among a small set of possible values. For such a purpose, he shall use the XOR-difference ΔL_{16} which satisfies

$$\Delta L_{16} = f_{K_{15}}(R_{14}) \oplus f_{K_{15}}(R_{14} \oplus \varepsilon). \tag{3.4}$$

From this equation, one can deduce the active S-boxes in the 15th round. If two S-boxes are active, then two solutions for ε are possible since every pair of S-boxes share at most two input bits (see Fig. 3.2). If a single S-box is active then two solutions are also possible that correspond to the input bits of the active S-box which does not enter another S-box.[2]

---

[2] Note that a one-bit differential in the input of a DES S-box cannot produce a zero differential in the output.

Once a few candidates for ε have been selected, (3.3) is involved to distinguish K_{16}. As in the previous attack, this is done for each S-box independently. From (3.3), we have

$$\mathsf{P}_i^{-1}(\Delta R_{16} \oplus \varepsilon) = \mathsf{S}_i(\mathsf{E}_i(L_{16}) \oplus K_{16,i}) \oplus \mathsf{S}_i(\mathsf{E}_i(L_{16}^{\not{}}) \oplus K_{16,i}). \tag{3.5}$$

This time, only the values in $\{0, 1\}^6$ for $K_{16,i}$ which do not satisfy (3.5) for any of the selected ε are discarded. Since for most ε (28 over 32) we have $\mathsf{P}_i^{-1}(\Delta R_{16} \oplus \varepsilon) = \mathsf{P}_i^{-1}(\Delta R_{16})$, the discrimination of $K_{16,i}$ is only slightly less efficient than in the previous attack. On the other hand, more S-boxes are active in the 16th round (since $\Delta L_{16} = \Delta f_{15}$ often has a Hamming weight greater than 1), which speeds up the overall discrimination of K_{16}.

In contrast to the attack on the 16th round, this one does not work for every ε. Indeed, the assumption that a single bit is flipped (or only a few bits are flipped) is necessary for the attacker to be able to isolate ε from among a small set of candidates and hence to obtain an efficient discrimination from (3.5).

3.3.3 Attack Results

The attack described in [49] assumes that one bit of the right half of the DES internal state is flipped at a random position during some random round. The attack first identifies whether the fault occurs in one of the last rounds by checking ΔL_{16}. For instance, if ΔL_{16} has a single bit set to 1, then it is likely that the fault occurred in the 16th round (since in that case $\Delta L_{16} = \varepsilon$). Also, if the bits of ΔL_{16} that equal 1 only output from one or two S-boxes, then it is likely that the fault occurred in the 15th round according to (3.4). Once the round where the fault occurred has been identified, the attacker can distinguish between the wrong key values as described above. In [49], it is also argued that the faults occurring in the 14th round can be exploited using counting methods, but very few details are given. According to the authors, their attack enables one to recover the full last round key using between 50 and 200 ciphertexts pairs when the fault occurs at a random position in the right half of the DES internal state in a random round. If the attacker is able to choose the exact fault position, then three ciphertexts pair are sufficient.

These results demonstrate the sensitivity of DES to fault analysis. It appears that one must protect at least the last three or four rounds of DES against DFA, for instance by computing them twice and comparing the outputs. However, some questions remain open: First, to what extent can DFA exploit faults occurring in the middle rounds of DES? Also, is DFA tolerant to more general fault models where several bits of the DES internal state are flipped? These issues are addressed in the next section.

3.4 Generalization and Extension to Middle Rounds

We present in this section a generalization of Differential Fault Analysis on DES that has been introduced by Rivain in [345]. The proposed attack exploits faults which occur in the middle rounds of DES and which corrupt possibly more that one bit of the internal state.

3.4.1 Generalized DFA: Basic Principle

The basic principle of the generalized DFA starts from the following relation, which holds for every target round and for every error vector ε

$$\Delta R_{16} = f_{K_{16}}(L_{16}) \oplus f_{K_{16}}(L_{16}^{\not{z}}) \oplus \Delta L_{15}. \qquad (3.6)$$

The attack presented in the previous section uses the fact that, when a fault occurs in one of the last rounds of the cipher, the adversary knows ΔL_{15} or a few candidates for it. For instance, $\Delta L_{15} = 0$ and $\Delta L_{15} = \varepsilon$ for the attacks on the 16th and 15th rounds respectively. Therefore (3.6) yields one or a few equations per subkey $K_{16,i}$, which enables an efficient recovery of the whole K_{16}. However, an attacker may not be able to retrieve ΔL_{15} when the fault occurs in a previous round and/or when it corrupts several bits of the DES internal state. In those situations, the attack presented in the previous section does not apply anymore.

In fact, one does not necessarily need to recover ΔL_{15} to mount an attack. The most elementary requirement is that the statistical distribution of ΔL_{15} be significantly biased. This distribution actually depends on two main factors: the average number of bits that are flipped by the fault injection and the number of times the f-function is crossed from the fault location to L_{15}. For instance, if an error ε is induced in the left half of the DES internal state at the end of the 13th round, then we have $\Delta L_{15} = \Delta L_{13} = \varepsilon$. If the error only corrupts a few bits (i.e. ε has a small Hamming weight), then the distribution of ΔL_{15} is strongly biased from the uniformity. More generally, a fault injection in the left half of the DES internal state skips one round before propagating through the function f. Besides, the error propagation path from L_r to L_{15} passes through the function f only once for $r = 12$, twice for $r = 11$ and so on (see Fig. 3.7). This is quite low considering the slow diffusion of the DES f-function. As a result, a fault induced in L_r may produce a differential ΔL_{15} with a distribution that is significantly biased. This bias enables the construction of wrong key distinguishers.

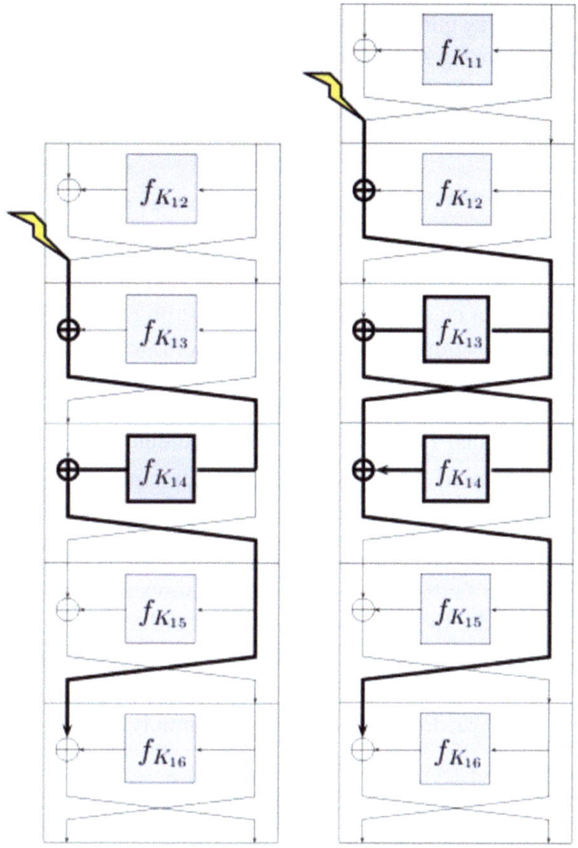

Fig. 3.7 Error propagation paths from L_r to L_{15} for $r = 12$ and $r = 11$

3.4.2 Wrong Key Distinguishers

As in the previous attacks, every subkey $K_{16,i}$ is retrieved independently. For every $i \in \{1, \ldots, 8\}$, (3.6) is equivalent to

$$\mathsf{P}_i^{-1}(\Delta R_{16}) = \mathsf{S}_i\big(\mathsf{E}_i(L_{16}) \oplus K_{16,i}\big) \oplus \mathsf{S}_i\big(\mathsf{E}_i(L_{16}^{\lightning}) \oplus K_{16,i}\big) \oplus \mathsf{P}_i^{-1}(\Delta L_{15}). \quad (3.7)$$

Let us define g_i as the function that predicts $\mathsf{P}_i^{-1}(\Delta L_{15})$ according to a pair $(C, C^{\lightning})$ and a guess k on the value of $K_{16,i}$:

$$g_i((C, C^{\lightning}), k) = \mathsf{S}_i\big(\mathsf{E}_i(L_{16}) \oplus k\big) \oplus \mathsf{S}_i\big(\mathsf{E}_i(L_{16}^{\lightning}) \oplus k\big) \oplus \mathsf{P}_i^{-1}(\Delta R_{16}).$$

From (3.7), we have $g_i(C, C^{\lightning}, k) = \mathsf{P}_i^{-1}(\Delta L_{15})$ for the correct key guess $k = K_{16,i}$. On the other hand, $g_i(C, C^{\lightning}, k)$ can be assumed to have a uniform distribution for a wrong key guess. This is a classical assumption in block cipher cryptanalysis known as the wrong key assumption.

Let $p_i(\delta)$ denote the probability that $\mathsf{P}_i^{-1}(\Delta L_{15})$ equals δ for every $i \in \{1, \ldots, 8\}$ and for every $\delta \in \{0, 1\}^4$. According to the wrong key assumption, we have

$$\Pr\left[g_i((C, C^{\lightning}, k) = \delta\right] = \begin{cases} p_i(\delta) & \text{if } k = K_{16,i}, \\ \frac{1}{16} & \text{otherwise.} \end{cases} \tag{3.8}$$

Provided that the distribution $p_i(\cdot)$ is significantly biased, (3.8) clearly shows that $K_{16,i}$ can be distinguished from a wrong guess. Such a distinction can be made in practice using classical estimators from statistics such as the likelihood or the squared Euclidean imbalance (SEI). The corresponding distinguishers are defined hereafter. Both of them compute a value $d(k)$ for every key candidate k which is expected to be maximal for the correct key candidate $k = K_{16,i}$. These distinguishers take as input a set of N pairs of correct-faulty ciphertexts $(C_n, C_n^{\lightning}, 1 \leq n \leq N$. The choice of the distinguisher to use depends on the attacker's knowledge of the fault model.

- **Likelihood distinguisher** The attacker is assumed to have an exact knowledge of the fault model; namely he knows the distribution of ε. In that case, he can estimate the distribution $p_i(\cdot)$ in order to use a maximum likelihood approach. The likelihood of a key candidate k is defined as the product of the probabilities $p_i\big(g_i(C_n, C_n^{\lightning}, k)\big)$ for $n = 1, \ldots, N$. For practical reasons, one usually computes the logarithm of the likelihood; namely $d(k)$ is defined as

$$d(k) = \sum_{n=1}^{N} \log\big(p_i\big(g_i(C_n, C_n^{\lightning}, k)\big)\big).$$

- **SEI distinguisher** The attacker does not have precise knowledge of the fault model and is hence not able to estimate the distribution $p_i(\cdot)$. In that case, an alternative strategy is to look for the strongest bias in the distribution of $g_i(C_n, C_n^{\lightning}, k)$. This is done by computing the squared Euclidean distance to the uniform distribution (i.e. the SEI); namely $d(k)$ is defined as

$$d(k) = \sum_{\delta \in \{0,1\}^4} \left(\frac{\#\{n;\, g_i(C_n, C_n^{\lightning}, k) = \delta\}}{N} - \frac{1}{16} \right)^2.$$

Remark 3.1 If an attacker can induce different kinds of faults (i.e. following different fault models), then a distinct distribution $p_i(\cdot)$ is induced per fault model. In that case, the SEI of $p_i(\cdot)$ shall be estimated for every distinct fault model independently. The SEI distinguisher is then defined as the sum of the SEIs for the different models.

Table 3.1 Number of faults to recover the 16th round key with a 99 % success rate

Round	Distinguisher	Bit error		Byte error	
		Chosen pos.	Random pos.	Chosen pos.	Random pos.
12	Likelihood	7	11	9	17
	SEI	14	12	17	21
11	Likelihood	11	44	210	460
	SEI	30	71	500	820
10	Likelihood	290	1500	13400	18500
	SEI	940	2700	26400	23400
9	Likelihood	3.4×10^5	2.2×10^7	$>10^8$	$>10^8$
	SEI	1.4×10^6	$>10^8$	$>10^8$	$>10^8$

3.4.3 Attack Results

The results of several attack simulations are reported in [345]. The attacker is assumed to be able to inject some fault in the left half of the DES internal state L_r at the end of some round $r \in \{9, 10, 11, 12\}$. Several fault models are considered: either a single bit is flipped or one byte is switched to a random value, and the fault position is either random (among the 32 bit positions or the four byte positions in L_r) or chosen by the attacker. In the latter case, the bit positions are chosen from among the ones entering a single S-box in order to slow down the error propagation and maximize the bias in the distribution of ΔL_{15} while the byte positions are all chosen successively. For every round number and every fault model, the likelihood distinguisher and the SEI distinguisher are both applied. For the likelihood distinguisher, the distributions $(p_i(\cdot))_i$ have been empirically computed based on several correct and faulty encryptions of random plaintexts under random keys.[3] Table 3.1 summarizes the numbers of correct-faulty ciphertexts pairs required for a 99% success rate in recovering the whole last round key.

The attacks on the 11th and 12th rounds are very efficient: less than 25 faults are sufficient on the 12th round, while on the 11th round less than 100 faults are sufficient in a bit error model and less than 1×10^3 faults are sufficient in a byte error model. On the tenth round, the attacks are still fairly efficient: the best attack (chosen position bit error model, likelihood distinguisher) requires 290 faults whereas the least efficient attack (chosen position byte error model, SEI distinguisher) requires 2.64×10^4 faults. It is on round 9 that the attacks become quite costly since the most efficient attack in the bit error model (chosen position, likelihood distinguisher) requires around 3.4×10^5 faults and all the attacks in the byte error model require more than 10^8 faults.

[3] Note that the $(p_i(\cdot))_i$ distributions are almost independent of the secret key (this fact is argued in [47]).

These results show that one should at least protect the last six rounds of DES against DFA. To resist a powerful adversary (precise fault model, high number of correct-faulty ciphertext pairs), it seems prudent to protect the last eight rounds.

3.4.4 Extension to Early Rounds Based on a Decryption Oracle

If an attacker has access to a decryption oracle then the attacks presented so far can be employed to exploit errors occurring in the early rounds of the cipher. In fact, the attacker may obtain a faulty ciphertext $C^{\lightning}$ from a plaintext P by inducing a fault at the end of the first round. The plaintext P can then be viewed as the faulty result of a decryption of $C^{\lightning}$ for which a fault has been induced at the beginning of the last round. The attacker then asks for the decryption of $C^{\lightning}$ that provides him with a plaintext $P^{\lightning}$, P). The pair $P^{\lightning}$ thus constitutes a pair of correct-faulty results of the decryption algorithm with respect to an error induced at the beginning of the last round. According to this principle, any fault attack on round r of an encryption can be transposed to a fault attack on round $16 - r$ of a decryption. For instance, the attack presented in this section, which exploits faults occurring in rounds $r \geq 9$, can be applied to exploit faults on rounds $r \leq 7$, provided that the attacker has access to a decryption oracle. In that case, the same number of rounds should be protected at the beginning and at the end of the cipher in order to obtain a homogenous security level.

When no decryption oracle is available, it is still possible to attack the early rounds of DES. This is the subject of the next section.

3.5 Attack on Early Rounds Based on Internal Collisions

In the previous section, we presented a DFA technique able to exploit faults occurring in the middle rounds of DES. We now present an attack against early rounds of DES that was introduced by Hemme in [178].

3.5.1 Notations and Definitions

In the following we shall denote by (P, P') a pair of plaintexts, and by (L_r, R_r) and (L'_r, R'_r) the underlying intermediate values of the DES internal state. We shall also denote by ΔL_r (or ΔR_r) the XOR-difference between L_r and L'_r (or R_r and R'_r). Note that this notation differs from the previous sections where it is used for the XOR-difference between L_r (or R_r) and its faulty counterpart $L_r^{\lightning}$ (or $R_r^{\lightning}$) for the same plaintext.

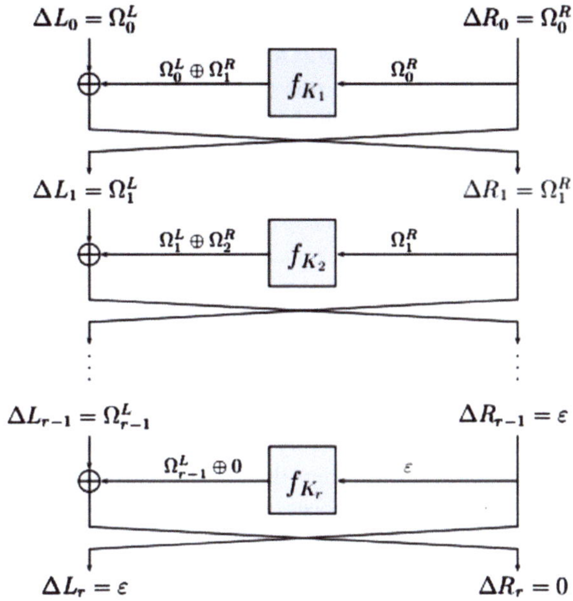

Fig. 3.8 Differential path of a correct pair with respect to a r-round ε-characteristic

We will call an *r-round characteristic*[4] an $(r+1)$-tuple $\Omega = (\Omega_0, \Omega_1, \ldots, \Omega_r)$ where $\Omega_i = (\Omega_i^L, \Omega_i^R) \in \{0,1\}^{32} \times \{0,1\}^{32}$ and $\Omega_{i+1}^L = \Omega_i^R$ for every $i \geq 0$. For $\varepsilon \in \{0,1\}^{32}$, an r-round characteristic Ω is called an r-round ε-characteristic if $(\Omega_r^L, \Omega_r^R) = (\varepsilon, 0)$. A correct pair with respect to a r-round characteristic Ω is a pair of plaintexts (P, P') such that $(\Omega_i^L, \Omega_i^R) = (\Delta L_i, \Delta R_i)$ for every $i \in \{1, \ldots, r\}$. Namely, a correct pair (P, P') is such that at round $i \leq r$ the XOR-differences in input and output of the f-function are Ω_{i-1}^R and $\Omega_{i-1}^L \oplus \Omega_i^R$ respectively. For instance, Fig. 3.8 represents the differential path of a correct pair with respect to an r-round ε-characteristic. Eventually, the *probability p_Ω of a characteristic Ω* is the probability that a pair of plaintexts $(P, P \oplus \Omega_0)$ is a correct pair with respect to Ω.

3.5.2 Attack Description

The attack assumes that the adversary can ask for the encryption of chosen plaintexts under a secret key K which he aims at recovering. The adversary is further able to induce some fault in the left half of the DES internal state at the end of some round r. More precisely, L_r is replaced by $L_r \oplus \varepsilon$ in the encryption of P which produces a faulty ciphertext $C^{\lightning}$. Note that, equivalently, the adversary could induce a fault in R_{r+1} or in the output of the f-function at round $r+1$. The error vector ε is randomly distributed among a set $\mathscr{E} \subseteq \{0,1\}^{32}$. For instance, in a single-bit error model, $\mathscr{E}$ is the set of 32-bit words whose Hamming weights equal 1. The attacker is

[4] This definition slightly differs from the definition in [178].

assumed to know the set of possible errors $\mathcal{E}$ as well as an associated set of r-round ε-characteristics $\{\Omega_\varepsilon; \varepsilon \in \mathcal{E}\}$.

Once a faulty ciphertext $C^{\lightning}$ is obtained from a plaintext P, the attacker searches for a plaintext P' such that (P, P') is a right pair with respect to an r-round ε-characteristic Ω_ε (where ε is the induced fault). Note that a correct pair with respect to Ω_ε satisfies

$$(L'_r, R'_r) = (L_r \oplus \varepsilon, R_r) = (L_r^{\lightning}, R_r^{\lightning}),$$

and consequently

$$C' = \bigcirc_{i=r+1}^{16} F_{K_i}(L_r \oplus \varepsilon, R_r) = C^{\lightning}.$$

Therefore, to determine a plaintext P' such that (P, P') is a right pair with respect to Ω_ε while the error vector ε is a priori unknown, the attacker asks for the encryption of $P' = P \oplus \Omega_{\hat{\varepsilon},0}$ for every $\hat{\varepsilon} \in \mathcal{E}$. If one of the obtained ciphertexts C' equals $C^{\lightning}$, then the attacker considers that (P, P') may be a correct pair with respect to $\Omega_{\hat{\varepsilon}}$. The probability that this process successfully outputs a correct pair is

$$p_{\text{suc}} = \sum_{\hat{\varepsilon} \in \mathcal{E}} \Pr[\varepsilon = \hat{\varepsilon}] p_{\Omega_{\hat{\varepsilon}}}.$$

On the other hand, a pair (P, P') such that $C' = C^{\lightning}$ may be returned which is not a correct pair. Such a wrong pair can result from a plaintext $P' = P \oplus \Omega_{\hat{\varepsilon},0}$ such that $(\Delta L_r, \Delta R_r) = (\varepsilon, 0)$ but $\hat{\varepsilon} \neq \varepsilon$ or such that $(\Delta L_r, \Delta R_r) = (\hat{\varepsilon}, 0) = (\varepsilon, 0)$ but $(\Delta L_i, \Delta R_i) \neq \Omega_{\hat{\varepsilon},i}$ for some i. Such a wrong pair occurs with a probability

$$p_{\text{err}} = \sum_{\hat{\varepsilon} \in \mathcal{E}} \Pr[\mathsf{DES}_K(P \oplus \Omega_{\hat{\varepsilon},0}) = C^{\lightning}] - p_{\text{suc}}.$$

However, if the characteristics are well chosen, namely if they are chosen with high probabilities p_{Ω_ε}, then the error probability is negligible compared to the success probability (the choice of the characteristics is addressed later in Sect. 3.5.4).

Once a correct pair (P, P') with respect to an r-round ε-characteristic Ω_ε has been obtained, it is used to infer some information on the first round key. As illustrated in Fig. 3.8, such pair satisfies

$$f_{K_1}(R_0) \oplus f_{K_1}(R'_0) = \Omega^L_{\varepsilon,0} \oplus \Omega^R_{\varepsilon,1}. \tag{3.9}$$

This equation enables an attacker to distinguish the value of the first round key K_1. In practice, since wrong pairs may occur, one does not discard key guesses as in the original DFA presented in Sect. 3.3 but rather uses a counting strategy. More precisely, every subkey $K_{1,i}$ is recovered separately by updating 64 counters $c(k)$, $k \in \{0, 1\}^6$. For every selected pair (P, P') and associated r-round ε-characteristic Ω_ε, the attacker tests whether a guess k for $K_{1,i}$ is consistent with respect to (3.9). Namely, he tests whether k satisfies

$$\mathsf{S}_i(\mathsf{E}_i(R_0) \oplus k) \oplus \mathsf{S}_i(\mathsf{E}_i(R_0') \oplus k) = \mathsf{P}_i^{-1}(\Omega_{\varepsilon,0}^L \oplus \Omega_{\varepsilon,1}^R). \tag{3.10}$$

If the previous equation holds, then $c(k)$ is incremented. Note that if S_i is not active (i.e. $\mathsf{E}_i(R_0) = \mathsf{E}_i(R_0')$), all the counters are incremented and no information about $K_{1,i}$ is inferred (one may equivalently decide to increment no counter when S_i is not active). Therefore, a correct pair is useful for discriminating $K_{1,i}$ only if S_i is active, which occurs if and only if $\mathsf{E}_i(\Omega_{\varepsilon,0}^R) \neq 0$. Consequently, the ε-characteristics must be chosen such that $\Omega_{\varepsilon,0}^R$ is not zero, and, more generally, $\Omega_{\varepsilon,0}^R$ should activate the most S-boxes possible.

Let us now assume that S_i is active in the first round. If (P, P') is a correct pair, then a few counters are incremented (four on average) among which one corresponds to the value of $K_{1,i}$. If (P, P') is a wrong pair, then a few counters are incremented that correspond to random guesses. Therefore, the correct guess is counted more frequently on average. The maximal counter is hence expected to be for $k = K_{1,i}$ once enough pairs have been analyzed. Assuming that the error probability p_{err} is negligible compared to the success probability p_{suc}, the number of correct-faulty DES encryptions required to get an important success rate in recovering K_1 depends on the number of correct pairs, i.e. on the success probability p_{suc}. The next section presents a way to increase this probability.

3.5.3 Attack Improvement

A possible improvement of the attack is to use several r-round ε-characteristics per error vector $\varepsilon \in \mathscr{E}$. For every ε, let $\mathscr{C}_\varepsilon$ denote a set of several r-round ε-characteristics. Given a ciphertext $C^{\lightning}$ obtained from the faulty encryption of a plaintext P, the attacker tries to encrypt $P' = P \oplus \Omega_{\hat{\varepsilon},0}$ for every $\Omega_{\hat{\varepsilon}} \in \mathscr{C}_{\hat{\varepsilon}}$, for every $\hat{\varepsilon} \in \mathscr{E}$. The resulting probability of getting a correct pair becomes

$$p_{\text{suc}} = \sum_{\hat{\varepsilon} \in \mathscr{E}} \Pr[\varepsilon = \hat{\varepsilon}] \sum_{\Omega_{\hat{\varepsilon}} \in \mathscr{C}_{\hat{\varepsilon}}} p_{\Omega_{\hat{\varepsilon}}},$$

which is clearly higher than in the original attack. However, the higher the number of characteristics per $\varepsilon \in \mathscr{E}$, the higher the number of correct DES encryptions that are required per faulty encryption. It thus appears that, for any given attack success rate, there is a trade-off between the number of faulty DES encryptions and the number of correct DES encryptions (cf. experiments in Sect. 3.5.5).

3.5.4 Choosing Good Characteristics

In order to maximize the efficiency of the attack described above, it is important to choose the r-round ε-characteristics with probabilities as high as possible. It is worth noting that the probability p_{Ω_ε} of a characteristic Ω_ε is the probability that for a random secret key K and for a random plaintext P, the pair $(P, P \oplus \Omega_{\varepsilon,0})$ is a correct pair with respect to Ω_ε. In the context of a fault attack, the secret key is unknown and the probability that a pair is a correct pair for a given secret key may differ from p_{Ω_ε}. However, according to [47], such a difference is only slight and p_{Ω_ε} is a good approximation of the real probability.

Let $\Omega = (\Omega_0, \Omega_1, \ldots, \Omega_r)$ be an r-round characteristic. The probability p_Ω of Ω satisfies

$$p_\Omega = \prod_{i=1}^{r} p_\Omega^{(i)},$$

where $p_\Omega^{(i)} = \Pr[(\Delta L_i, \Delta R_i) = \Omega_i | (\Delta L_{i-1}, \Delta R_{i-1}) = \Omega_{i-1}]$ is the probability that Ω propagates well through the ith round. The probabilities $p_\Omega^{(i)}$ can be in turn expressed as products of probabilities $p_\Omega^{(i)} = \prod_{i=1}^{8} p_\Omega^{(ij)}$ where $p_\Omega^{(ij)}$ is the probability that Ω propagates well through the jth S-box in the ith round which satisfies:

$$p_\Omega^{(ij)} = \Pr\left[\mathsf{S}_j(X) \oplus \mathsf{S}_j(X \oplus \mathsf{E}_j(\Omega_{i-1}^R)) = \mathsf{P}_j^{-1}(\Omega_{i-1}^L \oplus \Omega_i^R)\right],$$

where X is a uniform random variable over $\{0, 1\}^6$. This probability is given by the *XOR-difference distribution table* of S_j for the input differential $\mathsf{E}_j(\Omega_{i-1}^R)$ and the output differential $\mathsf{P}_j^{-1}(\Omega_{i-1}^L \oplus \Omega_i^R)$ (see [47]).

Based on these equations, one can easily compute the probability of a characteristic. It is even possible to calculate the r-round ε-characteristics that have the highest probabilities given r and ε using the search algorithm of Matsui [275]. The results of such a search are given in [178] for 1-, 2- and 3-round ε-characteristics for the set $\mathcal{E}_{1\text{-bit}}$ of error vectors whose Hamming weights equal 1 (i.e. corresponding to a single-bit fault model).

3.5.5 Attack Results

The results of several attack simulations are reported in [178], where the attacker is assumed to be able to induce a single-bit fault at a random position in L_r (or equivalently R_{r+1}). Table 3.2 summarizes these results which are given in terms of average number of found pairs (right and possibly wrong) and average number of extracted key bits (on average) for different numbers of characteristics and faulty/correct DES encryptions. For a given $\varepsilon \in \mathcal{E}_{1\text{-bit}}$, the r-round ε-characteristic Ω_ε is chosen such that

Table 3.2 Attack results for several simulations

r	Faulty + correct DES encryptions	No. characteristics	No. found pairs	No. extracted key bits	No. simulations
1	$100 + 1600$	16	9.16	17.46	10000
	$500 + 8000$	16	45.95	41.03	10000
	$500 + 16000$	32	55.43	46.74	10000
	$1000 + 32000$	32	110.92	47.94	10000
	$1500 + 48000$	32	166.46	48.00	10000
2	$5 \times 10^3 + 4 \times 10^4$	8	6.20	17.65	1000
	$1 \times 10^4 + 1.6 \times 10^5$	16	17.67	36.36	1000
	$5 \times 10^4 + 8 \times 10^5$	16	88.88	45.59	1000
	$5 \times 10^5 + 1.6 \times 10^7$	32	888.11	47.86	1000
	$1 \times 10^6 + 3.2 \times 10^7$	32	1779.18	48.00	1000
3	$5 \times 10^6 + 7 \times 10^7$	14	6.25	34.52	100
	$1 \times 10^7 + 1.4 \times 10^8$	14	13.40	42.42	100
	$5 \times 10^7 + 1 \times 10^9$	20	67.25	47.30	20

it has the highest probability p_{Ω_ε}. The attacks are also carried out with some subsets of these 32 characteristics. In that case, each subset is composed of the characteristics having the highest probabilities among the 32.

Table 3.3 summarizes the simulation results for the improved attack. The used characteristics are chosen as those having the highest probabilities among all the r-round ε-characteristics with $\varepsilon \in \mathscr{E}_{1\text{-bit}}$. Therefore, several characteristics are used for some ε's, while no characteristic is used for other ε's. Furthermore, the choice of $\Omega^R_{\varepsilon,0}$ is made taking care that the most S-boxes possible are active in the first round. From Table 3.3, we clearly see that the improved attack requires less faulty DES encryptions than the original attack. The total number of DES encryptions (faulty and correct) is also slightly lower.

These results show that about 4×10^4 (or 1×10^6, 1×10^8) DES encryptions are required for the recovery of the whole round key K_1 when L_1 (or L_2, L_3) is corrupted. Since a fault injection in L_r is equivalent to a fault injection in R_{r+1}, these attacks may be performed by corrupting the right half of DES's internal state at the beginning of round $r + 2$. In other words, one can retrieve the first round key in 4×10^4 (or 1×10^6, 1×10^8) DES encryptions by attacking the third (or fourth, or fifth) round.

These attacks may also apply with more general fault models, e.g. where one byte or one S-box output is randomly corrupted. In [178] it is argued that such attacks shall only use the ε-characteristics for single-bit errors $\varepsilon \in \mathscr{E}_{1\text{-bit}}$ since they have the highest probabilities p_{Ω_ε}. Therefore, we expect an increase in the number of DES encryptions roughly by a factor $1/p$ where p is the probability that a single-bit error occurs. For instance, in a byte error model $p = 8/256$ (and $1/p = 32$), and in an S-box error model $p = 4/16$ (and $1/p = 4$).

Table 3.3 Improved attack results for several simulations

r	Faulty + correct DES encryptions	No. characteristics	No. found Pairs	No. extracted key bits	No. simulations
1	$10 + 990$	99	3.96	10.55	10000
	$200 + 12800$	64	58.62	45.84	10000
	$400 + 40000$	100	94.21	48.00	10000
2	$100 + 28100$	281	2.09	10.23	1000
	$500 + 140500$	281	11.07	31.25	1000
	$1000 + 160000$	160	12.67	36.18	1000
	$2000 + 278000$	139	15.67	41.34	1000
	$5000 + 555000$	111	26.42	46.83	1000
	$10000 + 1110000$	111	53.42	47.95	1000
3	$1 \times 10^5 + 1.62 \times 10^7$	162	1.56	13.82	100
	$5 \times 10^5 + 8.10 \times 10^7$	162	6.52	36.21	100
	$1 \times 10^6 + 1.62 \times 10^8$	162	13.96	45.78	100
	$5 \times 10^6 + 8.10 \times 10^8$	162	66.68	48.00	100

According to these results, one should at least protect the first four rounds of DES. To resist a powerful adversary (precise fault model, high number of correct-faulty DES encryptions), it seems sensible to protect the first six rounds. Note that if a decryption oracle is available to the adversary, the trick described in Sect. 3.4.4 enables a more efficient attack, which implies that more rounds (six to eight) should be protected.

Chapter 4
Differential Fault Analysis of the Advanced Encryption Standard

Christophe Giraud

Abstract In October 2000, Rijndael was selected as the Advanced Encryption Standard (AES). Since then, this cryptosystem has been widely used to ensure the confidentiality of information stored in embedded devices. Therefore, over the last decade many researchers have studied this algorithm, leading to the publication of many Differential Fault Analyses (DFAs) on the AES. In this chapter, we present the state of the art of DFA of the AES. After describing the AES, we present in detail three of the most efficient DFAs on this cryptosystem. These attacks have different characteristics, allowing an attacker to recover the secret key from one faulty ciphertext or if faults have been induced in the middle rounds of the AES. We then present a table summarizing the characteristics of each and every DFA on the AES published so far. Finally, we present the main countermeasures proposed to counter fault injection attacks on the AES.

4.1 Introduction

Due to the very short DES key size to the increasing computational power of computers, NIST launched in September 1997 a call to find candidates for a successor to DES [310]. This algorithm, called the Advanced Encryption Standard (AES), had to be able to encrypt 128-bit blocks and be available in three different key sizes: 128, 192 and 256 bits. From amongst the fifteen submissions, five finalist algorithms were selected in August 1999: MARS [76], RC6 [348], Rijndael [112], Serpent [13] and Twofish [363]. In October 2000, the algorithm Rijndael proposed by Daemen and Rijmen was chosen to be the DES successor [209].

One of the particularities of the AES [142] is representing the intermediate cipher result, called the State, as a two-dimensional byte array with four rows and four

C. Giraud (✉)
Oberthur Technologies, Pessac, France
e-mail: c.giraud@oberthur.com

M. Joye and M. Tunstall (eds.), *Fault Analysis in Cryptography*,
Information Security and Cryptography, DOI: 10.1007/978-3-642-29656-7_4,

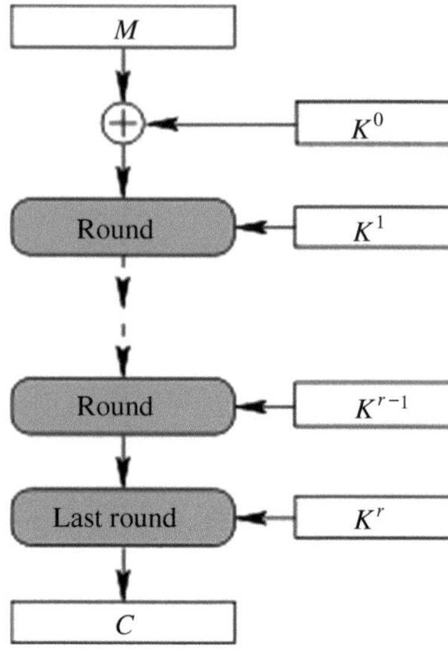

Fig. 4.1 Global structure of AES

Fig. 4.2 Representation of the state

S_0	S_4	S_8	S_{12}
S_1	S_5	S_9	S_{13}
S_2	S_6	S_{10}	S_{14}
S_3	S_7	S_{11}	S_{15}

Table 4.1 Number of rounds of AES depending on the key length

	Key length (in bits)	Number of rounds r
AES-128	128	10
AES-192	192	12
AES-256	256	14

columns. Such a State $S = (S_0, \ldots, S_{15})$ is represented by the array depicted in Fig. 4.2.

The AES is composed of r rounds (cf. Fig. 4.1), r depending on the key length, as shown in Table 4.1.

Each round is composed of four transformations:

1. *SubBytes* (SB), which is a non-linear substitution function (i.e. an S-Box) and applies independently to each byte of the State.
2. *ShiftRows* (SR), which performs a rotation on the last three rows of the State array. The second row is rotated on position to the left, the third (or fourth) row is rotated two (or three) positions to the left.
3. *MixColumns* (MC), in which the state array columns are considered as polynomials with coefficients in $\mathbb{F}_{2^8}$ and multiplied with the polynomial $\{\texttt{03}\} \cdot x^3 + \{\texttt{01}\} \cdot x^2 + \{\texttt{01}\} \cdot x + \{\texttt{02}\}$ modulo $x^4 + 1$.

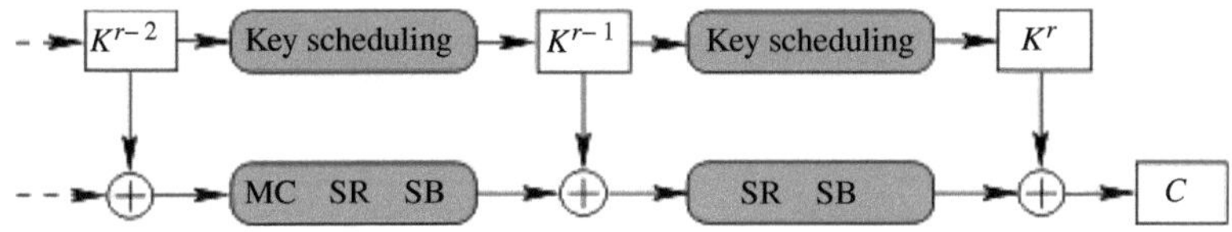

Fig. 4.3 The last two rounds of the AES

4. *AddRoundKey*, in which a round key is added to the State by using the bit-wise XOR operation.

The last round is composed of the same transformations as the other rounds, with the exception of the *MixColumns* transformation; cf. Fig. 4.3.

The AES key scheduling provides $r + 1$ round keys K^i of 128 bits, r depending on the AES key length (cf. Table 4.1). To do so, the AES key is firstly expanded into a $128(r + 1)$-bit buffer, then split into $r + 1$ round keys.

The key expansion is performed by operating on the columns of the array representing the AES key. Figure 4.4 depicts the beginning of the expansion where

- *SubWord* (RW) applies the AES S-Boxto each of the four bytes of its input,
- *RotWord* (RW) is a rotation such that a four-byte input (a, b, c, d) gives (b, c, d, a),
- $Rcon[i]$ is a four-byte array defined by $Rcon[i] = (x^{i-1}, \{00\}, \{00\}, \{00\})$, x^{i-1} being a power of x (x represented by $\{02\}$) in the field $\mathbb{F}_{2^8}$.

For more information about the AES, the reader can refer to [113].

Since May 2002, the AES has been the new standard for symmetric encryption. It has been therefore used more and more for secure applications, in particular in the embedded environment. In the latter case, the implementations of the AES must resist specific attacks such as side-channel analysis (e.g. SPA, DPA) and fault analysis (e.g. DFA). This chapter deals with the state of the art of Differential Fault Analysis of the AES and it is organized as follows. Section 4.2 presents the principle of DFA on the AES before describing in detail three of the most efficient attacks. These attacks allow an attacker to recover the AES key with a very relaxed fault model and/or a very small number of faulty ciphertexts. In Sect. 4.3 we compare the characteristics of the various DFAs on the AES published so far. Finally we present in Sect. 4.4 the different specific countermeasures which have been proposed to protect the AES against DFA.

4.2 Differential Fault Analyses of the AES

When Rijndael was selected as the AES, no DFA was known against this new cryptosystem. Indeed, in 2000 only the original differential fault attack of Biham and Shamir [49] was known on symmetric cryptosystems and it does not directly apply to Rijndael. Therefore, many researchers have worked on this subject and many DFAs on the AES have been published since 2000. In this section, after presenting

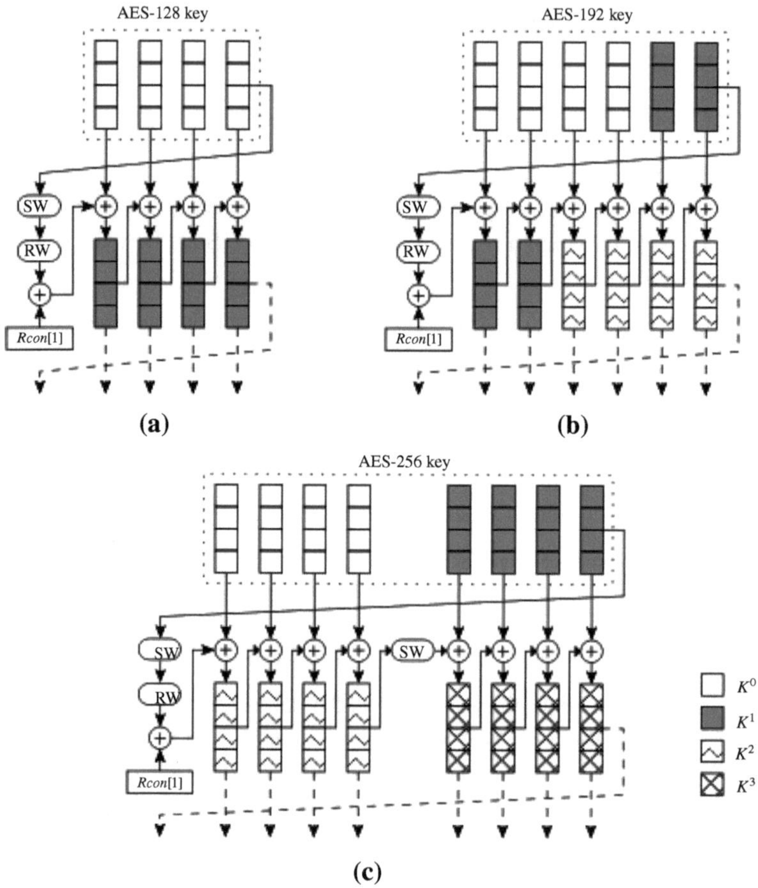

Fig. 4.4 Beginning of the AES key scheduling. **a** AES-128 Key expansion. **b** AES-192 key expansion. **c** AES-256 key expansion

the principle of DFA on the AES, we describe three of the most efficient attacks published so far. The first one is often cited as a reference by many variants published afterwards. It presents a very good compromise between the fault model and the required number of faulty ciphertexts. Secondly, we present an attack on the middle rounds of the AES. It allows an attacker to efficiently recover the secret key even if the last four rounds of the AES are protected against fault attacks. Thirdly, we describe an attack which is very efficient in terms of fault model since it allows the disturbance of three diagonals of the State array. Finally, we briefly present some attacks exploiting faults induced during the first rounds. These attacks can be very efficient in practice if only the last rounds are protected against fault injection.

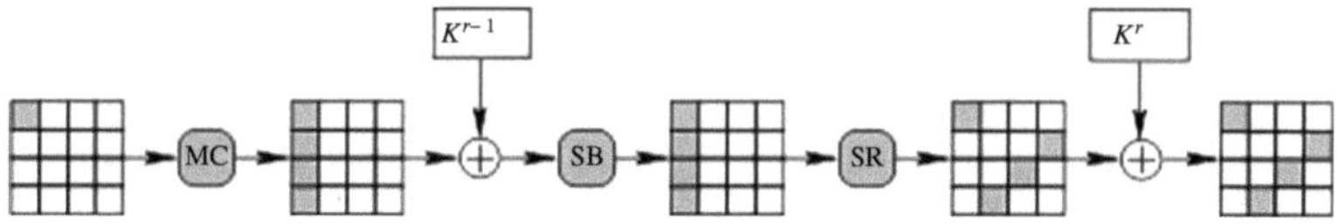

Fig. 4.5 Propagation of the differential by assuming a byte fault at the input of *MixColumns* of round $r-1$

4.2.1 Principle of DFA on the AES

The objective of an attacker performing DFA on the AES is to recover enough information on the last round key (or the last two round keys) to be able to mount an exhaustive search to recover the AES-128 key (or AES-192 or AES-256 key).

To obtain information on the last round key(s), the principle of DFA on the AES is to analyze the differential at some point between a correct and a faulty AES execution. Such a differential is computed from a guess on a small part of the last round key and from a pair $(C, C^{\lightning})$ of correct and faulty ciphertexts. If the guess is correct, the differential must satisfy some properties depending of the corresponding fault model, which gives us a key distinguisher.

4.2.2 The Standard DFA on the AES

At CHES 2003, Piret and Quisquater presented a new DFA on the AES which requires few faulty ciphertexts and a relaxed fault model [324]. Such a combination is rare and it makes this attack a reference. In the rest of this section, we first present their attack, before presenting its application to AES-192 and AES-256.

4.2.2.1 Basic Attack

The principle of this attack is to induce a fault resulting in a differential of one byte at the input of the last *MixColumns*. By guessing four bytes of the last round key, the attacker tests if the corresponding differential at the output of the last *MixColumns* corresponds to a one-byte differential at its input.

First of all, the attacker computes a list $\mathcal{D}$ of possible differences at the output of one column of the *MixColumns* transformation, assuming a one-byte difference at its input. The list $\mathcal{D}$ thus contains 4×255 four-byte elements. This operation is done only once and can be used for future attacks.

Secondly, the attacker obtains a pair $(C, C^{\lightning})$ of correct and faulty ciphertext, $C^{\lightning}$ obtained from a random byte fault injected into the input of the *MixColumns* transformation of round $r-1$. If the fault is injected into the first column of this State then $C^{\lightning}$ will differ from C in bytes at position 0, 7, 10 and 13. If the fault is

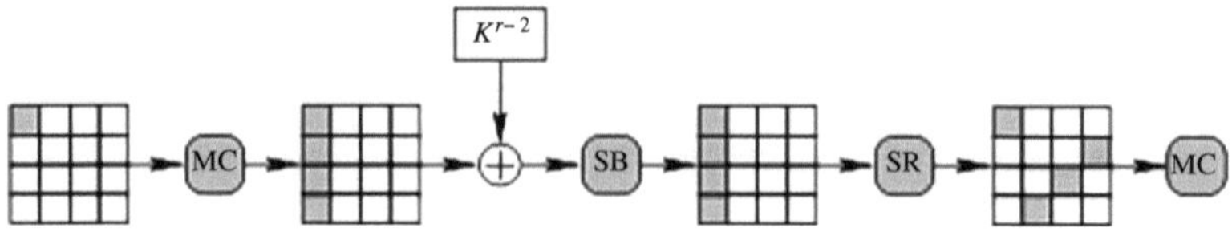

Fig. 4.6 Propagation of the differential by assuming a random byte fault at the input of *MixColumns* of round $r-2$

injected into the second column, then $C^{\lightning}$ will differ from C in bytes at positions 1, 4, 11 and 14. Similarly for the next two columns. For the sake of simplicity, we assume that the fault implies a difference between C and $C^{\lightning}$ on the four bytes at positions i, j, k, l (one per row). The attacker then considers the four key bytes K_i^r, K_j^r, K_k^r, K_l^r of the last round key, and for each of the 2^{32} candidates he computes

$$\begin{aligned}\Delta_i &= SB^{-1}(C_i \oplus K_i^r) \oplus SB^{-1}(C_i^{\lightning} \oplus K_i^r),\\ \Delta_j &= SB^{-1}(C_j \oplus K_j^r) \oplus SB^{-1}(C_j^{\lightning} \oplus K_j^r),\\ \Delta_k &= SB^{-1}(C_k \oplus K_k^r) \oplus SB^{-1}(C_k^{\lightning} \oplus K_k^r),\\ \Delta_l &= SB^{-1}(C_l \oplus K_l^r) \oplus SB^{-1}(C_l^{\lightning} \oplus K_l^r).\end{aligned} \tag{4.1}$$

The four-byte result $(\Delta_i, \Delta_j, \Delta_k, \Delta_l)$ is then compared with the 1,020 elements contained in the list $\mathscr{D}$. The candidates $(K_i^r, K_j^r, K_k^r, K_l^r)$ for which a match is found are gathered in a list $\mathscr{L}$.

With one pair $(C, C^{\lightning})$, the list $\mathscr{L}$ contains 1,036 elements on average. By using another pair $(C, C^{\lightning})$ with a fault injected into the same column, the corresponding four bytes of the last round key are uniquely determined with a 98 % probability.

Therefore the last round key can be recovered by using eight faulty ciphertexts with faults induced at chosen locations.

4.2.2.2 Improved Attack

To extend the attack described in Sect. 4.2.2.1, Piret and Quisquater noticed that if a random byte fault is induced between the *MixColumns* of rounds $r-3$ and $r-2$, the corresponding differential at the input of the *MixColumns* of round $r-1$ has four non-zero bytes, one per column of the State array (cf. Fig. 4.6) [324]. Each of them can thus be exploited by using the method described in Sect. 4.2.2.1 and releases information about different parts of the last round key. By using such a fault, one pair of correct and faulty ciphertexts allows the attacker to reduce the number of possible values for the last round key to $1036^4 \approx 2^{40}$, from which an exhaustive search to recover a 128-bit AES key can be done. With two pairs of correct and faulty ciphertexts, the last round key is uniquely identified with a 92 % probability.

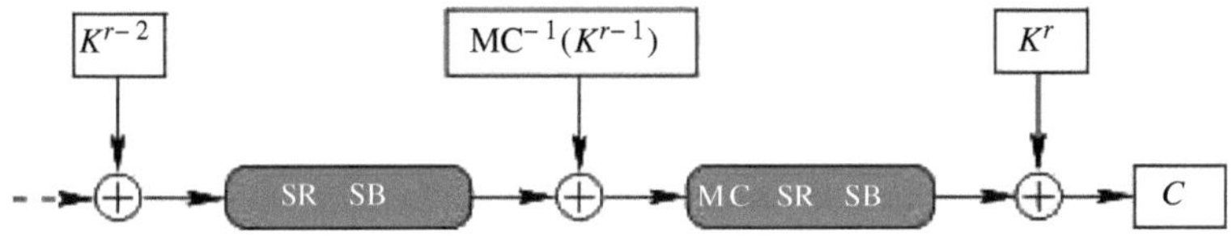

Fig. 4.7 Another view of the last two rounds of the AES

4.2.2.3 A Remark on the Complexity

To improve the timings of the exhaustive search amongst the 2^{32} candidates (K_i^r, K_j^r, K_k^r, K_l^r) in (4.1), Piret and Quisquater noticed that the four equations of (4.1) are independent. Therefore they suggest testing the first two key bytes first in order to make a list $\mathscr{L}$ of the candidates which match with the two left most bytes of the elements of list $\mathscr{D}$. Once this very fast exhaustive search is done, the attacker tries to extend the value of each element of $\mathscr{L}$ by one byte by testing the second and the third key bytes in the same way, the possible values for the second key byte being taken from $\mathscr{L}$. Then the same method is applied by testing the third and fourth key bytes. Finally we perform the test presented in Sect. 4.2.2.1 by considering the candidates (K_i^r, K_j^r, K_k^r, K_l^r) in $\mathscr{L}$ only.

4.2.2.4 Application to AES-192 and AES-256

When attacking the AES-192 or the AES-256, the goal of an attacker is to recover the last two round keys, from which the entire AES key can be easily computed.

To do so, the attacker first applies the method presented in Sect. 4.2.2.2 to obtain the last round key K^r by using two pairs of correct and faulty ciphertexts where the faults have been injected between the *MixColumns* of rounds $r-3$ and $r-2$.

Second, the attacker obtains two other pairs (C, $C^{\lightning}$) of correct and faulty ciphertexts by injecting random byte faults between the *MixColumns* of rounds $r-4$ and $r-3$. Since *MixColumns* is linear, one can rewrite the last two rounds of the AES as depicted in Fig. 4.7.

Therefore for each pair (C, $C^{\lightning}$), the attacker computes

$$\begin{cases} A &= MC^{-1}(SR^{-1}(SB^{-1}(C \oplus K^r))) \\ A^{\lightning} &= MC^{-1}(SR^{-1}(SB^{-1}(C^{\lightning} \oplus K^r))) \end{cases} \tag{4.2}$$

and applies the method described in Sect. 4.2.2.2 with A (or $A^{\lightning}$)) instead of C (or $C^{\lightning}$) to recover $MC^{-1}(K^{r-1})$. The penultimate round key is then obtained by computing the image of $MC^{-1}(K^{r-1})$ through *MixColumns*. Finally the whole AES key is computed from the last two round keys.

To conclude, the AES key is obtained by using four faulty ciphertexts in the 192- and 256-bit cases.

4.2.2.5 Conclusion

This attack combines a very relaxed fault model with very good efficiency in terms of faulty ciphertexts. Indeed, by using a random byte fault on any temporary variable between the *MixColumns* of rounds $r-3$ and $r-2$, a whole 128-bit AES key van be uniquely identified using two faulty ciphertexts. The same result can be achieved with only one faulty ciphertext and an exhaustive search amongst 2^{40} candidates. Moreover, this attack can be easily extended to the 192- and 256-bit key length cases. These different points make this attack a reference, used as a basis by many variants published afterwards.

4.2.3 A DFA on the Middle Rounds of the AES

At CARDIS 2006, Phan and Yen presented new DFAs on the AES which exploit techniques from block cipher cryptanalysis [323]. In this paper, they show in particular how by combining fault attacks and the Square distinguisher presented by Daemon et al. in [111], one can recover the AES key by using faults induced in the AES middle rounds. In this section, we first recall an AES property, before presenting the corresponding DFA on the AES.

4.2.3.1 Square Distinguisher

In the paper introducing the block cipher Square [111], a dedicated attack on reduced versions of Square is described which also applies to reduced versions of Rijndael. It exploits the following property of AES:

Property 1 *By taking a set of 256 plaintexts identical to each other except for one byte in which they take all the possible values, after three rounds the XOR of the 256 values results in a State containing 0 in all byte positions.*

Phan and Yen exploited this property to mount a DFA on the AES middle rounds. Let us present their attack in the following sections.

4.2.3.2 Basic Square-DFA

The principle of this attack is the following: the attacker always encrypts the same plaintext and he injects 255 uniformly distributed random byte faults into the same byte of the input of round $r-3$. Let us denote by $C^{\not{\iota}\, j}$, $j \in \{1, \cdots, 255\}$, the 255 corresponding faulty ciphertexts and by C the correct ciphertext. From Property 1, the XOR of the 256 corresponding inputs of the last round must be equal to zero. Therefore, to recover the ith byte of the last round key, the attacker tests for each possible value for K_i^r if the following equality holds:

$$SB^{-1}(C_i \oplus K_i^r) \oplus \bigoplus_{j=1}^{255} SB^{-1}(C_i^{\not{\,}j} \oplus K_i^r) = 0. \tag{4.3}$$

If so, the attacker has recovered the value of the ith byte of the last round key with high probability. He then repeats this search for the remaining 15 bytes of K^r by using the same 256 ciphertexts.

One may note that this attack applies not only if the fault is injected into the input of round $r-3$ but also if the fault is injected into any temporary variable between the *MixColumns* of rounds $r-4$ and $r-3$.

4.2.3.3 Extended Square-DFA

Phan and Yen noticed that it is possible to attack one round before, i.e. between the *MixColumns* of rounds $r-5$ and $r-4$. In this case, Property 1 implies that the XOR of the inputs of round $r-1$ must result in zero. To exploit this property, the attacker must decrypt the ciphertexts by the last two rounds and perform an exhaustive search similar to the one described in Sect. 4.2.3.2. To do so, he has to guess a column of K^{r-1} and the corresponding four bytes of K^r.

By repeating this four times, the attacker obtains both K^{r-1} and K^r, i.e. he recovers the AES key whatever its length. However, the complexity of such an attack is much higher that the one of Sect. 4.2.3.2 since we need to guess eight four-byte variables and not only 16 times a one-byte variable (i.e. we need to perform an exhaustive search amongst 2^{66} candidates instead of 2^{12}). Therefore, such an attack is very difficult to put into practice.

4.2.3.4 Application of Basic Square-DFA to AES-192 and AES-256

In the case of AES-192 and AES-256, the attack of Sect. 4.2.3.2 can be extended to recover the penultimate round key K^{r-1} by using 255 extra faulty ciphertexts obtained from a uniformly distributed random byte fault induced on the State between *MixColumns* of rounds $r-5$ and $r-4$. By guessing a column of K^{r-1} and by using the knowledge of K^r, the attacker decrypts the corresponding four bytes of the faulty ciphertexts by the last two rounds. Then he tests whether the corresponding four bytes of the XOR of the corresponding 256 States are equal to zero or not. Finally, the attacker iterates this search on the three other columns with the same ciphertexts until the penultimate round key is recovered.

4.2.3.5 Conclusion

We have presented here a DFA on the middle rounds of the AES which exploits a property used by traditional cryptanalysis. This attack is very powerful in the sense that it efficiently applies even if the last four rounds of the AES are protected against fault attacks. Moreover this attack can also be mounted if the last five rounds are protected but it requires four exhaustive searches amongst 2^{64} candidates.

4.2.4 A Very Effective DFA on the AES

In 2009, Saha et al. published on ePrint [352] an improved version of Piret and Quisquater's attack. This very efficient DFA allows the attacker to uniquely identify a 128-bit AES key by using only four faulty ciphertexts with faults disturbing up to three diagonals of a State. In the rest of this section we first present an observation on the AES which is then used in Sect. 4.2.4.2 to mount a one-shot DFA on the AES by assuming that a single diagonal of the State is disturbed. This fault model is then relaxed in Sect. 4.2.4.3.

4.2.4.1 Preliminary Observation

To mount their attack, Saha et al. made the following observation. If a diagonal of the input of round $r-2$ is disturbed then the fault impacts a column at the end of this round and spreads out to the entire State at the end of round $r-1$. However, the bytes of each of the four columns of the matrix which represents the differential between a correct and a faulty State at the output of round $r-1$ have some relationship which can be used to obtain information about the last round key. Such a propagation and the corresponding relationships are depicted in Fig. 4.8.

4.2.4.2 Description

Let us now explain how an attacker can recover information on the last round key from a pair $(C, C^{\lightning})$ where the fault has been induced on the first diagonal of the input of round $r-2$. If such a fault is induced, one can observe from Fig. 4.8 that the following set of equations is valid for a pair $(C, C^{\lightning})$:

$$\begin{aligned} SB^{-1}(C_0 \oplus K_0^r) \oplus SB^{-1}(C_0^{\lightning} \oplus K_0^r) &= 2(SB^{-1}(C_{13} \oplus K_{13}^r) \oplus SB^{-1}(C_{13}^{\lightning} \oplus K_{13}^r)) \\ SB^{-1}(C_{13} \oplus K_{13}^r) \oplus SB^{-1}(C_{13}^{\lightning} \oplus K_{13}^r) &= SB^{-1}(C_{10} \oplus K_{10}^r) \oplus SB^{-1}(C_{10}^{\lightning} \oplus K_{10}^r) \\ SB^{-1}(C_7 \oplus K_7^r) \oplus SB^{-1}(C_7^{\lightning} \oplus K_7^r) &= 3(SB^{-1}(C_{13} \oplus K_{13}^r) \oplus SB^{-1}(C_{13}^{\lightning} \oplus K_{13}^r)) \end{aligned} \tag{4.4}$$

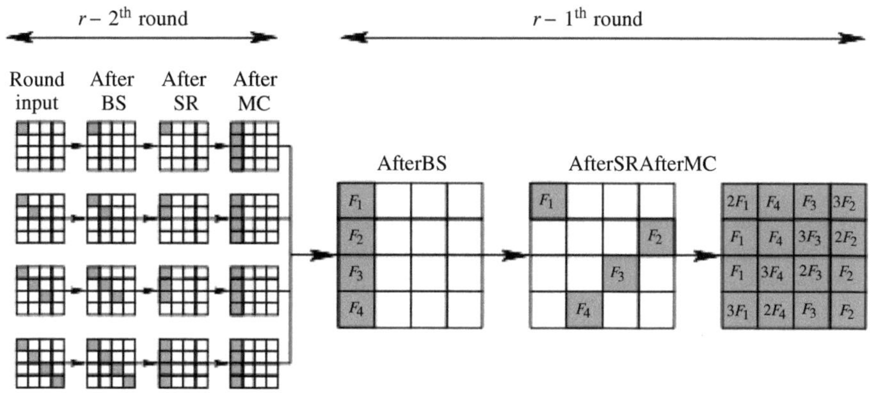

Fig. 4.8 Equivalence of the propagation of different kinds of fault induced on the first diagonal of the input of round $r-2$

First of all the attacker reduces the space of possible candidates for key bytes K_0^r and K_{13}^r by guessing all possible values for them in the first equation and filtering out all pairs that do not satisfy it. He now has a reduced key space for K_{13}^r which is used to perform a similar key space reduction with the second equation. Then the same methodology is applied to the third equation. After this three-step key space reduction, the number of possible values for $(K_0^r, K_7^r, K_{10}^r, K_{13}^r)$ is about 2^8 on average.

By observing the differential at the output of the *MixColumns* of round $r-1$ (cf. Fig. 4.8), one can see that the attacker can obtain similar equations with the last three columns. These other three sets of three equations are then used to reduce the last round key space to $(2^8)^4 = 2^{32}$ possible values.

In the description presented above, we assumed that the position of the faulty diagonal was known to the attacker. To relax this assumption, the attacker has to guess the position of the faulty diagonal and repeat the attack four times. In such a case, the last round key space will be reduced to 2^{34} possible values.

In the case of AES-128, the AES key is recovered by using one faulty ciphertext and by performing an exhaustive search amongst the 2^{34} candidates.

4.2.4.3 A Relaxed Fault Model

In the last part of their paper, Saha et al. extend the fault model used in Sect. 4.2.4.2 by assuming that three diagonals of the state array between the *MixColumns* of round $r-3$ and the *ShiftRows* of round $r-2$ are disturbed by the fault. In such a case, the attacker can obtain a set of equations similar to (4.4) for each column of the state array after the *MixColumns* of round $r-1$. If the position of the fault-free diagonal is known, such a set allows the attacker to reduce the number of candidates for the corresponding four bytes of the last round key to 2^{24} on average. By using such a fault model and four faulty ciphertexts, the last round key is uniquely identified.

4.2.4.4 An Improvement for AES-128

In [403], Tunstall and Mukhopadhyay published a one-shot attack specific to AES-128. The first step of their attack is very similar to the one presented in Sect. 4.2.4.2, with a more restricted fault model (i.e. a random byte fault between the *MixColumns* of the seventh and eighth round). In the second step of their attack, they exploit the relationships between the bytes of K^9 and K^{10} to analyse the differential at the output of the *MixColumns* of round 8. Their attack reduces the number of candidates for the last round key to 2^{12} by using only one faulty ciphertext.

4.2.4.5 Conclusion

The fault attack presented in this section is one of the most efficient published so far since only one random diagonal fault between the *MixColumns* of round $r-3$ and the *ShiftRows* of round $r-2$ reduces the number of possible values of the last round key to 2^{34}. Moreover, if three out of four diagonals of the State are disturbed, this attack makes it possible to uniquely identify the last round key by using four faulty ciphertexts.

4.2.5 Attacking the First Rounds

Previous sections dealt with attacks exploiting faults induced on the last rounds. However, it is also possible to mount DFA on the AES when faults have been induced on the first rounds. We present in this section two such attacks.

4.2.5.1 Attacking the Key Addition

The first attack exploiting faults induced on the first round was presented by Blömer and Seifert in 2003 [55]. Even if the fault model is rather strong, it allows one to attack the beginning of the AES.

The principle of their attack is quite simple. The attacker encrypts a null message twice, the first time without any disturbance; the second time the attacker forces the first bit of the State to 0 after the first *AddRoundsKey* transformation. If the two ciphertexts are equals then it means that the first bit of the first round key is equal to 0; otherwise it means that this bit is equal to 1. Therefore, the attacker can recover the value of the first round key by iterating this attack on the other 127 bits.

One may note that this attack can be easily adapted when the fault is induced after any *AddRoundsKey* transformation, i.e. this attack is effective at each round of the AES.

4.2.5.2 Adaptation of Attacks Exploiting Faults on the Last Rounds

Another way to attack the first rounds by using DFA is to adapt attacks exploiting faults on the last rounds. In this section we give an example by using Piret and Quisquater's attack (cf. Sect. 4.2.2).

Let us assume that a fault has disturbed one byte of the State at the beginning of the second round when encrypting a message M, leading to a faulty ciphertext $C^{\lightning}$. If we denote by $M^{\lightning}$ the message whose encryption corresponds to $C^{\lightning}$, then $M^{\lightning}$ differs from M on four bytes since the fault has been induced after the first *MixColumns* transformation. Therefore, the attacker can recover the message $M^{\lightning}$ by performing a fast 32-bit exhaustive search. In a second step the attacker applies Piret and Quisquater's basic attack (cf. Sect. 4.2.2.1) to $(M, M^{\lightning})$ instead of $(C, C^{\lightning})$, allowing him to obtain information on four bytes of the first round key. Iterating this attack, the attacker can recover the first round key by using the same number of faulty ciphertexts as in Piret and Quisquater's attack, with an overhead of a 32-bit exhaustive search per faulty ciphertext.

4.3 Comparison of DFAs on the AES

Over the last ten years, many DFAs on the AES have been published. Due to the various fault models which have been used, it is difficult to compare the efficiency of these attacks. In this section, we present such a comparison. To do so, we first present a way of classifying the different fault models that have been used in previous publications, before presenting the characteristics of each and every DFA on the AES.

4.3.1 Fault Models

The efficiency of a fault attack does not depend only on the number of faulty ciphertexts required to recover the secret key; it depends also on the practicality of the corresponding fault model. In order to compare the various fault models used in the attacks published so far, we present a way of characterizing a fault which depends on its impact and its location. The impact of a fault corresponds to the effect a fault has on the variable which is affected and the location of a fault indicates where the fault must be performed during the execution of the algorithm. These two characteristics are detailed below.

4.3.1.1 Impact of the Fault

To classify the impact of a fault on a variable, we can use an array whose columns indicate the numbers of bits affected by the fault:

- *bit*: only one bit is affected by the fault,
- *byte*: a whole byte of the temporary result is disturbed,
- *word*: four bytes are modified by the fault;

and whose rows indicate the kind of modification of these bits:

- *stuck-at*: the bits are always set to the same value; typically in this case, the attacker assumes that the fault sets all the affected bits to 0 or to 1,
- *flip*: the bits are complemented,
- *random*: after the fault, the value of the disturbed bits is random,
- *uniformly distributed random*: after the fault, the value of the n affected bits is uniformly distributed between 0 and $2^n - 1$.

From a practical point of view, the number of bits affected by the fault mainly depends on the characteristics of the component under attack. For instance, if the device has an eight-bit CPU with a 32-bit cryptoprocessor, then if a fault occurs during a CPU operation, it is probable that a byte of the temporary result will be erroneous, whereas if a fault disturbs a cryptoprocessor operation, the impact of the fault will probably affect 32 bits.

Regarding the practicality of the row characteristics, it is quite difficult to produce a uniformly distributed error. Most of the time, attacks assuming a uniformly distributed random fault model can also be achieved by using a random fault model with more faulty ciphertexts.

4.3.1.2 Location of the Fault

This characteristic indicates on which (part of the) State the disturbance must occur. It mainly corresponds to the ability of the attacker to synchronize the disturbance during the execution of the algorithm. In the case of the AES, the main fault locations are the following:

- *Chosen*: the bits of the variable disturbed by the fault as well as the variable itself can be chosen by the attacker,
- *Random on a chosen temporary variable*: the attacker can choose only which variable is affected by the fault,
- *Random between MixColumns of round n and round $n + 1$*: the attacker knows only that the fault has occurred on a State between the *MixColumns* of rounds n and $n + 1$.

Of course, the more precise the attack must be, the more difficult it is to put into practice.

Table 4.2 Number of faulty ciphertexts required to mount a DFA attack on AES-192 depending on the impact and the location of the faults

Ref.	Fault impact	Fault location	Number of faulty ciphertexts	Number of candidates left
[324]	Random byte	Random between MC of rounds 9 and 10+ Random between MC of rounds 8 and 9	$2+2$	1
[393]	Random byte	Random on a chosen temporary variables (input of round 10+ input of round 9)	$2+1$	2^8
[224]	Random byte	Known between MC of rounds 9 and 10+ Random between MC of rounds 8 and 9	$1+1$	2^8

Table 4.3 Number of faulty ciphertexts required to mount a DFA attack on AES-256 depending on the impact and the location of the faults

Ref.	Fault impact	Fault location	Number of faulty ciphertexts	Number of candidates left
[324]	Random byte	Random between MC of rounds 11 and 12+ Random between MC of rounds 10 and 11	$2+2$	1
[393]	Random byte	Random on a chosen temporary variable (input of round 12) in encryption + in decryption	$2+2$	2^{13}
[224]	Random byte	Random between MC of rounds 11 and 12+ Random between MC of rounds 10 and 11	$2+1$	2^{32}
[164]	Random byte	Random between MC of rounds 11 and 12+ Random between MC of rounds 10 and 11	$2+1$	2^{18}

4.3.2 Comparative Tables

Based on the classification of the fault models in the previous section, we present below a table for each AES key size which compares the various DFAs on the AES published so far. For each attack, we indicate the kind of fault model, the number of faulty ciphertexts and the number of candidates left for the secret key.

In Tables 4.2 and 4.3, we present the characteristics of the more efficient DFAs on AES-192 and AES-256 published so far.

Table 4.4 Number of faulty ciphertexts required to successfully mount a DFA attack on AES-128 depending on the impact and the location of the faults

Ref.	Fault impact	Fault location	Number of faulty ciphertexts	Number of candidates left
[426]	Bit flip	Random on a chosen temporary variable (input of round 10)	32	1
[55]	Bit stuck at 0	Chosen (after the first round key addition)	128	1
[160]	Uniformly distributed random byte	Chosen (K^9, K^8 and M^8)	31	2^{16}
		Random on a chosen temporary variable (K^9, K^8 and M^8)	248	2^{16}
[127]	Uniformly distributed random byte	Random between *MixColumns* of rounds 8 and 9	50	1
[84]	Uniformly distributed random byte	Chosen (K^9 and K^8)	22	2^{24}
[324]	Random byte	Random between *MixColumns* of rounds 7 and 8	1	2^{40}
			2	1
[323]	Uniformly distributed random byte	Fixed between *MixColumns* of rounds 6 and 7	255	1
[51]	Bit flip	Chosen	32	1
		Random on a chosen byte	4,096	1
[296]	At most three bytes per column	Random on a chosen temporary variable (input of *MC* of round 9)	6	1
	Each and every byte	Random on a chosen temporary variable (input of *MC* of round 9)	1,495	1
[322]	Uniformly distributed random word	Chosen (K^9)	12	1
[392]	Random word (column)	Random on a chosen temporary variable (K^8)	2	2^{40}
			4	2^{16}
[229]	Three random bytes	Chosen (K^9)	2	2^{32}
			4	1
[148]	Random byte	Random on a chosen temporary variable (input of round 8)	1	2^{40}
			2	1
[229]	Random byte	Random on a chosen temporary variable (input of round 8)	1	2^{36}
			2	1
[403]	Random byte	Random between *MixColumns* of rounds 7 and 8	1	2^{12}
[352]	Random word (diagonal)	Random between *MixColumns* of round 7 and *ShiftRows* of round 8	1	2^{34}
			2	1
	Three random words (three diagonals)	Random between *MixColumns* of round 7 and *ShiftRows* of round 8	4	1

4.4 Countermeasures

To counter the attacks presented above, a naïve countermeasure, called duplication, consists in performing the AES twice and comparing the corresponding outputs. However, if an attacker injects the same fault twice, such an attack will be undetected. An improved version of this countermeasure consists in performing the AES, decrypting the output and checking if the corresponding result equals the original input. Once again, double fault attacks can still bypass this countermeasure, even if they are more complex to mount. Moreover, the main problem with these countermeasures is the overhead in terms of performances. Indeed such countermeasures double the time required to compute an AES. Therefore, one could have thought that many efficient ways of protecting the AES against DFA would have been proposed over the last decade. However, very few papers dealing with countermeasures for AES have been published.

Concerning the countermeasures published before 2005, Malkin et al. [266] list and discuss security/efficiency trade-offs of security methods based on space and time redundancies. First of all they list the countermeasures using either parity bits or nonlinear codes. Regarding the first group, Malkin et al. concludes that the corresponding security is weak and that a malicious attacker can always bypass such countermeasures. Concerning the countermeasures based on nonlinear codes, for instance [211], which presents a hardware countermeasure with a good detection rate but with a hardware overhead comparable to duplication. In their conclusion space redundancy countermeasures, Malkin et al. underline that they are as expensive as duplication if realistic attackers are considered.

Since 2005, two other such methods have been published. In [158], Genelle et al. propose a way of protecting AES software implementations. They propose several appropriate digest values for each AES transformation. Compared to the doubling countermeasure, their method achieves roughly the same performance, but their main achievement is when one wants to combine FA and DPA countermeasures (which is always the case in practice). In such a case, their method is 28 % faster than the doubling method. In [281], a similar method to [211] has been proposed in which the authors add an eight-byte redundancy to the State and perform the AES transformations in a polynomial ring of degree 2 over $\mathbb{F}_{256}$. Regarding the countermeasures based on time redundancies, their cost is also close to that of duplication, except in certain particular implementation contexts such as feedback encryption modes.

In 2007, Kermani and Reyhani-Masoleh propose a structure-independent fault detection scheme which can be applied to any hardware implementation of the *SubBytes* transformation [220]. Although the suggested test detects every kind of error, their implementation is very costly since it induces a 93 % area overhead.

4.5 Conclusion

In this chapter we presented the state of the art of DFA on the AES and we described in detail three attacks which can be used to recover the AES key with a minimal number of faulty ciphertexts and a very practical fault model. However, for all the attacks presented in this chapter to succeed, the attacker must be able to cipher the same message twice (at least). However, such a condition is not always true in practice. Therefore, DFA on the AES does not always apply, but when it does, the various DFAs published on the AES are extremely efficient. This implies that the corresponding implementations must be protected against fault attacks. However, as shown in Sect. 4.4, very few efficient countermeasures have been proposed so far. Since the last published attacks are very efficient in terms of both fault model and number of faulty ciphertexts, the proposal of efficient DFA countermeasures for the AES in terms of performance, memory consumption and detection rate is a future challenge for cryptologists.

Chapter 5
Countermeasures for Symmetric Key Ciphers

Jörn-Marc Schmidt and Marcel Medwed

Abstract Since a single fault can lead to a recovery of the whole secret key of an AES-128 implementation, protection against fault attacks is vital for security-related devices. Moreover, the fatal impact of undetected faults implies high requirements for such devices: no erroneous result must be revealed with its correct counterpart. Given the fact that secret-key algorithms are not usually based on continuous algebraic structures complicates incorporating redundancy. Designing countermeasures that guarantee this property is a challenging task. As a result, a large number of different countermeasures have been developed. Each of them employs redundancy in a different way, which makes their efficiency heavily dependent on the application scenario and on the assumed adversary. This chapter presents a comprehensive study of fault countermeasures for symmetric key algorithms. It discusses the different levels where countermeasures can be deployed, points out the strengths and weaknesses of the different countermeasures and finally identifies their optimal field of usage.

5.1 Introduction

Compared to algorithms like RSA or elliptic curve-based algorithms, block ciphers usually have little or no algebraic structure. This makes their protection against fault attacks much harder as it is difficult to introduce redundancy in an efficient and yet secure way. Several approaches to tackle this problem have been proposed. Figure 5.1 categorizes the different approaches to protect symmetric ciphers from malicious adversaries. The two main topics of this chapter are Dual Modular Redundancy (DMR) and coding-theoretical approaches. These countermeasures can

J.-M. Schmidt
Institute for Applied Information Processing and Communications (IAIK)
Graz University of Technology, Graz, Austria

M. Medwed
Crypto Group, Université Catholique de Louvain, Louvain-la-Neuve, Belgium

M. Joye and M. Tunstall (eds.), *Fault Analysis in Cryptography*,
Information Security and Cryptography, DOI: 10.1007/978-3-642-29656-7_5,

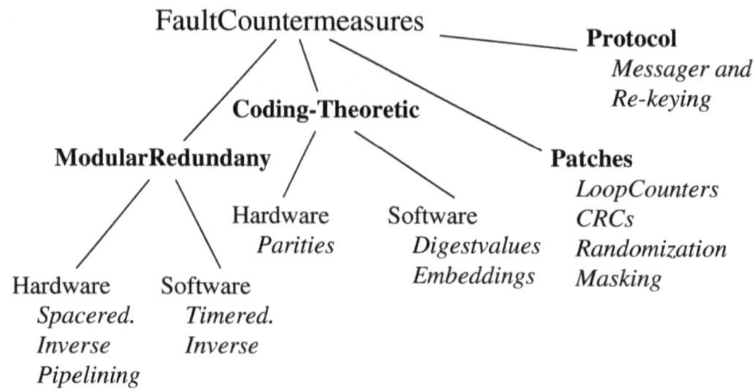

Fig. 5.1 Countermeasures for symmetric key primitives

provide (almost) complete protection for the algorithm. Besides that, more general building blocks that patch only a specific attack scenario or offer only limited protection can be used. They are usually applied to repair weaknesses of other countermeasures. Finally, in scenarios where the algorithm is not fixed, the protection can be implemented at protocol level.

We start our discussion in Sect. 5.2 with the general building blocks and techniques that are not tailored to a special algorithm. Afterwards, Sect. 5.3 shows how DMR schemes can be optimized using block cipher-specific properties. We explain the different approaches for hardware and software implementations. Next, we look at the second main topic: coding-theoretical countermeasures. Section 5.4 focuses on possibilities of using error detection codes to protect AES implementations. These solutions are directly tailored to the algorithm. Finally, we discuss protocol-level countermeasures, which do not protect the cipher itself, but embed it into a protocol that makes fault attacks impractical. Section 5.5 presents the different protocol-level approaches and discusses their strengths and drawbacks. The chapter is concluded in Sect. 5.6.

5.2 General Building Blocks to Protect Implementations

In order to protect implementations against fault attacks, several methods, which can be applied to any algorithm, are available. In the following, we discuss some of these building blocks.

5.2.1 Protecting Loops

A very powerful method to attack a device is to tamper with the counter of a loop. An adversary can try to reduce the number of rounds performed during the computation of a symmetric cipher and apply cryptanalytic methods afterwards to retrieve the secret key [89]. Another way an adversary can benefit from modifying the counter is during an output function. An output loop that does not terminate correctly and runs longer than expected may output memory regions that are not intended to leak out. These can include the secret key storage [14].

In order to detect manipulations of the loop counter, a loop invariant can be used. For this purpose a second variable j is used which counts in the opposite direction of the loop counter i. The second variable is initialized with the final loop value. Thus, the sum of i and j stays constant by construction if no fault is present. This method requires a second variable and an additional incrementation of a counter.

Another way to protect the integrity of a loop is to add a signature that is updated in every run of the loop. In [149], a checksum protects a Montgomery powering ladder, but the same approach can be applied to any loop. Before outputting any value, the signature is checked against a previously stored value. The overhead of the method depends on the checksum used and its size.

These techniques are used to provide high-level program flow protection for countermeasures that ensure data integrity. The drawback of both methods is that they only protect the loop structure itself; modifications of the operations inside the loop are impossible to detect.

5.2.2 Cyclic Redundancy Checks

Attacks are not limited to the time the device is actually powered and processes data. An adversary can also aim at injecting faults into nonvolatile memory to modify parameters that are required for subsequent computations. A manipulation of the parameters can lead to powerful attacks: If an adversary is able to modify bits of the secret key, the information about whether this leads to errors in the computation, so-called safe-errors, is sufficient [15]. For example, an adversary can tell that a key bit is 1 by setting it to 1 and observing that this does not lead to an erroneous output. Furthermore, a permanent fault in the S-Box of a secret key algorithm can be exploited [361].

In order to ensure the integrity of the stored data, an additional Cyclic Redundancy Check (CRC) can be added. Therefore, a checksum is stored together with the data that allows one to ensure that no one has tampered with it when it is loaded from the memory. Subsequent computations involving the data have to be secured by other means, since CRCs are not suited to efficiently calculating with encoded data.

5.2.3 Modular Redundancy

A straightforward way to protect implementations is to use modular redundancy. That is, the algorithm is executed several times. Either in parallel or sequentially. If the results are the equal, the computation is assumed to be correct. Thus, in order to successfully inject a fault, an adversary has to induce the same fault several times. On the one hand, this requires a strong adversary. On the other hand, the costs of the countermeasures are quite high.

The first method, repeating the computation, increases the time a computation takes; therefore it is referred to as time redundancy. The second method, cloning the function, is called space redundancy. It implies a hardware overhead of at least 100 %.

Space redundancy has the advantage of using different circuits. Therefore, also permanent errors can be detected, which tend to remain unnoticed when time redundancy is applied. Furthermore, if the other circuits have different layouts than the original one, this frustrates the efforts of an adversary who tries to inject the same fault in all circuits.

The most common form of modular redundancy for error detection is double modular redundancy (DMR).

5.2.4 Dual Rail Implementations

Besides using the traditional approaches to protect an algorithm, it is also possible to use a different way to implement the circuit in hardware to protect it from malicious adversaries. One approach is to use an *(m-of-n) encoding*. Each bit is represented by n wires, from which exactly m carry a 1. In [189], Ishai et al. discussed the security of bits represented by multiple wires.

A special form of this encoding, i.e. (1-of-2) encoding, is a dual rail implementation, which is also used for asynchronous circuits and for DPA-resistant logic styles. One wire always carries the inverted value of the other one. Thus, an adversary has to target two wires to flip one bit. If an undefined state is induced by the adversary, it spreads throughout the circuit due to early propagation detection mechanisms [327].

5.2.5 Randomization and Masking

In order to make an attack as difficult as possible to conduct, an implementation can make use of random delays during its execution. While this does not directly prevent an adversary from injecting a fault, it limits the precision an adversary can hope to achieve. Thus, attacks that can only exploit a special set of faults become harder.

Masking the data and operations has a similar effect. In [188] masking is realized by linear secret sharing and prevents an adversary from gathering information by probing wires of the device. Also, due to the mask, an adversary can hardly predict the effect of the injected fault, e.g. setting the value that passes by to 0 does not imply that the processed value is zeroed, but only that it is set to a random value that depends on the mask. However, Boscher et al. have shown that masking is not suited to preventing fault attacks [60].

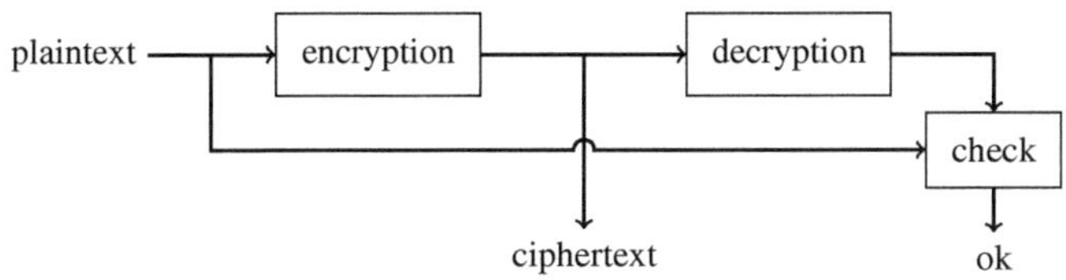

Fig. 5.2 Correctness check using the decryption

5.3 DMR Approaches Specific to Block Ciphers

Dual modular redundant approaches are the most straightforward way to implement error detection but they usually double the area costs or halve the performance of hardware implementations. However, for block ciphers the situation is slightly different as they have a structure which often allows optimizations. Depending on the block cipher and on the mode of operation it is sometimes even possible to keep the overhead close to zero. Furthermore, studies have shown that such approaches are the most efficient for hardware implementations if a realistic fault model is assumed [266].

5.3.1 Using the Inverse

In the previous section, we argued that injecting twice the same fault is always sufficient to overcome space-redundant approaches and that permanent errors defeat time-redundant countermeasures. In contrast to algorithms in general, block ciphers present a bijection. Thus, it is possible to feed the result into the algorithm's inverse, in our case the decryption. The decrypted ciphertext can then be checked against the original plaintext. This is depicted in Fig. 5.2. For time-redundant detection schemes this approach has a significant advantage over encrypting twice and comparing the ciphertexts, namely that permanent errors are detected with high probability.

For space-redundant detection schemes, it depends on the fault model whether the scheme provides higher security than standard DMR schemes. This is because, for block ciphers, the decryption is usually composed of the inverse round functions in reverse order. Thus, during the decryption, at a certain point the only present error is the very same as that induced during the encryption. If it is possible to reverse the fault injection, the error will not be detected. Note that this is specific to block ciphers and is not typically true for public key cryptographic algorithms, e.g. RSA. In general, reversing an error is expected to be more difficult than injecting the same fault twice. However, for an adversary who can induce random byte faults, the probability of correcting the error is the same as that of injecting the same error again.

For time-redundant approaches (as usually pursued in software implementations), the performance reduction for encrypting twice or encrypting and decrypting is nearly

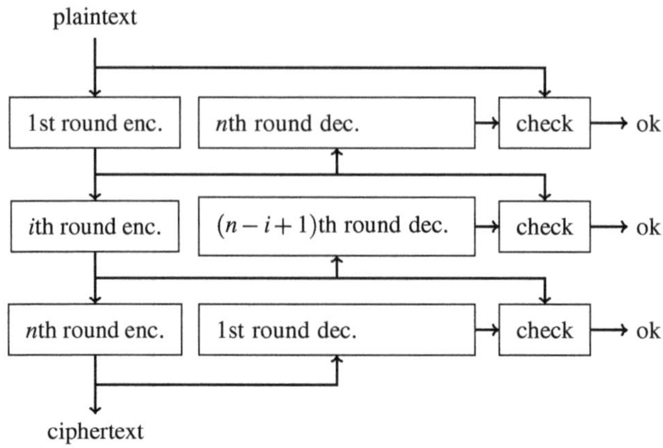

Fig. 5.3 Round-based inverse check

the same. At first glance, it looks as if the case is even worse for space-redundant approaches. There, in addition to area for the decryption hardware, it is not even possible to run the two algorithms in parallel as in the case of encrypting twice. Thus, a naïve implementation, as depicted in Fig. 5.2, decreases the performance by 50 % and increases the area by 100 %. If more than one block is processed, the throughput penalty is reduced as the ith block can be encrypted, while the $(i-1)$th one is checked. Nevertheless, the latency increase stays 100 %. However, it is possible to significantly improve the scheme in terms of area and latency.

The latency can be improved by implementing the countermeasure in a finely grained manner as proposed in [217]. A block cipher consists of several rounds. For each round its inverse round can be immediately applied to the round's output. Thus, while the ith round of the encryption is performed, the $(i-1)$th round can be checked. This reduces the latency overhead from 100 % to $(1/n) \times 100$ % for an n-round block cipher. Figure 5.3 illustrates this interleaved approach.

The area overhead of this scheme strongly depends on the cipher but in general it can be expected to be less than that for a standard DMR scheme. Usually, hardware realizations of block ciphers implement both, an encryption and a decryption circuit. If encryption and decryption are realized by two independent circuits, then the above scheme requires only little additional hardware. In such a case the overhead is caused by the additional registers, multiplexers and comparators for the check process. However, most implementations share resources between encryption and decryption. The most extreme case is presented by involution ciphers, where encryption and decryption are performed by the same building blocks and only the control logic differs. An extensive study on inverse-based error detection for the AES finalists can be found in [217].

Fig. 5.4 Correctness check using the involution property

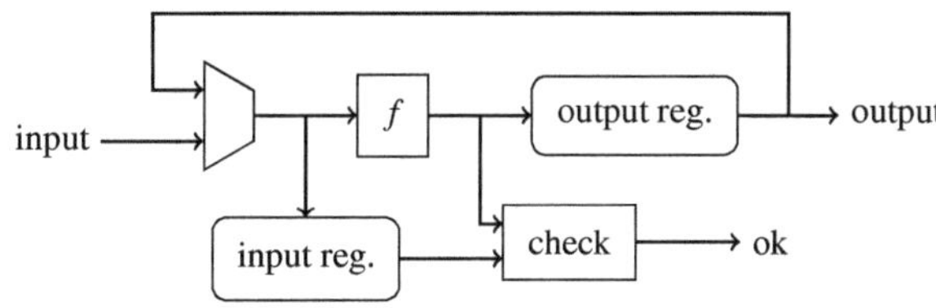

5.3.2 Involution Ciphers

For involution ciphers the substitution-permutation network consists of self-inverse operations. Thus, encryption and decryption can be performed with almost the same circuit. In fact, it is usually sufficient to extend the control logic, as for the operations itself no inverses are needed. This property can be used for error detection as shown by Joshi et al. [194]. The basic idea of their technique is depicted in Fig. 5.4 for one such round operation. In the first clock cycle, the multiplexer forwards the input to f. Thus, after the first cycle, the input register holds the input and the output register holds the output of f. In the second clock cycle the multiplexer forwards the feedback path. As f is an involution, $x = f(f(x))$ and the output of f is now the original input again. The advantage is that it is always possible to pursue a time-redundant detection strategy which detects permanent errors but does not require much additional hardware. The hardware overhead and the latency depend on the granularity of the scheme. On the down side, the scheme comes with a throughput decrease of 50 %. In general, the throughput can only be improved by adding hardware.

5.3.3 Feedback Modes

Another peculiarity of block ciphers is the way they handle large amounts of data. As the block length of the cipher is fixed, the data has to be split into several blocks. A mode of operation defines how these blocks are handled. The simplest mode of operation is the electronic code book (ECB) mode. In this mode, each data block is encrypted independently. However, in order to achieve stronger dependency between the blocks, often so-called feedback modes are used. For such modes, the output of the last encryption is somehow incorporated into the next encryption. In the cipher block chaining (CBC) mode, for instance, the first plaintext block is XORed with an initialization vector and every consecutive plaintext block is XORed with the previous ciphertext block before being encrypted. As a result one encryption has to be finished before the next can start.

A common technique to increase the throughput of a circuit is pipelining. To pipeline a design, the combinatorial circuit is split into n parts, and registers are inserted between those parts. As a result, the circuit takes $n - 1$ clock cycles longer to produce the result. On the other hand, as the combinatorial parts between the registers are less complex, the clock frequency can be increased and thus the latency stays nearly the same. However, now a result is produced in every clock cycle and this increases the throughput by a factor of n.

Feedback modes of operation do not benefit from pipelining as each input depends on the previous output. Nevertheless, pipelining can still be advantageous for error detection. That is, in a pipelined design of an involution cipher, it is possible to implement each pipeline stage according to Fig. 5.4. This allows each pipeline stage to work on data in one cycle and to check the data in the successive cycle. Hence, for involution ciphers operating in a feedback mode, it is possible to implement time-redundant error detection without throughput decrease and at the same time detect permanent errors.

A similar technique might also make sense for noninvolution ciphers in feedback mode if decryption is not available. In such a case, the same input is encrypted twice. For instance, at the end of the second cycle, the input and the output of the first pipeline register are equal. This technique does not decrease the throughput. However, permanent errors are not detected.

5.4 Coding-Theoretical Countermeasures for AES

In order to implement fault detection based on coding theory, it is necessary to tailor the countermeasure to the algorithm used. Since it is the current NIST standard for symmetric encryption, we focus on AES in this section.

Even before fault attacks were proposed, circuits were equipped with parity bits to detect common hardware faults. Such schemes show very little hardware overhead but are often not very robust in the presence of a strong adversary. This is because the parities usually handle only small parts of data when it comes to nonlinear operations. As a result, local data fragments are often only protected by a single bit.

In order to provide strong data integrity at little cost, it is necessary to find parities or check symbols which handle the whole AES state. As the AES does not use a continuous algebra, no straightforward solution is possible. Therefore, digest values have been proposed. The idea is to use different suitable digest values for the different operations.

Another approach which is based on coding theory embeds the AES field into a larger ring. This approach was motivated by similar techniques for public key cryptography. The advantage of this scheme is that it provides a continuous protection, a constant error detection rate can be provided, and automatically protects the program flow of the algorithm.

Finally, also an infective computation-based countermeasure has been proposed. The idea here is to render an error check obsolete by automatically randomizing the output in the case of a fault.

5.4.1 Parity-Based Countermeasures

Parity-based schemes use linear codes to ensure the integrity of the AES state. Thus, the appended bits are a linear sum (over GF(2)) of the state. For instance, one option would be to XOR all bytes of a column together and to store them as parity. Thus, the new, protected state would consist of 16 data bytes and four parity bytes.

As actually three of the four AES round operations are also linear over GF(2), they are compatible with linear codes. In [212], the authors use this property to achieve a high detection rate for these operations. However, the AES also consists of the SubBytes operation, which is nonlinear. As a consequence, it is impossible to efficiently protect the whole state with such parities during all four operations.

In principle, there are two solutions to this problem. The first one protects each byte with a single parity during all four operations. The second solution is to use different protection schemes for linear and nonlinear operations.

The single parity approach was pursued in [34, 424]. There, the authors implement the SubBytes operation with lookup tables which consist of 256 nine-bit entries. The parity of each entry is the sum of the input and the output parities. We denote x as the input and $p(x)$ as the input parity. Analogously, y denotes the output, which can be derived as $y = \texttt{SubBytes}(x)$, and $p(y)$ is the corresponding output parity. The parity which is stored as the ninth bit is thus $(p(x) + p(y))$. Hence, during the SubBytes operation, x is replaced by y from the lookup table and the new parity is calculated as $p(x)+(p(x)+p(y)) = p(y)$. It can be checked that only a valid input can result in a valid output. Alternatively, it is possible to use a lookup table with 512 entries where only 256 lead to a valid output.

A similar approach can be pursued for hardware implementations that do not use lookup tables. In [220], it is shown how the S-Box can be realized as a pure matrix operation. Such an implementation is more expensive than for instance a composite field realization, but it allows us to reduce the circuit to one output bit. This is then equal to calculating the parity for the S-Box.

Although the last two single-bit approaches seem to be very similar, there is a big difference, namely that the first one incorporates the old parity in the calculation of the new one. If this is not the case, that is, if a new parity is calculated from the plain data only and the old parity is discarded, it has to be ensured that the operation is preceded by a check which is still active during the operation. Otherwise it would be possible to manipulate the input data without violating any checks as the parities would be calculated from the very same erroneous input.

The second solution to the problem is to split the protection between the linear and the nonlinear parts. In order to provide continuous protection, the methods for both parts have to be interleaved. Otherwise, the interconnection between the parts would be unprotected and an easy target for the adversary. For the protection of the nonlinear part, i.e. the S-Box, several approaches have been proposed. A possible solution is to split a tower field-based S-Box into five parts [221]. Each part can now be equipped with a parity. Another possibility is to protect the affine transformation of SubBytes together with the linear part. This leaves only the inversion to be protected, which can be achieved by checking whether the input times the output results in the neutral element [221].

However, the protection a parity bit offers is quite limited. Every fault that changes an odd number of bits in a value that is protected by a single parity will remain undetected. Thus, these methods offer cheap protection against adversaries that have only little control over the fault and tend to influence large parts of the device at once. Furthermore, they can significantly increase the effort for attacks that require the injection of several faults, like those on stream ciphers [183].

5.4.2 Operation-Specific Digest Values

In order to protect AES implementations, each operation can be analyzed and secured separately. The dedicated protection mechanisms have to be interleaved afterwards to provide protection for the whole algorithm. The advantage of this approach is that the redundancy can be tailored specifically to each operation. Genelle et al. proposed digest values to protect the whole AES State [158]:

5.4.2.1 SubBytes

The SubBytes operation is split up into the inversion and the affine transformation. The inversion is protected by computing the Multiplicative Digest Value (MDV) of all nonzero values of the state, i.e.

$$\mathrm{MDV}(x) = \prod_{\{i=0,\ldots,15 | x_i \neq 0\}} x_i .$$

If no error occurred during the inversion, the inverted MDV before the operations equals the MDV after it. Since the zero values are not modified by the inversion by definition, Zero Test (ZT) Values are compared before and after the operation:

$$\mathrm{ZT}(x) = \sum_{\{i=0,\ldots,15 | x_i = 0\}} 2^i .$$

In order to increase the detection probability for multiple-byte faults, the Generalized Multiplicative Digest Value (GMDV) can be added:

$$\mathrm{GMDV}(x) = \prod_{\{i=0,\ldots,15 | x_i \neq 0\}} x_i^{i+1} .$$

The affine transformation (AT) is protected by an Additive Digest Value (ADV), i.e.

$$\mathrm{ADV}(x) = \sum_{i=0}^{15} x_i \,.$$

Since $\mathrm{AT}(x_1 + x_2) = \mathrm{AT}(x_1) + \mathrm{AT}(x_2) + c$ for $x_1, x_2 \in \mathrm{GF}(2)^8$ and a constant c, we get $\mathrm{ADV}(AT(s)) = \mathrm{AT}(\mathrm{ADV}(s)) + c$. A possible extension for a multibyte fault scenario is the digest

$$\mathrm{ADV}_\lambda(x) = \sum_{i=0}^{15} \lambda_i x_i$$

for nonzero constants $\lambda_i \in \mathrm{GF}(2^8)$. Naturally, these constants differ before and after the computation of the affine transformation and have to be adjusted accordingly.

5.4.2.2 ShiftRows

The ShiftRows operation is protected by the ADV, since it does not modify it. For the multibyte variants, a Generalized ADV can be used, i.e. $\sum_{i=0}^{15} x_i^3$.

5.4.2.3 MixColumns

Since the MixColumns operation does not change the sum of a column, the ADV provides protection. The multibyte case is addressed by the ADV_λ value.

5.4.2.4 AddRoundKey

For the AddRoundKey operation, the ADV and the ADV_λ are again the natural choice. However, the corresponding values of the round key must be computed and added to the ADV/ADV_λ of the state before the operations to compute the estimated value to compare them with.

An implementation protected by these methods provides perfect security assuming that one byte is manipulated. For a model that allows two errors, an upper bound over all transformations can be given by $14/255^2$. The authors state furthermore that the computation of the additive check values can be combined for the one-byte fault mode to make the method more efficient.

5.4.3 Ring Embedding

In contrast to protecting every operation of the AES separately, extending all AES operations to work on a larger redundant data structure was proposed in [281]. The idea is to embed the AES field into a larger ring. This new ring has the form $(\mathbf{F}_D \times \mathbf{F}_C)$, where $\mathbf{F}_D$ denotes the data algebra and $\mathbf{F}_C$ the check algebra. The countermeasure exploits the property that every operation applied to such a ring element affects the data element and the check element (or signature) independently. On the other hand, altering either of the embedded elements is difficult without altering the other one. Similar techniques have been used for public key cryptography [155, 280, 329].

To protect the AES, the plaintext and the key are embedded into the ring together with defined signature values. Since the signature values after the encryption depend neither on the key nor on the plaintext, they can be checked against pre-calculated values. The main advantage of this scheme is that every operations modifies the signature. Thus, not only are data modifications detected, but also every manipulation of the program flow results in an invalid signature.

Although every AES operation can be expressed as a function which works on the AES field, such implementations are not very efficient. However, to implement an AES operating on a larger ring, such implementations would be needed. To avoid them the authors propose a ring which is especially suited for eight-bit arithmetic. This way, ring multiplications can be implemented using logarithm tables and the S-Box is implemented using redundant table lookups.

The countermeasure detects all first-order bit faults, byte faults and program flow manipulations with certainty. For all other faults the upper success bound is $1/256$. This probability holds for the entire algorithm.

5.4.4 Infective Computation

Instead of trying to detect faults, a possible strategy to prevent fault attacks is to make the output look random to the adversary in the case of an error. This is referred to as infective computation [432]. Thereby, the exploitable relation between the fault and the output is cut, making a fault analysis attack impossible. The advantage of these methods over detection is that they require no check procedures, which can be the target of a second-order attack [226].

Joye et al. presented a method to use infective computation for AES [200]: The main idea is to double the algorithm, while scrambling the data paths of the two implementations. The scrambling can be done byte-wise or bit-wise, while interleaving the bits provides better protection. A possible method to implement bit scrambling is depicted for a single operation in Fig. 5.5.

This method increases the effort of an adversary, while creating no extra cost for an already doubled hardware implementation. Thus, it could be implemented in addition to DMR approaches, providing diffusion in case an adversary is able to

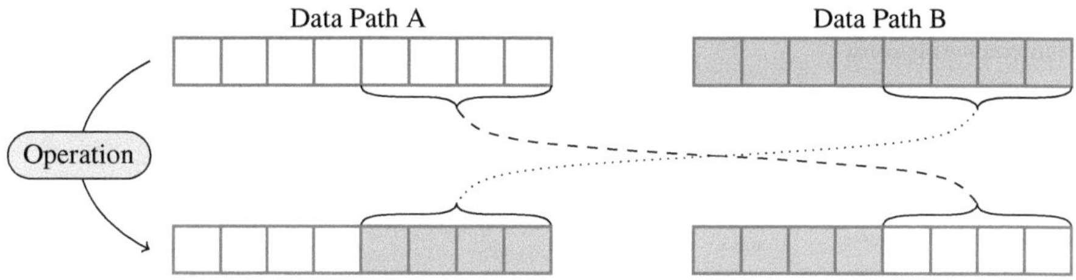

Fig. 5.5 Example of bit-wise scrambling for a byte of one operation

bypass the check routine. However, an adversary who manages to induce the same fault in both circuits circumvents checking and scrambling at once.

5.5 Protocol-Level Countermeasures

Protocol-level countermeasures are based on the fact that differential attacks need at least two encryptions with the same key and plaintext. Thus, they either randomize the key or the plaintext. Such countermeasures are not only very effective but also low in cost when compared to the previously discussed methods. Additionally, some of them even provide unified protection against DPA and fault attacks. Their main disadvantages are that (1) changes in the protocol are needed, which is often difficult in widely established systems, and (2) that only one party within a two-party communication can be protected.

In particular, we will look at three different approaches. The first one is based on all-or-nothing transforms and was actually intended to protect against DPA attacks. However, it automatically protects against fault attacks. The second technique is also based on message randomization and is the cheapest of all approaches. The disadvantage is that it only provides fault protection. Finally, the third countermeasure is based on re-keying and thus provides a unified countermeasure against DPA and fault attacks.

All three approaches can only protect one party in a two-party communication. This is because only the encryption is probabilistic and an adversary could always attack the decryption. The first approach can protect two parties against DPA but requires a second pre-shared key to achieve this. However, the fact that only one party can be protected is often not a concern in low-cost applications. This is because if only one party has to be low-cost (this could be an RFID tag or a smart card), the other party (e.g. an RFID or smart card reader) can be protected by different, more expensive, means.

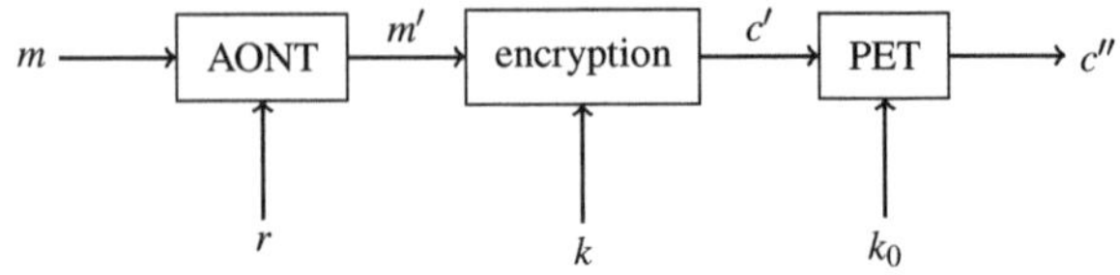

Fig. 5.6 AONT as a DPA and fault countermeasure

5.5.1 All-or-Nothing Transforms

In [278], all-or-nothing transforms (AONTs) were introduced as a DPA countermeasure. The basic idea is depicted in Fig. 5.6. The AONT takes the input message m and uses the value r to generate the randomized message m'. The value r is random and unknown to the adversary. Nor is it actually needed to reconstruct m from m'. After applying the AONT, an ordinary encryption of m' takes place. This would be already sufficient to counter fault attacks. However, DPA attacks can also be performed on the ciphertext and therefore a post-encryption transformation (PET) is necessary. The PET takes the value c' and uses another pre-shared key k_0 to generate c''. A possible realization of an AONT is optimal asymmetric encryption padding (OAEP) [30].

5.5.2 The Message Modifying Approach

Probably the most efficient approach to protect block ciphers against fault attacks is described in [171]. It encrypts $m' = m \oplus r$ instead of m. Here, r is random but made public together with the ciphertext. As a result, an adversary has no control over the message to encrypt. However, m' is known to the attacker and thus DPA attacks are possible.

5.5.3 Fresh Re-keying

The third protocol-level countermeasure we look at is fresh re-keying [282]. As the name already suggests, fault attacks are prevented by randomizing the key before every encryption. The idea, as illustrated in Fig. 5.7, is to use a function g to derive a session key $k' = g(k, r)$. Afterwards k' is used to encrypt m. The value r is random, but made public together with the ciphertext. The scheme is more expensive than the previous one because g must be more complex than a simple XOR in order to provide DPA security. However, in contrast to the AONT approach this scheme does not need an additional pre-shared key.

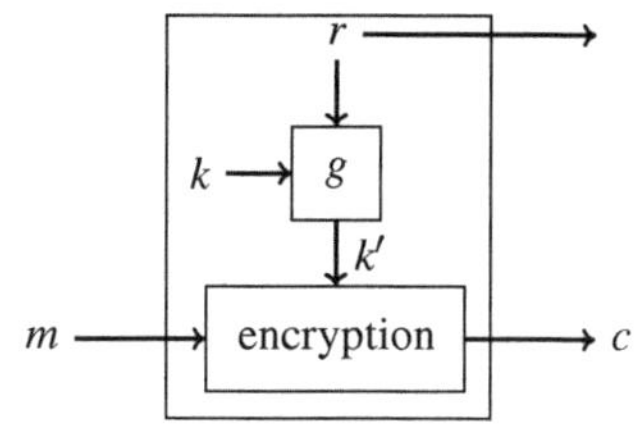

Fig. 5.7 Fresh re-keying

5.6 Conclusion

This chapter discusses different ways to protect structures of block ciphers against malicious adversaries. We presented general methods that can be applied to an arbitrary algorithm. These methods are very expensive and/or provide only limited protection against strong adversaries. Therefore, we looked at ways to optimize DMR approaches for block ciphers. Thereby, specific structures of the ciphers are used to decrease the costs, and to increase the effort of an adversary to inject an undetected fault.

A different approach to protect ciphers is to use coding-theoretic methods. They have to be tailored to the algorithm used. Since it is the current NIST standard for symmetric encryption, we put the emphasis for countermeasures based on coding theory on AES. On the hardware side we looked at parity schemes. It can be observed that most parity schemes are cheaper than DMR approaches but often provide less security when assuming a strong adversary. The discussed software countermeasures on the other hand can thwart rather strong adversaries. The cost for this benefit is the significantly higher execution time.

Another way to protect ciphers is to shift the protection mechanisms from the algorithmic level to the protocol level. By randomizing the message or the key prior to each encryption, an adversary is not able to obtain the same ciphertext twice, which is (at the least) required for the known fault attacks on block ciphers.

We can conclude that fault countermeasures for symmetric encryption are much more expensive than those for public key algorithms. Therefore, low-cost devices should be protected by protocol-level countermeasures whenever possible. For all other applications strong coding-theoretic countermeasures should be implemented if sufficient computational power is available. For a good security/performance trade-off we suggest DMR schemes. Finally, parity schemes can be used if precise attacks are not a threat and performance is important.

Chapter 6
On Countermeasures Against Fault Attacks on the Advanced Encryption Standard

Kaouthar Bousselam, Giorgio Di Natale, Marie-Lise Flottes and Bruno Rouzeyre

Abstract This chapter presents redundancy-based error detection mechanisms deployed in devices implementing the Advanced Encryption Standard for preventing fault-based attacks. Different forms of redundancy are examined, highlighting strengths and weaknesses with regard to cost, global error detection capabilities, and ability to detect errors.

6.1 Introduction

Cipher algorithms are often integrated as coprocessors for performance reasons. Hopefully, cryptanalysis on most recent algorithms is not practical.

Unfortunately, numerous types of attacks against secure devices rely on the hardware implementation of the ciphers. These attacks take advantage of logical or physical information naturally processed or leaked by the physical component. Invasive attacks, which use probes and irreversible modifications of the chip, are very powerful but destroy the package, and require the time of experts in well-equipped laboratories and a large budget.

Conversely, non-invasive side-channel attacks use leakage information related to the processed data such as the operational timing, the power consumption of the chip, the electromagnetic emanations of signals. In the middle, active but semi- or non-invasive fault-based attacks rely on perturbation of the circuit and use the (expected) production of erroneous results for inferring secret information. Formally, the faults reflect the physical conditions that cause a circuit to fail to perform in a required manner. The error is the visible aspect of the fault, i.e. a wrong observable signal

K. Bousselam · B. Rouzeyre
Université de Montpellier II, Montpellier, France

G. Di Natale · M.-L. Flottes
LIRMM / CNRS UMR 5506, Montpellier, France

M. Joye and M. Tunstall (eds.), *Fault Analysis in Cryptography*,
Information Security and Cryptography, DOI: 10.1007/978-3-642-29656-7_6,

produced by the defective device. The fault set induced by fault-based attacks includes transient faults on combinational gates and bit-flips on memory elements [253].

Several techniques can be used for fault injection. They rely either on perturbation of the environmental conditions (e.g. power supply, clock), or, with higher cost but better precision, on injection of transients or bit-flips into targeted signals (e.g. faults injected through laser beams).

There are two forms of countermeasures against fault-based attacks: sensor-based and error detection-based countermeasures. The former aim at detecting inappropriate environmental conditions (e.g. unexpected light, clock glitches). This chapter focuses on the latter and shows how potential faults can be detected through identification of erroneous signals during execution. Note that transient single-event upsets (SEUs) caused by various types of cosmic or terrestrial radiation also manifest themselves as bit-flips on storage elements and transient faults on logic gates. As a result, error detection schemes implemented for preventing fault attacks also detect natural transient and bit-flip faults.

Section 6.2 details an example of a symmetric block cipher, the Advanced Encryption Standard (AES), which will be used throughout this chapter as an application example. Several hardware implementations of this algorithm are presented, allowing us to trade off area for performance. Section 6.2.2 gives an overview of fault attacks on AES cryptoprocessors and explains how errors can be exploited for retrieving the secret encryption key. Section 6.3 presents solutions from the literature for online error detection on AES. These solutions are compared using common evaluation criteria such as implementation cost, error detection latency, and error detection rate. Finally, the quality of these protection mechanisms is also evaluated in terms of their capacity to detect the most frequent errors (Sect. 6.4.2).

6.2 Advanced Encryption Standard

Even if the error detection principles described in this chapter are general enough to be implemented for secure devices implementing many cryptographic algorithms, we use the AES as a support example. Section 6.2.1 details the AES algorithm, which was adopted as a symmetric key cryptographic standard by the US Government in 2001, and Sect. 6.2.2 focuses on its hardware implementations.

6.2.1 Algorithm Description

The AES algorithm is a symmetric block cipher using cryptographic keys of 128, 192, and 256 bits to encrypt and decrypt data in blocks of 128 bits. Symmetric encryption and decryption are performed by means of the same cryptographic secret key. This section focuses on the encryption/decryption algorithm for 128-bit cryptographic keys (details on others key lengths are fully given in [142]).

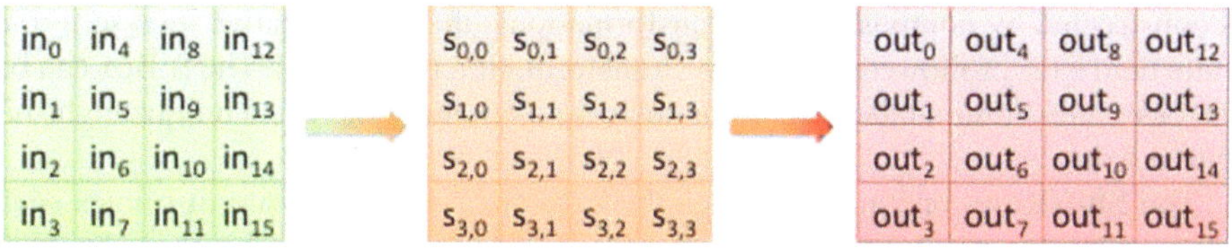

Fig. 6.1 Input, state and output arrays

Fig. 6.2 AES algorithm

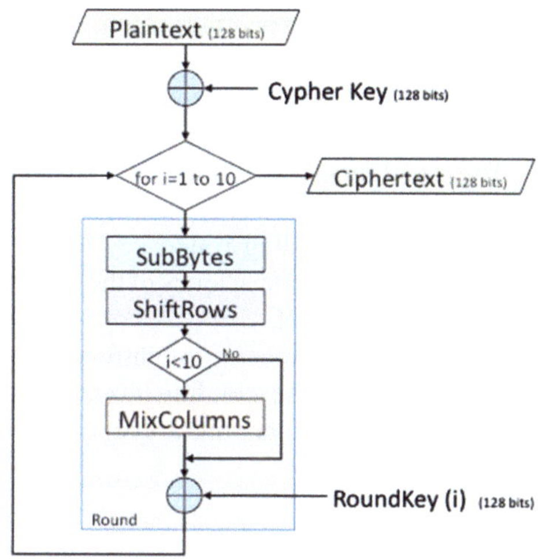

The basic unit for processing in the AES algorithm is a byte. Input, output and secret key bit sequences are internally processed on a two-dimensional array of bytes called the State. The State consists of four rows of four bytes. In the State array, denoted by the symbol s, each individual byte has two indices, with its row number r in the range $0 \leq r \leq 3$ and its column number c in the range $0 \leq c \leq 3$.

The AES algorithm is based on permutations and substitutions. Permutations are rearrangements of data, while substitutions replace one byte of data with another. AES performs permutations and substitutions using four functions, as explained later.

At the beginning of the encryption, the input is copied to the State array using the conventions described in Fig. 6.1. After an initial XOR operation between the State array and the cipher key, the State array is transformed by implementing a round function that is repeated ten times, with the final round differing slightly from the first nine rounds. The final State is then copied to the output as shown in Fig. 6.1. The round function is composed of four operations: SubBytes, ShiftRows, MixColumns, and AddRoundKey. These functions operate and modify the value of the State. As shown in Fig. 6.2, all rounds are identical with the exception of the final round, which does not include the MixColumns transformation.

The **SubBytes** transformation is a nonlinear byte substitution that operates independently on each byte of the State using a substitution table (S-Box). The S-Box is constructed by composing two transformations, the multiplicative inverse in the finite field $\mathbb{F}(2^8)$ followed by an affine transformation (over $\mathbb{F}(2)$). The S-Box can be seen as a substitution table where for each eight-bit input there is a unique eight-bit output value.

The **ShiftRows** operation changes the byte position in the state. It rotates each row by different offsets to obtain a new state. In particular, the first row is unchanged; the second row is rotated one byte position to the left, the third row is rotated two positions, and the fourth row three positions. ShiftRows is a linear transformation.

The **MixColumns** operates column-wise altering all the bytes of the same column, and combining the four bytes in each column. It treats a column as a third-degree polynomial with coefficients in $\mathbb{F}(2^8)$ and produces the new column by multiplying it with a constant matrix.

The round function is parameterized using a key expansion function that generates a variation of the original cipher key at each round repetition. The **AddRoundKey** adds this corresponding round key to the current State using bit-wise XOR. Not relevant for the self-consistency of the chapter, details on the key expansion procedure can be found in [142].

The decryption process straightforwardly follows the encryption process by using at each round the corresponding inverse functions, i.e. InvSubBytes, InvMixColumns and InvShiftRows. Functions are executed in the opposite order as that of round keys.

6.2.2 AES Hardware Implementations

Since the standardization of the AES in 2001, several papers have been published proposing different hardware implementations, to cope with different requirements and goals such as throughput, speed, target (i.e., FPGA or ASIC), power consumption, area, and resistance to side-channel attacks. This section presents the main classes of solutions, from the architectural point of view to the optimization of each block.

At a high level, three architectures are possible: iterated, loop-unrolled, and pipelined. In the iterated architecture the circuit's data path implements the basic functions of one round, and the data is iterated several times in order to obtain the final result. This architecture leads to the smallest (and the slowest) implementation. Loop-unrolled architectures implement two (or more) rounds per clock cycle, and the execution of the algorithm is iterated. They achieve the highest speed at the cost of greater area. Finally, pipelined architectures increase the throughput of encryption/decryption by processing multiple blocks of data simultaneously. They are realized by inserting some registers so that each pipeline stage can be one round function (or sub-function), the whole round, or more than one round (in the case of a loop-unrolled architecture).

Since area is one of the main concerns when dealing with embedded systems, such as smart cards, instead of implementing a data path that contains the whole round functions (i.e., with 16 instances of S-Boxes and four MixColumns as in [267, 357]) several works propose reducing the width of the data path in iterated-based solutions, thus increasing the overall number of clock cycles required to complete the encryption process. The smallest architectures proposed in literature are based on eight-bit data paths and contain only one S-Box. Very compact solutions for ASIC implementations are proposed in [137, 138, 174], while solutions targeting FPGA implementations are proposed in [135, 166].

Scalable solutions with varying data path widths are proposed in [267, 357]. In all these architectures, the designer can trade off performance for design cost by selecting the number of parallel S-Boxes that are implemented. The smallest version they propose is a 32-bit architecture. An optimized 32-bit architecture targeting FPGA implementations is proposed in [88], where round keys are precomputed and S-Boxes are optimized using lookup tables.

Concerning the hardware implementation of the round functions, first it must be noted that no optimizations can be performed on ShiftRows and AddRoundKey transformations, since no logic gates are needed for the former transformation and only one XOR layer is needed for the latter. However, several optimizations can be found for S-Boxes and MixColumns.

The MixColumns function can be implemented either in a straightforward manner by instancing all the XOR gates required to perform the matrix multiplication, or in an iterated (and smallest) way, where the overall multiplication is completed in four clock cycles (one for each byte). However, several approaches are proposed for S-Boxes:

1. ROM-based [433].
2. Truth table implementation (combinational logic): the S-Box is written in any hardware description language as a truth table and it is converted into a gate-level circuit using a logic synthesizer [119]. This solution allows the best performance in terms of speed and dynamic power consumption.
3. Lookup Tables, suited for FPGA implementations [88, 135, 175, 328];
4. Mathematical implementation: the arithmetic operation on $\mathbb{F}(2^8)$ of the SubBytes function is mapped to the isomorphic field $\mathbb{F}((2^4)^2)$. This alternative can provide better solutions in terms of area occupation and power consumption, at the cost of a slower solution [351, 421, 433].

To provide an idea of the differences between truth table-based implementations and mathematical implementations, we implemented the solutions proposed in [119] (truth table) and [421] (mathematical) in VHDL and synthesized the two circuits. Table 6.1 shows the comparison of area, speed and power consumption obtained using a 90 nm technology library provided by ST Microelectronics.

A compact solution is proposed in [357] for implementation of both encryption and decryption in the same device. Encryption and decryption hardware are merged by sharing the AddRoundKey XOR gates, GF(2^8) inverters in S-Boxes, and common terms between the permutation functions MixColumns and inverse MixColumns. This optimization focuses on applications that do not require duplex modes where encryption and decryption are performed at the same time.

Table 6.1 Comparison of area, delay and power between truth-table implementation and mathematical implementation (ASIC)

	Truth table	Mathematical
Area [μm^2]	1993	1122
Delay of critical path [ns]	1.25 (14 logic levels)	2.45 (25 logic levels)
Average Dynamic Power [mW]	0.136	0.907

6.3 Fault Attacks

As mentioned in the introduction, faults can be intentionally injected into the circuit in order to retrieve the secret key [55]. Given certain errors resulting from a fault, an attacker can deduce the key by comparing the result of a normal encryption with the faulty one.

Faults can be injected into the circuit by different means, such as temperature variation, clock frequency modification, glitches in power supply, and exposure to radiation or light [227]. The main advantage of laser-based fault injection is the localization of the fault in the timing and the spatial domains. Nevertheless, whatever the means used to inject the fault, the induced errors must satisfy certain conditions in order to be successfully exploited. This section briefly reviews some of the published fault attacks on AES and analyzes their conditions on the injected errors. All the reported attacks have it in common to assume that the error affects the state matrix at a given instant and presents a particular characteristic in terms of the number and the locations of erroneous bits.

One of the first cryptanalysis methods using faults on AES was published in [227]. The considered error is a single faulty bit in any of the 128 State bits at the input of the SubBytes operation during execution of the last encryption round (tenth round). This error of multiplicity 1 (only one bit is affected) is equivalent to an error of multiplicity 1 appearing on the round key or at the input of the previous AddRoundKey operation. This error spreads, affects the output of the SubBytes operation and, since the last round does not include the MixColumns operation, affects one (and only one) byte of the final output array.

In [55] the authors proposed attacks based on the "safe-error" principle, i.e. "the error affects the result or not". The considered fault is a stuck-at-0 affecting one bit of the key. If the result is faulty, it can be deduced that the actual value of the key bit is 1. In [51] the authors report another attack based on the injection of an error of multiplicity 1 which exploits collision effects, i.e. the fact that two messages (one without, the other with an error) give the same result.

Giraud also proposes a more complex attack with the advantage of considering a less restrictive error model than errors on a single bit [160]. Here, errors of multiplicity x ($x \geq 1$) affecting one byte are taken into account. Some other attacks relying on several fault injections on bytes are reported in [84, 127, 324].

The attack in [55], based on the "safe-error" principle, can be extended to the case of an attack on one byte. In this case all the input bits of an S-Box must be stuck-at-0. Then, by applying all 256 values at the input of the S-Box, a collision appears: when the byte has the same value as the key, the result equals the one obtained with the stuck-at fault.

In [296] the authors propose a generalization of the attack on a single byte, focusing on three or four bytes in the first State array column. The faults can be injected at any step of the AES process. For the attack to be successful, it is necessary to know whether the error affects three or four bytes or not. Furthermore, six and 1,500 error injections have to be performed for three and four faulty bytes respectively.

In this chapter, we do not discuss the actual capabilities of injecting faults according to the hypotheses underlying the attacks. Nevertheless, from this overview, the following can be concluded:

- Errors affecting one byte (or even four in [296]) of the State are easily exploitable. Thus it is of prime importance to detect errors located within a byte of the State. All error multiplicities (1 to 8) have to be considered.
- According to the variety of reported attacks, all rounds of the AES are prone to fault injections. Thus, the data protection mechanism must span the whole AES process.
- While errors affecting more than one byte are not typically exploitable, their detection is of interest since it helps in detecting an attack (for instance laser-based attacks need in practice many shots before succeeding in flipping bits within a single byte). Second, the ingenuity of hackers may make these attacks efficient in the future.

6.4 Error Detection Methods

Online testing methods aim at detecting errors in data processed by a device in mission mode. These errors are characterized by their multiplicity, i.e. the number of bits in the erroneous data that differ from the excepted data value (e.g. single- or multiple-bit errors). Error multiplicity is strongly correlated to the hardware implementation of the device and the faults likely to affect its structure (effect, duration and location in space and time dimensions).

Error detection exploits different forms of redundancy, namely hardware, temporal and information redundancies. The methods are generally evaluated in terms of error detection rate, error detection latency, performance, and area cost, and less frequently in terms of the capacity to detect the most likely errors. The following subsections review several error detection mechanisms and evaluate these mechanisms with regard to classical criteria. A discussion on the error detection rate is deferred to Sect. 6.5.

6.4.1 Hardware and Temporal Redundancies

Temporal redundancy consists of performing the same process (operation, round, or encryption) twice, or following it by the inverse process (operation, round, or decryption of the result). In the former case, the results of the two processes are compared in order to check for a possible mismatch. In the latter, the result of the inverse process is compared with the initial data. The temporal redundancy mainly affects the error detection latency (>100 %), but provides a very high error detection rate except in the case of two faults affecting both processes. This case is rather unlikely in attacks where the temporal redundancy relies on the execution of two different functions (e.g. a process followed by its inverse).

Conversely, simple duplication of the device (hardware redundancy) allows encrypting (or decrypting) the same data twice at the same time. It provides a very high error detection rate since any fault affecting one copy of the circuit and producing an error on an output is detected by comparison with the output of the other instance. Comparison of the two results can be performed during the operation of the device, at round or algorithm levels for different fault detection latencies.

The cost in terms of area overhead (>100 %) limits this approach to some operations in the round (perhaps at the exception of systems requiring high reliability against maliciously injected faults). However, duplicated circuits may increase the correlation between power consumption and confidential data under processing and thus may contribute to side-channel attacks such as Differential Power Analysis [240].

Karri, Wu, Mishra and Kim [218] propose a temporal redundancy solution for systems containing both encryption and decryption modules. The error detection scheme takes advantage of the inherent redundancy in the structure and relies on the inverse relationships that exist between encryption and decryption at the algorithm level, round level, and operation level. After encryption, the resulting data is processed by a corresponding decryption module; its output is then compared with the input of the preceding encryption process to check for an eventual mismatch. The fault detection latency decreases with the duplication level (algorithm level > round level > operation level). Maximum performance degradation and area overhead are respectively 61 % (when temporal redundancy is performed at the algorithm level) and 38 % (when temporal redundancy is performed at the operation level). A similar approach has been suggested in [35] for the key schedule module where an inverse key schedule module is used to reconstruct the previous round key from the current one.

Satoh, Sugawara, Homman and Aoki [358] propose a temporal redundancy scheme for the compact encryption/decryption implementation further described in [357]. Round block B_r is divided into two sub-blocks B_{r_1} and B_{r_2} for serial execution of two sub-operations op_{r_1} and op_{r_2}. The two sub-blocks are used alternatively for encryption (op_{r_i}) or decryption and error detection (inverse-op_{r_i} and comparison). At cycle n, for instance, op_{r_1} is executed on B_{r_1}. At the same time, inverse-op_{r_2} is executed on B_{r_2} for cross-checking with the op_{r_2} performed at cycle $n-1$.

At cycle $n + 1$, op_{r_2} is executed on B_{r_2} for completion of the round operation op_r, while inverse-op_{r_1} is executed in parallel for cross-checking with the op_{r_1} executed at cycle n. Taking advantage of the optimized hardware of the compact implementation, (encryption and decryption operations share the same hardware when possible), execution of inverse-op_{r_1} does not require extra hardware compared to op_{r_1}. Error latency is short (half-round level), and the error detection scheme does not affect much the performance of the initial compact implementation (at most 14.5 %).

Maistri, Vanhauwaert and Leveugle [265] propose a design solution that exploits temporal redundancy with a Double-Data-Rate (DDR) mechanism. The pipelined AES data path logic is partitioned into two classes, where nonadjacent stages in the pipeline are driven by two opposite clock signals. The DDR architecture allows us to halve the number of clock cycles per round, though maybe with a light impact on clock frequency compared with a design without protection. It takes advantage of the free cycles for recomputing the round on the same hardware. Two successive round operation outputs obtained from two copies of the same input data are checked for possible mismatches. A fault injection campaign performed on the low-area AES architecture from [267] shows an almost maximal error detection rate on the data-path at the cost of 36 % area overhead.

Temporal redundancy provides high error detection rates as long as both data under comparison are not affected by the same error. It assumes the attacker is able to induce two faults with identical affect on the structure at different times, i.e. during the first and the second operations to compare. The implementation of such attacks is all the more tricky as the two operations are different, for instance when the data input is first encrypted (first operation) and then decrypted (second operation) for cross-checking [218]. Conversely, the protection scheme based on hardware redundancy (simple duplication) can be fooled by the injection of two identical faults (same location) at the same time into the two versions of the circuit. Again, the attack implementation becomes more complex when the two circuits under comparison have different structural implementations.

6.4.2 Information Redundancy

Error detection codes check for a possible mismatch between a code predicted for an output from the current input and the code of the actual output of the process. Code prediction is performed in parallel with the main process (algorithm, round, operation) in such a way that predicted and actual codes can be compared at the end.

Wu et al. [424] compared predicted and actual codes at the round level on a 128-bit iterated AES hardware implementation with ROM-based S-Boxes (Fig. 6.3). For the prediction, a parity bit is computed for the 128-bit input of the round $P(x)$. $P(x)$ is then updated according to the round operations. For each S-Box S_i, a parity bit is predicted for words composed by its input (x_i) and the expected output (y_i). The 16 parity bits so calculated for the 16 S-Boxes are then added to obtain a single parity bit ($P(x) \oplus P(y)$). This parity bit is added again to the input parity bit $P(x)$ so that

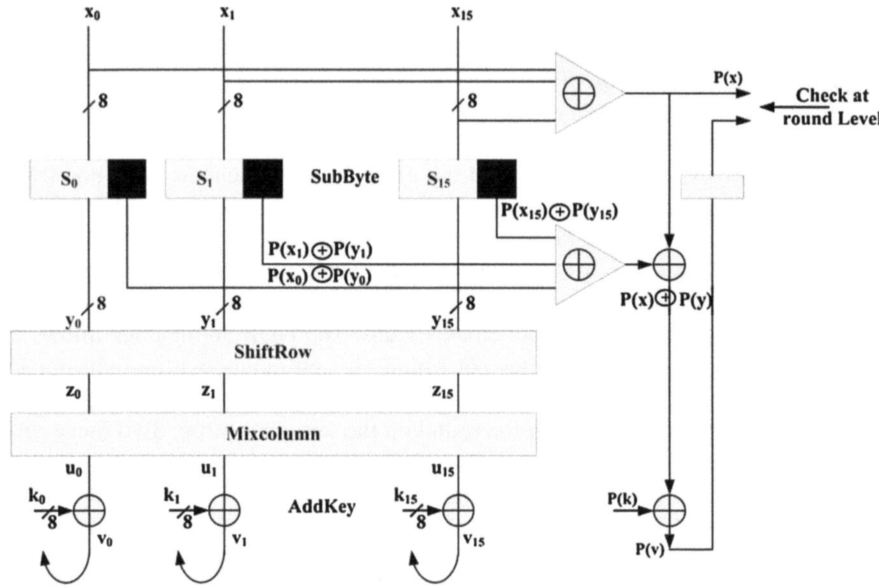

Fig. 6.3 Parity for state [424]

the parity $P(y)$ is determined. $P(y)$ is affected neither by the ShiftRows operation, which only changes the byte position in the State array, nor by the MixColumns operations, as demonstrated by Wu et al. [424]. This parity bit is XORed with the parity of the key k, $P(k)$, and stored in a one-bit register. Errors are detected when this predicted parity bit differs from the input parity computed at the beginning of the next round, which corresponds to the actual parity bit of the preceding round's output. Implemented on a Xilinx Virtex 1000 FPGA, this solution results in 8 % of hardware overhead and 5 % of performance degradation. We will refer to this architecture as "Parity for State".

Bertoni et al. [34] propose the use of a parity bit that is associated with each byte of the State matrix of a 128-bit iterated hardware implementation with ROM-based S-Boxes (see Fig. 6.4). Predicted parity bits on S-Box outputs are stored as additional bits in the ROMs (nine bits instead of eight in the original S-Boxes). In order to detect errors on input parities and in the memory, the authors propose increasing each ROM to 512×9 bits in such a way that all the ROM words addressed with a wrong input address (i.e. S-Boxes input with a wrong associated parity) deliberately store values with a wrong output parity so that the detection mechanism will detect the fault. As before, the parity bit associated with each byte is not affected by the ShiftRows operation. Conversely to the preceding scheme, where the global parity bit on the 128 bits remains unchanged after the MixColumns operation, when working at the byte level, the parity after the MixColumns operation is affected, therefore requiring the implementation of prediction functions. Finally, the parity bits after the AddRoundKey operation are computed as before by adding the current parity bits to

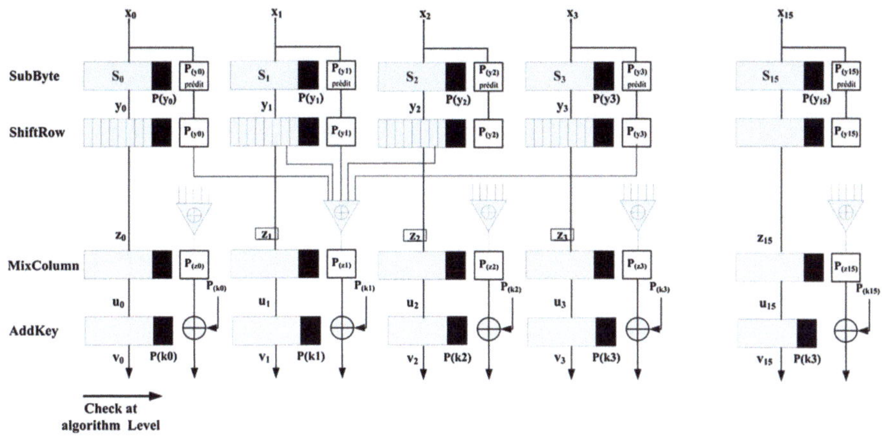

Fig. 6.4 Parity per byte [34]

the parity bits of the corresponding round key. We will refer to this architecture as "Parity per byte".

Yen and Wu [425] propose the use of an $(n+1, n)$ Cyclic Redundancy Check (CRC) over $\mathbb{F}(2^8)$ to detect errors during encryption, with $n = 4, 8, 16$. The generator polynomial is $g(x) = 1 + x$. For instance, for $n = 4$, 32 parity bits are added to the State, each parity bit being the column-wise sum of State bits. The error detection can be generated for each four-byte column of the State ($n = 4$ bytes, CRC(5,4)), for two columns ($n = 8$ bytes, CRC(9,8)), or for the whole 16-byte State ($n = 16$ bytes, CRC(17,16)). The CRC is adopted for most operations in the round, except Sboxes where two implementation types are explored: truth-table or $\mathbb{F}(2^8)$ inversion and affine transformation. In the first case, it is assumed that both SubBytes and InvSubBytes are implemented, allowing temporal redundancy at the operation level as in [218]. In the second implementation, error detection is applied separately to the $\mathbb{F}(2^8)$ inversion and the affine transformation. The error detection for the $\mathbb{F}(2^8)$ inversion is achieved by either temporal redundancy or CRC; the error detection for the affine transformation is achieved by CRC. This solution allows a very high error detection level at the cost of high area overhead. We will refer to this architecture as "CRC for State".

Kulikowski et al. [242] propose a protection scheme similar to the previous one, but based on a class of nonlinear systematic error-detecting codes called robust codes instead of a CRC. Motivations come from the inherent difference between the fault and the subsequent error model when considering protection for computation channels or for cryptographic devices possibly subject to active fault attacks. A protection based on nonlinear systematic robust codes performed on 32-bit iterated AES architectures provides for uniform protection against all errors without making any assumption about these errors, at the cost of 28 additional code bits, leading to a 77 % area overhead.

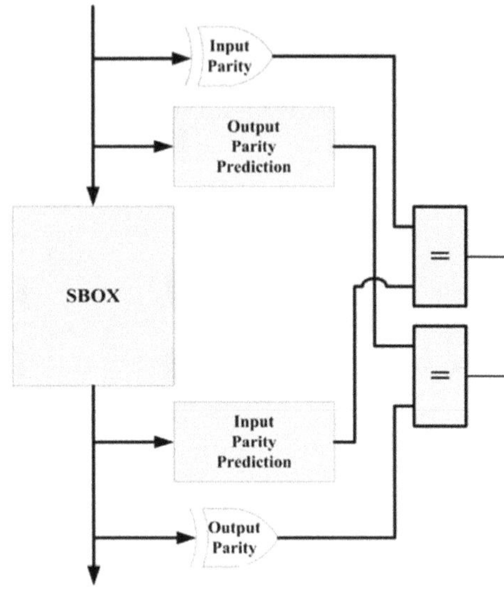

Fig. 6.5 Double parity for S-Box [120]

Di Natale, Flottes and Rouzeyre propose another solution that focuses on S-Boxes. The principle is to add two parity bits per S-Box, one parity bit for the input byte and one for the output byte, and therefore the logic for predicting such information (see Fig. 6.5). The actual output parity is compared with the predicted output parity bit, and the actual input parity bit is compared with the predicted one [120]. When the S-Box and the prediction circuits are synthesized as combinational logic, the area overhead is 38.33 % with respect to the original S-Box. With respect to the solution given in [34], where the authors also proposed the use of a double parity bit for the S-Box, this solution allows detecting an additional 27 % of errors.

In Fig. 6.6, the double parity checking on the S-Box is applied to the scheme proposed in [424]. For each S-Box output, the prediction of the input parity bit is calculated. This predicted input parity bit is then compared with the actual input parity of each S-Box. Finally, the error indication flag of the SubByte operation is obtained as the result of the logical OR of the sixteen indication flags issued from the S-Boxes. We will refer to this architecture as "Parity for State + double parity for S-Box".

In Fig. 6.7, the double parity checking on the S-Box is applied to the scheme proposed in [34]. In the original solution there is a parity bit for each byte of the State matrix. The principle of the proposed solution is that in the case of error detection at the S-Box level, the corresponding parity bit, which is then propagated to the ShiftRows operation, is deliberately forced to have the opposite value, so that the detection mechanism will detect the error. We will refer to this architecture as "Parity per byte + double parity for S-Box".

Another approach that focuses on S-Boxes is presented in [219]. Using composite field arithmetic, the S-Box is divided into five cascaded blocks, each of them designed

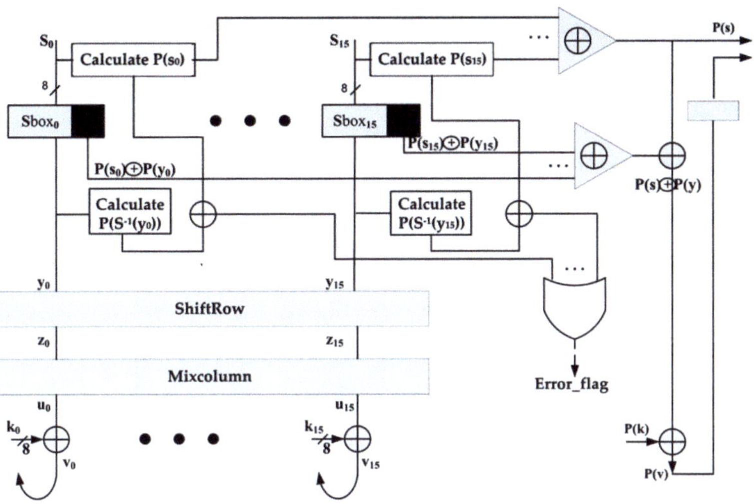

Fig. 6.6 Parity for State + double parity for S-Box [120, 424]

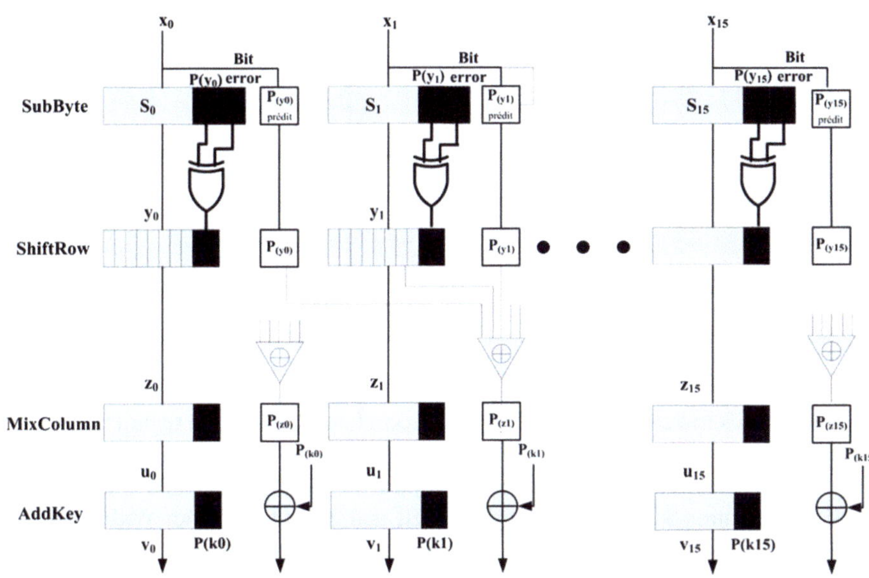

Fig. 6.7 Parity per byte + double parity for S-Box [34, 120]

so that single faults lead to an odd parity error at the output of the related block. Parity-based concurrent fault detection is associated with each block, i.e. five parity bits are used for each S-Box. This detection scheme is capable of detecting all the single faults occurring in an S-Box. The use of composite field arithmetic for decomposition of

Table 6.2 Synthesis results

	Area (μm)	Power (mW)	Timing (ns)	Check bits
Orig. AES	814083	607.15	14.4	
Parity for State	16.34 %	23.28 %	0 %	1
Parity for State + double parity for S-Box	30.13 %	70.07 %	15.5 %	2
Parity per byte	21.73 %	12.79 %	15.41 %	16
Parity per byte + double parity for S-Box	49.09 %	71.45 %	34.51 %	16
CRC for State	92.26 %	104.92 %	147.22 %	32

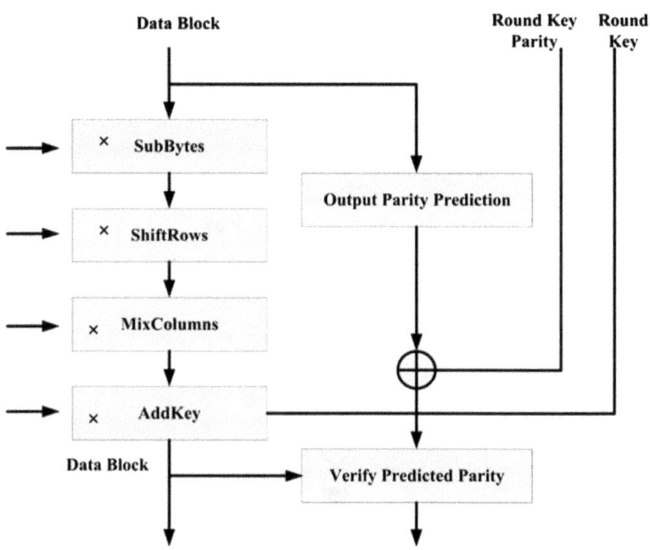

Fig. 6.8 Error injection model

the S-Box leads to a structure with lower area and larger delay with regard to standard implementations.

All the protection schemes under comparison are implemented on a 128-bit AES architecture. S-Boxes are implemented using random logic. With regard to the scheme presented in [425], we used CRC (5,4), i.e. four CRC bytes, one for each column of the State array (32 bits). Table 6.2 summarizes the main overheads entailed by those techniques compared to an original AES design without any error detection mechanism. The same synthesis options are used for all designs.

With regard to the error detection capability, we first analyzed the effect of errors affecting a single byte of the State matrix since known fault attacks rely on errors affecting a single byte of the State array. All the 1 to 8 multiplicity errors have been exhaustively injected into each byte and at each step before any round operation.

Table 6.3 reports the error multiplicity (one to eight erroneous bits), the number of simulated errors for each error multiplicity (e.g. $16 \times 40 \times 28 = 17,920$ error

Table 6.3 Detection capability for errors in a single byte

Error multiplicity	# Errors	Percentage of the undetected errors				
		Parity for State	Parity for State + parity for S-Box	Parity per byte	Parity per byte + double parity for S-Box	CRC for State
1	5120	13.01	0	0	0	0
2	17920	86.31	86.31	94.38	81.84	0
3	35840	12.23	0	0	0	0
4	44800	86.01	86.01	88.75	76.10	0
5	35840	12.77	0	0	0	0
6	17920	86.22	86.22	83.13	70.5	0
7	5120	12.01	0	0	0	0
8	640	86.88	86.88	77.5	26.72	0

instances for two erroneous bits among eight bits), affecting one byte over 16 in the State array, after one of the $4 \times 10 = 40$ operations executed during the encryption), and the percentage of the undetected errors for each technique under evaluation. As expected, errors with odd multiplicity are more easily detected with detection schemes based on parity codes. Only solution "Parity for State" does not guarantee a high error detection level since it uses a single parity bit for the whole 128-bit State matrix. In contrast, the CRC-based scheme outperforms other techniques by detecting even-multiplicity errors too, but at the cost of large penalties in terms of area overhead, power consumption, and performance degradation (see Table 6.2).

We also injected random errors affecting not only a single byte but bytes of any subset of the state matrix with error multiplicity ranging from 1 to 64 faulty bits. Note that even if errors affecting several bytes are not easily exploitable during DFA (only one laser-based fault injection over 10,000 leads to an exploitable error), their detection is of prime interest for detecting the attack itself. For each error multiplicity, 1,000 randomly chosen injection positions in the State matrix have been randomly selected, and this after every encryption operation (four operations per round, for ten rounds). Finally, a total of 40,000 injections have been performed for each error multiplicity, except for the case of multiplicity 1, where only 128 positions are available, and then 128×40 injections have been performed. A sample of simulation results showing the number of undetected errors is reported in Table 6.4.

Techniques "Parity per byte" [34] and "CRC for State" [425] detect all simulated errors with multiplicity larger than six bits (and therefore solutions based on "Parity per byte + double parity for S-Box" [34, 120]). In contrast, a large percentage of even multiplicity errors are not detected by the "Parity for State" solution [424], mainly due to errors occurring before the ShiftRows, MixColumns and AddRoundKey operations, since a single parity bit over the 128-bit State is used for error detection. However, the use of a double parity bit for each S-Box ("Parity for State +

Table 6.4 Detection capability for random errors (multiplicity 1 to 64 bits) injected into the 128-bit state matrix

Error multiplicity	Percentage of the undetected errors				
	Parity for State	Parity for State + double parity for S-Box	Parity per byte	Parity per byte + double parity for S-Box	CRC for State
1	12.85	0	0	0	0
2	86.09	85.93	5.41	4.71	1.37
3	12.37	0	0	0	0.01
4	86.58	86.25	0.80	0.66	0.18
5	12.69	0	0	0	0.003
6	86.43	86.42	0.20	0.15	0
7	12.40	0	0	0	0
8	86.26	86.18	0	0	0
9	12.37	0	0	0	0
10	85.95	86.05	0	0	0
20	86.19	86.3925	0	0	0
21	12.4425	0	0	0	0
30	86.095	86.275	0	0	0
31	12.685	0	0	0	0
40	86.23	86.14	0	0	0
41	12.41	0	0	0	0
50	86.1925	86.2325	0	0	0
51	12.4425	0	0	0	0
60	86.1575	86.255	0	0	0
61	12.4	0	0	0	0
64	86.3475	86.2775	0	0	0

double parity for S-Box" [120,424]) allows all the errors of odd multiplicity to be detected at the price of larger area overhead.

6.5 Faults and Errors

In the previous section, the code-based detection schemes were evaluated in terms of error detection capabilities. The errors were injected at different spatial and temporal positions in the State matrix. However, this information is not sufficient for evaluating the real strength of the detection schemes when a physical-induced fault is actually disturbing the circuit. Such faults may affect logic signals during execution of any encryption operation, thus resulting in different error profiles (i.e. different error multiplicities on the State matrix).

Table 6.5 S-Boximplementations

	Description	Tool	Area (#cells)
SB1	Truth Table	RTL Compiler©v06.10-s007	553
SB2	Truth table	Design Compiler©D-2010.03-SP5	474
SB3	Inversion truth table, affine transform. as combinational logic	Design Compiler©D-2010.03-SP5	477
SB4	Mathematical expressions [421]	Design Compiler©D-2010.03-SP5	193
SB5	Mathematical expressions [433]	Design Compiler©D-2010.03-SP5	176

This section presents the results of some experiments conducted to show the error profiles resulting from the injection of a single transient bit-flip fault (representative of faults injected by means of a laser beam), and discusses how this information can be used for choosing an appropriate code-based detection scheme. Bit flips are injected into the logic gates description of a circuit implementation instead of errors in the state being considered as before (see Tables 6.3 and 6.4).

- The ShiftRow module is not a concern for faults attacks since it does not involve any logic gate.
- The AddRoundKey module resumes on a single layer of XOR gates; therefore only errors of multiplicity 1 may occur at the output module, under the selected fault model.
- The MixColumns transformation operates on 32 input bits (four bytes in the same column of the State matrix) and it consists of multiplications of each byte by constant coefficients $\{02\}$, $\{03\}$ and $\{01\}$ in $\mathbb{F}(2^8)$ (see before). A possible implementation of the MixColumns operation is shown in Fig. 6.4. In this implementation a redundant layer of XOR gates is added to perform the multiplications by $\{02\}$ and $\{03\}$ independently of each other. Conversely, a "classical" implementation uses the result of the multiplication by $\{02\}$ and the data to compute the $\{03\}$ multiplication. With this slight modification, any single fault propagates to a single-bit error in the State matrix (possible fault sites are depicted in Fig. 6.9).

The case of the SubBytes module is far less favorable. Several hardware implementations can be designed, and the error propagation strongly depends on the implementation of the S-Box. For ROM-based designs, any fault in the decoder changes the expected address to another one. Thus the error multiplicity ranges from 1 to 8 bits at the S-Box output. For standard cell designs, we implemented different versions of the S-Box using different synthesis parameters and implementation styles. In particular, we implemented the following designs using the AMS 0.35 μm technology library. Table 6.5 reports the characteristics of these designs.

Faults have been exhaustively simulated for each S-Box implementation. All possible input values (256 values) are considered and the device is fault simulated for each possible inversion of the logic value at each input and output of each gate by using an in-house fault simulator [62]. The simulator allows collecting the number of erroneous bits at the output of the S-Box for each possible pair P composed by input/fault.

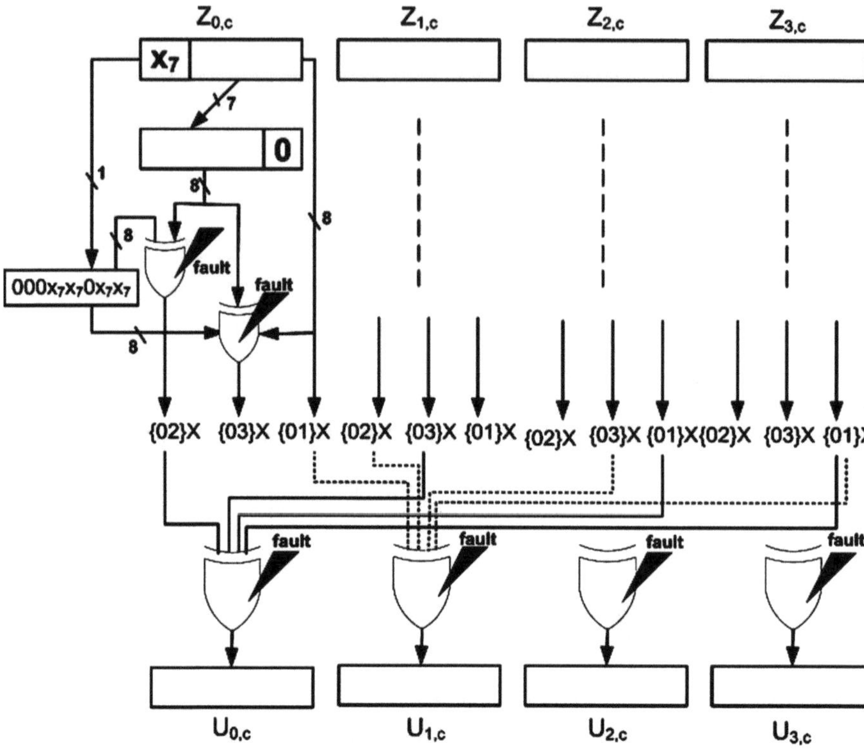

Fig. 6.9 MixColumns implementation

Table 6.6 reports the number of pairs P leading to error multiplicities ranging from 0 to 8 for each S-Box implementation. The first column (error multiplicity = 0) corresponds to all the cases where the faults produce no errors. In all other columns, the table cell reports, besides the number of pairs resulting in the corresponding error multiplicity, the percentage that represents these pairs with respect to the overall number of pairs and, in bold, the percentage that represents these pairs with respect to the number of pairs that lead to at least one error on the S-Box output (i.e., pairs falling in column 0 are discarded).

For instance, among the 19 % of fault simulations for which at least one output bit is erroneous at the output of the SB1 (being that 81 % of experiments result in no error), 78 % result in only one erroneous output bit, justifying therefore the use of a code-based solution that exploits a simple parity bit. Moreover, a deep study of the fault dictionary for the 14 pairs {input/fault} resulting in eight output errors allowed us to find that they correspond to faults injected into the input of the S-Box. Therefore, it might not be necessary to protect the S-Box with detection schemes able to detect eight error bits in this case if these errors are detected at the input of the S-Box.

Obviously, different S-Box implementations lead the series of fault injection to different profiles in terms of error multiplicity. The presented implementations differ

Table 6.6 Fault versus error multiplicity on S-Boxes output

	0	1	2	3	4	5	6	7	8
SB1	201806	36551	6448	2033	953	474	235	62	14
	81.2 %	14.7 %/**78.2 %**	2.6 %/**13.8 %**	0.8 %/**4.4 %**	0.4 %/**2.0 %**	0.2 %/**1.0 %**	0.1 %/**0.5 %**	0.0 %/**0.1 %**	0.0 %/**0.0 %**
SB2	191854	23849	16609	8212	3801	1544	619	145	23
	77.8 %	9.7 %/**43.5 %**	6.7 %/**30.3 %**	3.3 %/**15.0 %**	1.5 %/**6.9 %**	0.6 %/**2.8 %**	0.3 %/**1.1 %**	0.1 %/**0.3 %**	0.0 %/**0.0 %**
SB3	175674	7311	8973	4808	9949	27511	8486	212	20
	72.3 %	3.0 %/**10.9 %**	3.7 %/**13.3 %**	2.0 %/**7.2 %**	4.1 %/**14.8 %**	11.3 %/**40.9 %**	3.5 %/**12.6 %**	0.1 %/**0.3 %**	0.0 %/**0.0 %**
SB4	31632	16289	9688	14484	16418	13639	8641	1517	2252
	27.6 %	14.2 %/**19.6 %**	8.5 %/**11.7 %**	12.6 %/**17.5 %**	14.3 %/**19.8 %**	11.9 %/**16.5 %**	7.5 %/**10.4 %**	1.3 %/**1.8 %**	2.0 %/**2.7 %**
SB5	12323	7060	15431	18079	23306	23705	10120	1868	236
	11.0 %	6.3 %/**7.1 %**	13.8 %/**15.5 %**	16.1 %/**18.1 %**	20.8 %/**23.4 %**	21.1 %/**23.8 %**	9.0 %/**10.1 %**	1.7 %/**1.9 %**	0.2 %/**0.2 %**
		A							
		B %/**C %**							

A = number of pairs {input/fault}, B % = percentage over the total number of pairs, C % = percentage without considering pairs leading to 0 errors

in terms of power consumption, performance and area according to the chosen synthesis parameters. However, the synthesis tool generates also quite different implementations depending on whether the initial VHDL description is based on a truth table or a mathematical expression. When starting from a truth table, the number of errors at the output of the S-Box is concentrated around one or two erroneous bits. In contrast, mathematical-based architectures (SB3, SB4, SB5) are composed of several blocks that operate in cascade, and an error in a particular element is spread over several output bits. Therefore this type of implementation, first is, more sensitive to faults (72.4 % of pairs generate an error at the output for SB4 and 89.0 % for SB5, while only 18.8 % do so for SB1), and, second, generates a higher number of output errors (the highest average is between 4 and 5).

Countermeasures based on the use of single parity bit suffer from a low detection rate even for multiple errors. A possible solution is presented in [219, 222] where the authors propose decomposing the SubBytes operation in $\mathbb{F}((2^2)^2)$, leading to an implementation of five cascaded blocks. Each block is implemented in such a way that any single fault propagates to an odd number of bits on the block's output. A parity prediction block and a parity calculation block are associated with each block, guaranteeing thus that any single fault will be detected. The synthesis of this S-Box (without considering the detection scheme) leads to 258 cells by Synopsys Design Compiler© and high map effort. Compared to other mathematical descriptions of the S-Box (like SB4 and SB5) the extra area cost is justified by the increased detection capability (all single faults are detected).

To summarize, for protection against single transient faults, a simple parity scheme (over 32 and 128 bits) is sufficient for the MixColumns and AddRoundKey modules. For the protection of S-Boxes, the error detection scheme has to be selected conjunctly with the actual implementation of the S-Box. For instance, a simple parity scheme would be sufficient for the SB1 or SB2 designs whereas a more sophisticated error code detecting up to eight erroneous outputs has to be considered if the SB3, SB4, or SB5 is selected.

Furthermore, the type of fault against which to protect the circuit has also to be considered. For instance, if a laser beam is used to attack the circuit, it is very likely that the injected fault will not be a single bit-flip fault, as considered above, but a multiple fault due to the width of the laser beam. Such experiments as the ones described above have to be performed for deriving the error profile and selecting the appropriate error detection code.

In contrast to code-based protection schemes, time redundancy-based solutions are more robust against injected faults in the sense that their error detection capability does not depend on the error profile (as long as the duration of the fault does not span over the double computation).

6.6 Conclusion

This chapter presented a study on mechanisms deployed for detecting fault-based attacks on devices implementing the Advanced Encryption Standard. Fault-based attacks were discussed with respect to their requirements in terms of error multiplicity (spatial and temporal characteristics). We analyzed more precisely some error detection schemes based on information redundancy, with respect to their cost and error detection capability.

We also studied the correlation between the effect of a physical fault (modeled as a bit flip) and the error profile at the output of the circuit. It turns out that the error profile strongly depends on the implementation style. For some particular implementations, most of the faults result in only one or two erroneous output bits, while there is no internal fault affecting all the output bits, thus justifying the use of simple and inexpensive codes. On the other hand, some other designs optimized in terms of area and/or speed lead to far less favorable error profiles, thus requiring more expensive coding styles. As a consequence, the circuit synthesis and the choice of the protection scheme must be performed conjunctly in order to reduce the overall circuit cost.

Something to further investigate would be the outcome of correlating the fault model with a physical effect induced by the attack technique. For instance, it is very likely that a voltage glitch attack will produce a multiple fault at the logic level, while a laser-based attack will produce a more localized fault. Therefore the type of detection mechanism has to be customized according to the fault attack, which the designer wants to prevent.

Another thing to check is the conditions under which the detection method does, or does not, enable backdoors for other types of attacks. For instance, information redundancy-based solutions tend to increase the correlation of the circuit power consumption, with the processed data increasing the side-channel leakage.

Part III
Fault Analysis in Public Key Cryptography

Chapter 7
A Survey of Differential Fault Analysis Against Classical RSA Implementations

Alexandre Berzati, Cécile Canovas-Dumas and Louis Goubin

Abstract Since its first introduction by Bellcore researchers, fault injection has been considered as a powerful and practical way to attack cryptosystems, especially when they are implemented on embedded devices. In this chapter, we will review how fault injection has been practically and efficiently exploited to attack some implementations of the celebrated RSA. The first attacks were based on perturbing execution flow or a private key; powerful attacks exploiting modifications in the public key have recently appeared. These new attacks are particularly relevant since they highlights the need for also protecting public key elements against faults.

7.1 Introduction

Since the advent of side-channel attacks, resistance to classical cryptanalysis is no longer sufficient to ensure the security of cryptographic algorithms. In practice, implementations of algorithms on electronic devices are potential sources of leakage that an attacker can use to completely break a system [74, 153, 240]. The injection of faults during the execution of cryptographic algorithms is considered as an intrusive side-channel method because secret information may leak from malicious modifications of a device's behavior [49, 56]. In this context, the security of public key cryptosystems [56] and symmetric ciphers in both block [49] and stream [182] modes have been challenged. Recently, some interesting results have been obtained by attacking public key cryptosystems. More precisely, several papers have demonstrated that the perturbation of public elements may induce critical flaws in implementations of public key cryptosystems [44, 70, 225].

A. Berzati · C. Canovas-Dumas (✉)
CEA Leti Grenoble, France
e-mail: cécile.dumas@cea.fr

L. Goubin
Université de Versailles Saint-Quentin-en-Yvelines, Versailles, France

M. Joye and M. Tunstall (eds.), *Fault Analysis in Cryptography*,
Information Security and Cryptography, DOI: 10.1007/978-3-642-29656-7_7,

In this chapter, we present a survey of the application of fault attacks to different RSA implementations, classical and more sophisticated. After giving a presentation of the RSA cryptosystem and its classical implementation, we chronologically review these attacks and some of the proposed countermeasures. Although the first attacks focused their efforts on exploiting perturbations of secret elements or related operations (see Sect. 7.3), recent works have addressed the security of public elements (see Sect. 7.4). This new trend is all the more interesting since public elements are usually handled in a less secure way than private ones. Furthermore, it may lead to powerful attacks regardless of the perturbation induced.

7.2 RSA Implementations

The celebrated RSA was invented in 1977 by Rivest et al. [349]. This method was the first cryptographic algorithm instantiating the principle of public key cryptography, introduced earlier, as a concept, by Diffie and Hellman [121]. Nowadays, RSA is certainly the most widely used algorithm to ensure the security of communications or transactions.

The RSA security relies on the Integer Factorization Problem (IFP). The best known method for solving an instance of this problem is the Number Field Sieve, whose complexity is sub-exponential with respect to the size of the RSA field [252]. To the best of our knowledge, the larger RSA modulus ever factored by this method has a bit length of 768 [232]. As a consequence, this method cannot be used today to factor practically sized moduli (i.e. 1,024 or 2,048 bits).

In practice, the RSA can be derived in both standard and CRT modes. The standard mode is the straightforward way for implementing RSA. Its principle is recalled below. The CRT mode of RSA, where CRT stands for Chinese Remainder Theorem, is usually used to perform efficient signatures in embedded systems. Further details about CRT-RSA implementations can be found in [283, 330].

7.2.1 Standard RSA

7.2.1.1 Key Generation

Let N, the public modulus, be the product of two large prime numbers p and q. The bit length of N, denoted by n, also stands for the RSA length. Let e be the public exponent, coprime to $\varphi(N) = (p-1) \cdot (q-1)$, where $\varphi(\cdot)$ denotes Euler's totient function. Then the pair $K_p = (N, e)$ is the RSA public key, which is spread over a network. The public key exponent e is linked to the private key exponent d by the following equation:

$$e \cdot d \equiv 1 \ (\mathrm{mod}\varphi(N)).$$

This exponent is also called private key K_s, which is kept secret by its owner.

7.2.1.2 RSA Encryption

Let m be the plaintext that will be ciphered with the RSA algorithm. Then, the cipher operation relies on the following modular exponentiation involving the public exponent e:

$$C = m^e \bmod N.$$

Here C denotes the ciphertext. After sending this ciphertext throughout a network, the expected receiver of the message may want to recover the original message by decrypting C. Then, this operation consists of computing

$$\tilde{m} = C^d \bmod N.$$

If no error occurs during computation, transmission or decryption of C, then we expect to get

$$\tilde{m} = m \bmod N.$$

One can notice that performing a decryption implies knowledge of the private key. This ensures that only its owner, and so the expected receiver, is able to recover m.

7.2.1.3 RSA Signature Scheme

As for the decryption, the RSA signature S of a message m consists of a modular exponentiation to the power d

$$S = m^d \bmod N.$$

To check the validity of the received signature (S, m), the associated public key is used to ascertain that

$$m \stackrel{?}{=} S^e \bmod N.$$

In general, this is not really the plaintext m that is signed but a hash $\dot{m}$. This operation is performed by using a hash function $\mathscr{H}$ such that $\dot{m} = \mathscr{H}(m)$. Then, the sent signature is $(S, \dot{m})$ where $S = \dot{m}^d \bmod N$. Furthermore, the usage of a hash function is generally combined with a padding scheme (e.g. RSA-OAEP [30] for the RSA encryption and RSA-PSS [31] for the signature). For the different fault attacks presented in this paper, we assume that all RSA decryptions or signatures are performed with some hash and/or deterministic padding function.

7.2.2 *Modular Exponentiation Methods*

Modular exponentiation is one of the core operations for implementing asymmetric cryptography. Indeed for both RSA decryption and RSA signature schemes, one has to compute a modular exponentiation of the input message as efficiently as possible.

For a naïve implementation of this operation, we may write

$$m^d \bmod N = \underbrace{m \cdot m \cdot \ldots \cdot m}_{\text{d times}} \bmod N.$$

The computational complexity of the algorithm above is exponential with respect to the exponent length, which is not acceptable for implementing a modular exponentiation. However, a method as intuitive as the previous one, but much smarter, consists in expressing the exponent in a binary basis: $d = \sum_{i=0}^{n-1} 2^i \cdot d_i$. Then, we can perform a modular exponentiation by scanning the bits of the exponent:

$$m^d \bmod N = \prod_{i=0}^{n-1} m^{2^i d_i} \bmod N. \tag{7.1}$$

As a consequence, the power is no longer performed by multiplying d times a message m by itself but by multiplying, at most $\log_2(d)$ times, square powers of m. Hence the cost of the exponentiation methods derived from this principle, also called binary exponentiation, is linear with respect to the exponent length, which is much more efficient than the naïve approach. The next section briefly introduces common implementations of binary exponentiations

7.2.2.1 "Right-to-Left" Exponentiation

The first method to perform a modular exponentiation is to scan the exponent bits from the least to the most significant ones. That is why it is also referred as the "right-to-left" method. In practice, this method is the most intuitive since it consists of computing consecutive square powers of the input message and, depending on the current exponent bit value, multiplying these powers to the accumulation register. As a consequence, this algorithm exactly transcribes (7.1). The complete algorithm is detailed below.

Algorithm 7.1: "Right-to-left" modular exponentiation

Input: $m, d = \sum_{i=0}^{n-1} 2^i \cdot d_i, N$
Output: $S = m^d \bmod N$

$A \leftarrow 1$;
$B \leftarrow m$;
for $i = 0$ **to** $n-1$ **do**
  if $d_i = 1$ **then**
    $A \leftarrow A \cdot B \bmod N$;
  end
  $B \leftarrow B^2 \bmod N$;
end
return A

7.2.2.2 "Left-to-Right" Exponentiation

It is also possible to scan the exponent bits from the most to the least significant bits. The associated method is not surprisingly called "left-to-right" exponentiation. This method differs from the dual one by an accumulation register withdrawal and a different execution flow. Indeed, each iteration begins with the execution of a square that can be followed, depending on the current exponent bit value, by a multiplication. This method is detailed in Algorithm 7.2.

Algorithm 7.2: "Left-to-right" modular exponentiation

Input: $m, d = \sum_{i=0}^{n-1} 2^i \cdot d_i, N$
Output: $S = m^d \bmod N$

$A \leftarrow 1$;
for $i = (n-1)$ **to** 0 **do**
    $A \leftarrow A^2 \bmod N$ **if** $d_i = 1$ **then**
        $A \leftarrow A \cdot m \bmod N$;
    end
end
return A

7.2.2.3 Other Variants

Although both the previously presented algorithms are quite efficient methods of computing a modular exponentiation, their implementation may leak information on the exponent. Indeed, the execution of a multiplication directly depends on the current value of the exponent bits. This makes smart card implementations of modular exponentiation algorithms potentially vulnerable to side-channel attacks, such as DPA (see Sect. 1.3). A basic solution to thwarting this flaw is to make the execution independent of the exponent value. Such a variant was proposed by Coron [102] and is known as the Square and Multiply Always algorithm. In this case, one square and one multiplication are sequentially executed at each iteration, whatever the value the current exponent bit may take.

The performance of the exponentiation algorithms can also be improved. For example, we can derive from the previous algorithms the Sliding Window Exponentiation used in the OpenSSL library. The principle of this variant is to scan the exponent by, at most, k-bit windows. Hence, after precomputing some odd powers of the input message m, this method significantly reduces the number of iterations for a modular exponentiation, and so the performance.

Finally, some variants allow one to improve both security and performance. For example, this is the case with the Montgomery exponentiation algorithm [294]. Obviously, there are other efficient implementations of modular exponentiation, but we advise an interested reader to take a look at [234] for more details.

7.3 Classical Fault Analysis of Standard RSA Implementations

Since the introduction of fault attacks at the end of the 1990s, the security against perturbation of CRT-based implementations of RSA has not been the sole implementation mode targeted [56]. Indeed, the security of standard RSA implementations has been also challenged, leading to various attack methodologies. Among these fault attacks, we have chosen to distinguish between two main categories. The first one deals with the perturbation of intermediate computations. The fault attacks that belong in this category take advantage of the perturbation of intermediate values or of the execution flow. The second category includes attacks based on exploiting modifications of RSA public elements. This trend is quite recent but has ever led to successful applications against various standard RSA implementations.

The following details the different attacks that we have identified from our state of the art study.

7.3.1 Perturbation of Intermediate Computations

7.3.1.1 Register Faults

Bellcore researchers not only introduced the concept of fault attacks [32] but also showed it could be applied to many public key cryptosystems, including standard RSA, and their various implementations. They explained in [56] how to exploit fault injections during the execution of a standard RSA signature to recover the private exponent. The fault model they considered, the so-called register fault, is a transient or permanent bit-flip induced in the memory area containing the current value of the exponentiation algorithm. Using this model, they showed that the perturbation of standard RSA signature, implemented with a "right-to-left" exponentiation, may leak some secret information. The principle of the attack is described next.

General Methodology

The fault attack against a standard RSA signature proposed by Bellcore researchers can be split into two parts. The first one is online and consists in gathering sufficiently many message/faulty signature pairs $(m_i, \hat{S}_i)$ by inducing permanent faults, one per faulty signature, on the register that contains an intermediate value. Then, in the second part, the attacker tries to analyze the collected faulty signatures to recover the whole secret key. Hence, this part of the attack is completely off-line. The principle of the analysis is recalled below. Let S_i be the correct signature and let $\varepsilon(m_i) = \pm 2^b$ be the mathematical representation of a bit-flip with $0 \leqslant b < n$. Since this fault is supposed to be permanent, then corresponding faulty signature can be expressed as

$$\hat{S}_i = \left(\left(\prod_{j=0}^{t-1} m_i^{2^j d_j}\right) \pm 2^b\right) \cdot \prod_{j=t}^{n-1} m_i^{2^j d_j} \tag{7.2}$$

$$= S_i \pm 2^b \cdot \prod_{j=t}^{n-1} m_i^{2^j d_j} \bmod N \tag{7.3}$$

$$\text{or } S_i = \hat{S}_i \pm 2^b \cdot m_i^{d_{[t]}} \bmod N \tag{7.4}$$

with $d_{[t]} = \sum_{j=t}^{n-1} 2^j d_j$. If we use the signature's verification operations, the previous equation becomes

$$m_i = (\hat{S}_i \pm 2^b . m_i^{d_{[t]}})^e \bmod N. \tag{7.5}$$

One can notice that this equation is interesting because it only depends on the message m_i and the corresponding faulty signature $\hat{S}_i$ (i.e. the knowledge of the correct signature is no longer required). Moreover, the fault injection has isolated a part $d_{[t]}$ of the private exponent d. This is precisely this part of the exponent the attacker has to determine with a guess and determine approach.

The whole exponent is gradually recovered, from the most to the least significant bits, by repeating the previous analysis on different faulty signatures. In each analysis, w bits of exponent are retrieved. In more detail, for each analysis, the attacker has to simultaneously guess the values of the part of the exponent isolated $d_{[t]}$ and the fault induced $\pm 2^b$ such that (7.5) is satisfied. According to [56], the whole private exponent d can be determined in this way, with probability greater than $\frac{1}{2}$, from $(n/w) \log(2n)$ message-signature pairs. In this case, the attack complexity is about $O((2^w n^3 \log^2(n))/w^3)$ exponentiations. We can additionally remark that this fault attack was later generalized to "left-to-right"-based exponentiations by Blömer and Otto [315].

7.3.1.2 Faults on the Private Exponent

This attack was published by Bao et al. in [20] and then Bar et al. in [21]. The principle is to induce a transient error during the decryption that produces the same effect as a bit modification of the private exponent. In practice, such an effect can be obtained by flipping a bit of the private exponent, or by corrupting the evaluation made just before the conditional branch in the classical implementations of modular exponentiation (see Sect. 7.2.2). The following paragraph only describes the attack for a bit error on the exponent d.

General Methodology

Let m be a plaintext and C be the corresponding ciphertext obtained from an RSA encryption (see Sect. 7.2). In the case of a faulty computation, the deciphered text $\hat{m}$ is

$$\hat{m} = C^{\hat{d}} \bmod N.$$

The fault is exploited by dividing the erroneous result by a correct one: $\hat{m} \cdot m^{-1}$. The induced error can be modeled as a *bit-flip* of the t-th bit of d. Therefore, we have

$$\hat{m} = C^{\sum_{i=0,i\neq t}^{i=n-1} 2^i \cdot d_i + 2^t \bar{d}_t} \bmod N.$$

That implies that either $\hat{m} \cdot m^{-1} = C^{-2^t} \bmod N \Rightarrow d_t = 1$, or $\hat{m} \cdot m^{-1} = C^{2^t} \bmod N \Rightarrow d_t = 0$. This method can be repeated until we obtain enough information on the private exponent. Note that this attack is also suitable in the case of a multiple error model [20]. Moreover, the principle can be adapted to attacking cryptosystems based on discrete logarithms (e.g. DSA, El-Gamal). Finally, this attack strategy has been extended and generalized by Joye et al. [203], who describe an improved attack that relies on knowledge of faulty deciphered texts.

7.3.1.3 Exploiting Safe-Errors

Classical fault attacks often exploit the difference between a correct and a faulty output to deduce some secret information. However, Joye and Yen noticed that some secret information may leak even if a fault causes no effect on the final result of the computation [427]. This is why this particular kind of fault attacks is also called "safe-error" attack. The attacker must be able to perform some perturbation which has a significant probability of achieving a given effect. Among these attacks, two categories are usually distinguished between: C safe-error attacks that target dummy operations [429] and M-safe-error attacks that target register allocations [427]. In order to illustrate the principle of safe-error attacks in the context of RSA, we have chosen to detail the C safe-error attacks against the Square and Multiply Always exponentiation [102].

General Methodology

The purpose of the Square and Multiply Always exponentiation is to make its execution independent from the exponent value. Hence, the conditional branch is withdrawn (see Sect. 7.2.2) and an extra dummy multiplication is added such that, regardless of the value of the exponent, a square and a multiplication are always executed at each iteration. The principle of the C safe-error attacks proposed by Yen et al. [429] is to exploit faults induced while dummy multiplications are performed. To achieve this, the attacker has to inject a fault during the computation of an RSA signature and, if the signature remains "error-free", it means that a dummy multiplication has been infected. Therefore, the attacker can deduce that the exponent bit value handled while the perturbation was provoked is 0. Otherwise, the result would be incorrect and the exponent bit value equal to 1. The attacker has to repeat the attack at each iteration to gradually recover the whole private exponent. One can notice that this attack does not apply to the classical exponentiation algorithm but applies to a variant expected to defeat Simple Power Analysis. As a consequence, only checking the correctness of the output may not be enough to protect an implementation of a cryptographic algorithm against faults. Furthermore, designers of secure instantiations must be careful not to add new vulnerabilities while trying to defeat other ones.

7.4 Exploiting Perturbations of RSA Public Modulus

Although the issue of exploiting malicious modifications of public elements was addressed in the context of elliptic curve-based cryptosystems [90], it took half a decade to see the first application to RSA. Indeed, the first fault attack against public key elements was presented by Seifert with a method for corrupting RSA signature verification [298, 367]. This fault attack aims to corrupt a signature verification mechanism by modifying the value of the public modulus N. Nevertheless, no information about the private exponent d is revealed with this fault attack.

Whether it is necessary or not to protect RSA public elements was an open question until Brier et al. proposed an attack for recovering the whole private key. This attack, inspired by Seifert's [298, 367], was published in [70] and reviewed in [92]. It makes it possible to extract the private key using a modulus perturbation. As in Seifert's attack, the fault on the modulus is induced before executing the exponentiation. Hence, if the faulty modulus has a small divisor r, the attacker will be able to solve an instance of the Discrete Logarithm Problem from the corresponding faulty signature and obtain $d \bmod r$.

A new fault attack against "right-to-left"-based implementations of the core RSA exponentiation [39], complemented by the attack of the dual implementation [37], has recently appeared in the literature. Contrary to previous attacks, authors assumed that the fault is injected during the execution of an RSA signature. Then, from the knowledge of a correct and a corresponding faulty signature, the attacker guesses and determines simultaneously the faulty modulus and the part of the private exponent that has been isolated by the fault injection. To recover the whole exponent, the attacker has to repeat the analysis for a sufficient number of signatures with faults at different moments of the execution.

In the following we will detail the different fault attacks published against RSA public elements.

7.4.1 Modifying N Before a Signature to Solve a Small Discrete-Log Problem (DLP)

Although Seifert—with his attack proposal to corrupt an RSA signature verification mechanism [298, 367]—first addressed the issue of exploiting RSA signatures performed under a faulty public modulus N, the first analysis leading to a complete secret key recovery was described by Brier et al. [170]. The main idea behind their attack is to analyze faulty signatures performed under a faulty modulus N to recover small parts of the private exponent d (i.e. 20–30 bits from each faulty signature). All the parts of exponent extracted from different faulty RSA signatures are finally combined with the Chinese Remainder Theorem to build the full private exponent.

General Methodology

This fault attack can be split into two distinct phases. The first one is "online" and consists in gathering K pairs message/faulty signatures $(m_i, S_i)_{1 \leqslant i \leqslant K}$ computed with faulty moduli $\hat{N}_i \neq N$ such that

$$S_i = \dot{m}_i{}^d \bmod \hat{N}_i .$$

The authors assumed that the values of the different faulty $(\hat{N}_i)_{1 \leqslant i \leqslant K}$ are not known to the attacker. However, they have also supposed that theses values are uniformly distributed over the n-bit long integers [70, 92]. From this set of faulty signatures, the attacker performs an "off-line" phase which consists in recovering the private exponent by parts. The proposed analysis is derived in different variants depending on the decision of the attacker to generate, or not generate, a table of possible values for faulty moduli. But, generating such a table, also referred as the dictionary of moduli [70], requires one to choose a fault model. Finally, one can notice that the number of signatures to gather (i.e. the parameter K) depends on the method chosen to perform the analysis. Both methods are detailed below.

Analysis without a dictionary

This first method is used when the attacker is not able to identify a fault model from the set of gathered faulty signatures or, if the identified model induces a dictionary too large to be handled (e.g. more than 2^{32} entries). In this case, it is not possible to retrieve faulty moduli used to perform the faulty signatures. To overcome this difficulty, the authors give a means of finding some factors p^a of $\hat{N}_i$ thanks to the Eq. (7.6) that is satisfied under some conditions[1] with probability 1 if $p^a \mid \hat{N}_i$ and $\frac{1}{r}$ otherwise (where r is a small prime that divides the multiplicative order of $\dot{m}_i$),

$$S_i = \dot{m}_i{}^{d \bmod \varphi(p^a)} \bmod p^a . \tag{7.6}$$

Then, for such a p^a, if $\varphi(p^a)$ is divisible by a small enough prime r_k (i.e. 20–30-bit long), the attacker can take advantage of the bias (see [70, Proposition 1] for details) with a counting method that enables him to determine parts of the private exponent $d_k = d \bmod r_k$ by solving small discrete logarithms on the faulty signatures gathered. Hence, when $R = \prod_k r_k$ is bigger than N (and obviously bigger than $\varphi(N)$), it is possible to use the Chinese Remainder Theorem to build the whole private exponent d from the recovered parts.

---

[1] p is a prime number such that $p \nmid \dot{m}_i$ and $p \nmid S_i$.

One can notice that the advantage of this method is that it may lead to a full key recovery regardless of the faults injected into the different moduli. According to [70, 92], the implementation of this methodology enables one to recover 512-bit private exponents (1,024-bit in the case of a small public exponent e[2]) by gathering about 2.5×10^4 faulty signatures. On the other hand, 6×10^4 faulty signatures are enough to recover 1,024-bit secret exponents (2,048-bit in the case of a small public exponent e).

Analysis with a dictionary

The attack performance can be improved if the attacker is able to identify and validate a fault model from the faulty signatures collected during the "online" phase. From this fault model and the knowledge of N, the attacker can establish a list of possible values for the faulty moduli. This list is also called the dictionary of moduli. Once the dictionary is generated, the attacker tries to guess the pairs $(m_i, \hat{S}_i)$ that result from a computation with one of the dictionary entries. Each successful guess, or "hit", brings a certain amount of information on the private exponent d. In terms of performance, this approach is very interesting since, according to [70, 92], 28 "hits" and only 1,100 faulty signatures are enough for recovering a 1,024-bit RSA private exponent. Moreover, by finding these "hits" with a statistical approach, it is possible to extract the private exponent with a number of faults equal to the number of "hits" (which is proved to be optimal). In this case 28 faulty signatures may suffice to recover a 1,024-bit RSA private exponent.

Although fault attacks has been considered as a powerful way to attack implementations of cryptographic algorithms, the presented attacks highlight that even noncritical elements, such as public keys, have to be protected against fault injection. While public elements are not supposed to reveal secret information, their perturbation may be a source of leakage leading to the corruption of a signature verification mechanism [298, 367] or, worse, to a full private key recovery [70, 92]. However, the use of an exponent randomization technique may be an effective way of defeating such attacks. Furthermore, these attacks only apply to perturbations that occur before performing the core exponentiation of the RSA signature. The attack presented in the next section prove that this claim is no longer the case since the perturbation of public elements during the signature can also be exploited.

7.4.2 Exploiting Faults on N During the Computation of an RSA Signature

In Seifert's and Brier et al.'s proposals [70, 367], the authors exploited perturbations of the public modulus provoked before the core exponentiation so that the whole

[2] When e is small, the authors take advantage of the RSA equation $e \cdot d \equiv 1 \bmod \varphi(N)$ to determine the most significant part of d. Indeed, knowing that $\varphi(N) = N + 1 - p - q \approx N$ for its most significant part, then $d \approx \frac{1+k\cdot(N+1)}{e}$ with $k < e$. Hence, if e is small (e.g. $e = 2^16 + 1$), the most significant part of d can be directly deduced from the previous relation completed by an exhaustive search on k.

signature is performed with a faulty modulus $\hat{N}$. The attack presented by Berzati et al. [39] extended the fault model by enabling an attacker to exploit faults injected into a "right-to-left"-based exponentiation. The modification of N was supposed to be a transient random byte fault modification. It means that only one byte of N is changed to a random value. Moreover, they also assumed that the value of the faulty modulus $\hat{N}$ is not known to an attacker. However, the time location of the fault is a parameter known to an attacker and used to perform the cryptanalysis. This fault model has been chosen for its simplicity and practicability in the smart card context [52, 160]. Furthermore, it can be easily adapted to 16- or 32-bit architectures. This attack was later generalized to "left-to-right"-based implementations including Montgomery and Sliding Window exponentiations. It has also been proven that exponent randomization method—also known as exponent blinding—suggested by Kocher [239] may not be efficient for protecting RSA implementations against faults in public key elements [41].

General Methodology

In order to detail the general methodology, we assume that the implementation of the modular exponentiation is "right-to-left"-based (see Sect. 7.2.2.1) and that an attacker is able to modify one byte of the public modulus N at the tth step of the exponentiation. If we denote by A_t and B_t the internal register values, then for a fault occurring while B_t is computed, we have

$$\hat{S}_t = A_t \cdot \hat{B}_t^{d_t} \cdot \ldots \cdot \hat{B}_t^{2^{(n-1)-t} \cdot d_{n-1}} \bmod \hat{N} \tag{7.7}$$

$$= A_t \cdot \hat{B}_t^{\frac{d_{[t]}}{2^t}} \bmod \hat{N} \tag{7.8}$$

where $d_{[t]} = \sum_{i=t}^{n-1} 2^i \cdot d_i$ Hence, the fault injection splits the computation into a correct and a faulty part. The main consequence is that a part of the private exponent, namely $d_{[t]}$, is isolated by the fault injection. In practice this part of the exponent is composed of a known (already determined) part and a part to guess. So, if the part to guess is small enough, it is possible to guess and determine it from a faulty/correct signature pair $(\hat{S}_t, S_t)$. Therefore, since the faulty modulus is also unknown to the attacker, he has to choose a candidate value $\hat{N}'$ (built according to the fault model) and another candidate value for the part of $d_{[t]}$ to guess. Then he computes from the correct signature $A'_t = S_t \cdot \dot{m}^{-d'_{[t]}} \bmod N$ and

$$S'_{(d'_{[t]}, \hat{N}')} = A'_t \cdot \left(\dot{m}^{2^{t-1}} \bmod N\right)^{2 \cdot \frac{d'_{[t]}}{2^t}} \bmod \hat{N}'.$$

If this rebuilt faulty signature satisfies (7.9) then the pair of candidate values $(\hat{N}', d'_{[t]})$ is the expected one with high probability. Otherwise, the attacker has to choose another candidate pair and perform this test again.

$$S'_{(d'_{[t]},\hat{N}')} = \hat{S}_t \bmod \hat{N}'. \tag{7.9}$$

The whole exponent is gradually recovered by cascading the previous analysis with signatures faulted at different moments of the execution and using the knowledge about already found parts. The performance of this fault attack mainly depends on the number of exponent bits w that the attacker can extract from correct/signature pair. This parameter is a trade-off between fault number and performance, and so has to be carefully chosen. Authors estimated that for $w = 4$ (which ensures reasonable execution time), the number of faulty signatures to gather for recovering a 1,024-bit private exponent is about 250 [39]. Finally, one can notice that a similar analysis can be performed when the first operation infected by the fault is a multiplication (i.e. the computation of A_t is infected first). The details of this variant are provided in [39].

Application to "Left-to-Right"-Based Exponentiations

After exploiting public key perturbation during execution of "right-to-left" implementations of RSA signatures, Berzati et al. generalized their attack to left-to-right-based exponentiations. Although both algorithms are very similar, a generalization of the previous fault attack is not straightforward. To illustrate the difficulties induced by this generalization the authors consider a classical "left-to-right" exponentiation (see Sect. 7.2.2.2) and an attacker able to modify the modulus N during its execution, as in the previous fault model but t steps before the end of the execution. Denoting by A_{n-t} the internal register before the perturbation of N, the faulty signature $\hat{S}_t$ can be expressed as

$$\hat{S}_t = A_{n-t}{}^{2^t} \cdot \dot{m}^{d_{[t]}} \bmod \hat{N}, \tag{7.10}$$

and this time $d_{[t]} = \sum_{i=0}^{t-1} 2^i \cdot d_i$ denotes the t-bit least-significant part of d. By observing (7.10), one can notice that, in contrast to the "right-to-left" case, t cascaded squares are induced by the fault injection. Hence, extending the previous analysis supposes that the attacker is able to compute modular square roots (which is a difficult problem in RSA rings). Fortunately, since these additional squares are computed modulo $\hat{N}$, it is possible to efficiently compute square roots when $\hat{N}$ is prime (or B-smooth) using the probabilistic Tonelli and Shanks algorithm [97]. To validate the consistency of considering only prime faulty moduli, the authors provided a complete theoretical analysis based on number theory. Hence they estimated that, according to the fault model and for a 1,024-bit RSA, one faulty modulus over 305 will be prime [37]. Moreover, they observed that, although two square roots may be computed from a given quadratic residue, the number of the tth square root is not 2^t but is bounded by the bigger power of 2 dividing $\hat{N} - 1$. Since this power is usually quite small (i.e. smaller than 5 in their experiments), they concluded that, in practice, the computation of square roots does not prevent an attacker from using a guess and determine approach derived from "right-to-left" analysis and recovering parts of d when "left-to-right"-based exponentiations are targeted.

In the previous paragraph, we detailed a proposal for attacking the dual implementations of the modular exponentiation [37, 39]. In both cases, an attacker takes advantage of a fault that occurs during the computation of an RSA signature. However, the previous results show that "right-to-left"-based exponentiations seem to be easier to attack than "left-to-right" ones. Moreover, this attack methodology has been later reused to successfully defeat the randomized exponent countermeasure [41]. Although this countermeasure seems to be efficient against modifications of public elements that occur before the computation of signatures [70], Berzati et al. showed that signatures partially infected by a faulty public modulus are still exploitable when the private exponent is blinded [41]. However, the authors suggest the use of the Probabilistic Signature Scheme with RSA (RSA-PSS) [31] to defeat their attacks. Eventually, this work completes the state of the art and highlights the need to protect RSA public elements against perturbations, even during the computation of signatures.

7.5 Conclusion

The study of the injection of faults into RSA implementations shows that a large panel of different attacks exist. Of course the popularity of RSA is widely accountable for this, but the variety of the proposed implementations, even secured ones, leads to different fault exploitations. The first instance of fault attacks has led to very powerful applications, especially for RSA computed using the Chinese Remainder Theorem where one fault may suffice; standard RSA implementations seems to be more difficult to attack. Indeed, for such implementations, the goal of the attacker is not to factor the modulus but to gradually recover the private exponent. In both cases, conditional checks must be avoided or secured and public elements be protected in the same way as private ones. One can conclude that implementations that parcel the secret elements for the computation are by construction vulnerable to fault attacks. Countermeasures such as masking techniques lead to confusion regarding the isolated parts, but they may be not enough, as proven by the number of attacks that exploit the residual vulnerabilities.

Chapter 8
Fault Attacks Against RSA-CRT Implementation

Chong Hee Kim and Jean-Jacques Quisquater

Abstract RSA-CRT uses the Chinese Remainder Theorem to speed up the computation of an RSA decryption or a signature and reduces the size of the data stored in memory. This implementation is four times faster than the RSA standard implementation. This is why the CRT implementation of RSA is widely deployed in embedded systems. However, Boneh et al. showed that an error that occurred during the exponentiation could allow one break the implementation of RSA-CRT in 1997. This is a very powerful attack as one can easily find the key of RSA with only one faulty signature. Many countermeasures have been proposed to prevent this attack, but most of them have failed. In this chapter, we introduce a survey of the attacks and countermeasures against RSA-CRT implementations.

8.1 Introduction

Ciphertext indistinguishability against adaptive chosen-ciphertext attacks is an important security property of many encryption schemes [333]. To achieve indistinguishability, public key encryption schemes must be probabilistic [165]. Therefore encryption schemes used in practice introduce redundancy so that a random ciphertext will be valid with negligible probability. This includes the widely used RSA-OAEP encryption scheme [30]. There is therefore no need to add a fault detection mechanism: the correctness of the plaintext is explicitly checked by the decryption algorithm. This explains why fault attacks focus on RSA signature schemes.

RSA signatures can be computed in two ways: standard mode and CRT mode. Each mode is vulnerable to different fault attacks. In this chapter we describe fault attacks against RSA-CRT mode and countermeasures.

C. H. Kim · J.-J. Quisquater
Université Catholique de Louvain, Louvain-la-Neuve, Belgium

M. Joye and M. Tunstall (eds.), *Fault Analysis in Cryptography*,
Information Security and Cryptography, DOI: 10.1007/978-3-642-29656-7_8,

8.2 Fault Attacks Against RSA-CRT

RSA-CRT (RSA with CRT mode), first introduced by Quisquater and Couvreur [330], uses the Chinese remainder theorem to speed up the computation of an RSA decryption or a signature and reduces the size of the data stored in memory. In terms of binary operations, this implementation is four times faster than RSA standard mode. This is why the CRT implementation of RSA is widely deployed in embedded systems.

Here we describe RSA-CRT using Garner's reconstruction method [154]. Let $N = p \cdot q$ be RSA modulus and let $\varphi(N) = (p-1)(q-1)$, where p and q are large primes. We select a random integer e, $1 < e < \varphi(N)$, such that $\gcd(e, \varphi(N)) = 1$. Then we compute d such that $e \cdot d \equiv 1 \bmod \varphi(N)$. In the CRT mode, we compute $d_p = d \bmod (p-1)$, $d_q = d \bmod (q-1)$, and $i_q = q^{-1} \bmod p$. The public key is (N, e) and the private key is (p, q, d_p, d_q, i_q).

The RSA signature on a message M is computed as follows:

$$S_p = \mu(M)^{d_p} \bmod p, \quad S_q = \mu(M)^{d_q} \bmod q,$$
$$S = \mathrm{CRT}(S_p, S_q) = S_q + q(i_q(S_p - S_q) \bmod p),$$

where $\mu(\cdot)$ is an appropriate padding function. Without loss of generality we use M instead of $\mu(M)$ in the subsequent sections to simplify the explanation. Therefore we let $S = M^d \bmod N$. The signature is performed by computing two exponentiations modulo p and q (S_p, S_q) followed by a CRT recombination $(\mathrm{CRT}(S_p, S_q))$. This recombination relies on the Chinese remainder theorem.

Boneh et al. showed that if an error occurs in only one of the exponentiations (that is, during computation of S_p or S_q, but not during both), then the factorization of N is possible with the correct and faulty signatures, S and $\tilde{S}$ [56] (it is also referred to as the Bellcore attack as the authors were Bellcore researchers). For example, suppose that an error occurs during computation of S_p. Then a faulty $\tilde{S_p}$ will be used in the CRT recombination and $\tilde{S} = \mathrm{CRT}\ (\tilde{S_p}, S_q)$ will be returned. With the correct signature S and the faulty one $\tilde{S}$, the secret prime q can be computed by computing

$$q = \gcd((S - \tilde{S}) \bmod N,\ N).$$

It was shown that the attack is possible with only one execution of the algorithm [199]. That is, with only faulty signature the secret prime number q could be recovered by computing

$$q = \gcd((\tilde{S}^e - M) \bmod N,\ N).$$

8.3 Basic Countermeasures

The simplest way to prevent the Bellcore attack is to compute a signature twice and compare the two results. However, this doubles the computation time, and it cannot prevent permanent errors. Another way is to verify the signature with the public exponent e. That is, the device returns the signature S only when $S^e = m \bmod N$. However, this method is too costly if e is large. Furthermore, in some applications (e.g. Java card), one does not have access to the public exponent e during signature generation.

8.4 Shamir's Method and Variants

Most countermeasures against the Bellcore attack rely on the method first suggested by Shamir in 1997 [372]. In chronological order, they are [18, 53, 91, 103, 201, 228, 372, 409, 429]. Shamir's method and variants compute exponentiation with some redundancy as shown in Algorithm 8.1. A random integer r is first chosen and then modular exponentiation is computed based on $r.N$. If there is an error during the computation of the modular exponentiation, it is detected by evaluating if $S^* \equiv Z(\bmod\ r)$.

Algorithm 8.1: Basic concept of Shamir's method and variants

Input: $M \neq 0,\ d,\ N$.
Output: $M^d \bmod N$ or *error*.
begin
  Choose a (small) random integer r;
  Compute $S^* = M^d \bmod rN$ and $Z = M^d \bmod r$;
  if $S^* \equiv Z \bmod r$ **then**
    Output $S = S^* \bmod N$
  else
    Return *error*
  end
end

Shamir's method is shown in Algorithm 8.2. Both exponentiations are computed based on $p\,r$ and $q\,r$ respectively. Then there is a step to check an error. The disadvantage of Shamir's method is that it requires d, which is not known in CRT mode (only d_p and d_q are known). Joye et al. proposed a variant using only d_p and d_q as shown in Algorithm 8.3 [201].

However, the algorithm proposed by Joye et al. cannot detect an error in CRT recombination. Aumüller et al. exploited this security flaw [18]. If one of the S_p^*, S_q^* and i_q values is transiently modified during the recombination step, the fault is not detected. To thwart this attack, a countermeasure was also presented in [18]. However,

Algorithm 8.2: Shamir's CRT-RSA

Input: $M \neq 0, d, p, q, i_q$.
Output: $M^d \bmod N$ or *error*.

begin
  Choose a (small) random integer r;
  Compute $S_p^* = M^d \bmod (pr)$ and $S_q^* = M^d \bmod (qr)$;
  Compute $S_p = S_p^* \bmod p$ and $S_q = S_q^* \bmod q$;
  if $S_p^* \equiv S_q^* \bmod r$ **then**
    Output $S = \mathrm{CRT}(S_p, S_q)$
  else
    Return *error*
  end
end

Algorithm 8.3: Joye et al.'s CRT-RSA

Input: $M \neq 0, p, q, d_p, d_q, i_q$
Output: $M^d \bmod N$ or *error*

begin
  Choose a (small) random integer r_1 and r_2;
  Compute $S_p^* = M^{d_p} \bmod (pr_1)$, $s_1 = M^{d_p \bmod \varphi(r_1)} \bmod r_1$;
  $S_q^* = M^{d_q} \bmod (pr_2)$, $s_2 = M^{d_q \bmod \varphi(r_2)} \bmod r_2$;
  Compute $S_p = S_p^* \bmod p$ and $S_q = S_q^* \bmod q$;
  if $S_p^* \equiv s_1 \bmod r_1$ *and* $S_q^* \equiv s_2 \bmod r_2$ **then**
    Output $S = \mathrm{CRT}(S_p, S_q)$
  else
    Return *error*
  end
end

Yen et al. showed that this countermeasure introduced another vulnerability [432]. That is, the modified algorithm can be attacked by injecting a permanent fault during the CRT recombination.

8.4.1 Infective Computation

Yen and Joye introduced the concept of infective computation in 2000 [427] (see also [429]). The principle is that, if an error changes the result of one of the exponentiations (i.e., S_p or S_q), then it will also infect the result of the other one. Hence, even if $\tilde{S} \not\equiv S \pmod q$, $\tilde{S} \not\equiv S \pmod p$ and the faulty signature will not be exploitable with the classical gcd computation.

Moreover, the authors proposed using infective computation-based methods to avoid conditional checks. In Shamir's method, there is a conditional branch. Yen et

Algorithm 8.4: Basic concept of infective computation

Input: $M \neq 0, d, N$
Output: $M^d \bmod N$ or a random value in $\mathbb{Z}/N\mathbb{Z}$

begin
  Choose a (small) random integer r;
  Compute $S^* = M^d \bmod rN$ and $Z = M^d \bmod r$;
  Choose a random integer $\rho > r$;
  Compute $c = [\rho(S^* - Z) + 1] \bmod r$;
  Return $S = (S^*)^c \bmod N$
end

al. noted that error detection based on decisional tests should be avoided. From the viewpoint of low-level implementation of this decision procedure, it often totally relies on the status of the zero flag of a processor. The zero flag is a bit of the status register in a processor. So, if an attack can induce a random fault into the status register, a conditional jump instruction may perform incorrectly. As shown in Algorithm 8.4, the conditional check can be replaced with infective computation.

Unfortunately, the first two protocols using infective computation [429] were rapidly broken by Yen and Kim [428]. In spite of this, the concept of infective computation was later used in other protocols such as [53, 91].

8.4.2 BOS Algorithm and Attack

Blömer, Otto and Seifert suggested a countermeasure based on Shamir's method and infective computation [53]. The principle of the algorithm is recalled in Algorithm 8.5. At the beginning of the algorithm, two k-bit random integers, r_1 and r_2, have to be carefully chosen. From this the algorithm precomputes and stores in memory random public/private exponents:

$$d_1 = d \bmod \varphi(pr_1),\ e_1 = {d_1}^{-1} \bmod \varphi(pr_1),$$
$$\text{and } d_2 = d \bmod \varphi(pr_2),\ e_2 = {d_2}^{-1} \bmod \varphi(pr_2).$$

In the case of a correct execution, $c_1 = c_2 = 1$, and so $S = S^* \bmod N$. The length of the random value k is a trade-off between security and device capabilities.

However, this algorithm has two main drawbacks. First the knowledge of the private exponent d is required to compute d_1 and d_2, which is not a CRT secret key. The second one deals with the efficiency, because two modular inverses need to be performed in the precomputation phase (i.e., computation of e_1 and e_2).

Wagner found a way to exploit a vulnerability in Algorithm 8.5 [410]. The attack consists in injecting a random transient byte fault in M, just before computing S_p^*, in a way that subsequent accesses to M will be error-free. The returned signature will be faulty, since

Algorithm 8.5: BOS algorithm

Input: $M \neq 0, p, q, d, i_q$
Output: $M^d \bmod N$ or a random value in $\mathbb{Z}/N\mathbb{Z}$

```
begin
    Choose two k-bit random integers r1 and r2;
    Store in memory r1, r2, pr1, qr2, N, r1r2N, d1, e1, d2, e2;
    Compute S*_p = M^{d1} mod(pr1);
            S*_q = M^{d2} mod(pr2);
    Compute S* = CRT(S*_p, S*_q) mod(r1r2N) ;
    Compute c1 = (M − (S*)^{e1} + 1) mod r1;
            c2 = (M − (S*)^{e2} + 1) mod r2;
    Return S = (S*)^{c1c2} mod N
end
```

$$\tilde{S} = (\tilde{S^*}^{\tilde{c_1}}) \bmod N,$$

and $\tilde{S^*} = S^* \bmod q$ but $\tilde{S^*} \neq S^* \bmod p$. Then the attacker tries to guess the injected fault on M and compute the corresponding value for $\tilde{c_1}$ until she factorizes N by computing $\gcd(\tilde{S}^e - M^{\tilde{c_1}} \bmod N, N)$. Wagner claimed that the success probability of this attack depends on the byte location of the fault. For a 1,024-bit RSA and $k = 80$ bits, the success probability is around 4 %.

Blömer et al. proposed a variant that overcame the weakness by randomizing the computation of both c_1 and c_2 [52].

8.4.3 Ciet and Joye's Algorithm and Attack

Another countermeasure based on the infective computation and Shamir's method was proposed by Ciet and Joye [91].

This method, depicted in Algorithm 8.6, was designed to defeat all previously published attacks, even Wagner's attack [410]. However, its security has not been formally proved. The CRT parameters p, q, d_p, d_q, and i_q and the values computed during the first step are assumed to be error-free. This means that these values are implicitly protected against fault injection. As for Algorithm 8.5, the length of random values (l and k) can be tuned according to the application.

Berzati et al. exploited a flaw in the computation of the check values (i.e., c_1 or c_2) at FDTC 2008 [38]. Their attack was inspired by Wagner's attack against the BOS algorithm. They first assumed that an attacker was able to modify transiently one byte of the result of one of the exponentiations: $\tilde{S}_{p^*} = S_{p^*} + \varepsilon$, where ε stands for a byte value. They also assumed that no modular reduction occurred in the computation of the checks. Then $\tilde{c_1}$ directly depends on the fault: $\tilde{c_1} = 1 + \varepsilon$. Because of this particular form, the value of $\tilde{\gamma}$ will be easily found for 5.4 % of the cases (if the attacker limits the exhaustive search of $\tilde{\gamma}$ to 40 bits). As a result, the attacker will be

Algorithm 8.6: Ciet and Joye's algorithm

Input: $M \neq 0, p, q, d_p, d_q, i_q$
Output: $M^d \bmod N$ or a random value in $\mathbb{Z}/N\mathbb{Z}$

```
begin
    Choose two k-bit random integers r_1 and r_2;
    Store in memory p* = r_1 p, q* = r_2 q, i_{q*} = (q*)^{-1} mod p*, N = pq;
    Compute S_{p*} = M^{d_p} mod p*;
            s_2 = M^{d_q mod φ(r_2)} mod r_2;
    Compute S_{q*} = M^{d_q} mod q*;
            s_1 = M^{d_p mod φ(r_1)} mod r_1;
    Compute S* = S_{q*} + q i_{q*} (S_{p*} − S_{q*}) mod p* ;
    Compute c_1 = (S* − s_1 + 1) mod r_1;
            c_2 = (S* − s_2 + 1) mod r_2;
    Pick a l-bit random value r_3 and compute γ = ⌊(r_3 c_1 + (2^l − r_3) c_2) / 2^l⌋ ;
    Return S = (S*)^γ mod N
end
```

able to factor N by computing $\gcd(\tilde{S}^e - M^{\tilde{\gamma}} \bmod N, N)$. From their experiment, the authors evaluated that 83 signatures perturbed according to their model are enough to factor N with a success probability of about 99 %.

In order to fix this flaw, they suggested in [38] replacing the final exponentiation to the power of γ by the masking the operation already proposed, as a variant, by Ciet and Joye [91]:

$$S = (\gamma \cdot S^* \oplus (\gamma - 1) \cdot r) \bmod N .$$

8.4.4 *Vigilant's Algorithm and Attack*

Vigilant proposed an efficient method to protect a modular exponentiation against a fault attack and extended this result to the case of RSA-CRT [409]. The principle of Vigilant's secure exponentiation method consists in computing $M^d \bmod N$ in $\mathbb{Z}_{Nr^2}$ where r is a small random integer that is coprime with N. Then the base M is transformed into M^* such that

$$M^* \equiv \begin{cases} M \bmod N, \\ 1 + r \bmod r^2 \end{cases} .$$

This implies that

$$S^* = M^{*d} \bmod Nr^2 \equiv \begin{cases} M^d \bmod N, \\ 1 + d\,r \bmod r^2 \end{cases} .$$

Therefore a consistency check of the result S^* can be performed modulo r^2 from d and r. If the verification $S^* = 1 + d\,r \bmod r^2$ is successful, then the final result $S = S^* \bmod N$ is returned.

The application to RSA-CRT depicted in Algorithm 8.7 is derived from Shamir's method [372]. The principle is to perform both half exponentiations modulo $p\,r^2$ and $q\,r^2$. Therefore it is possible to perform a final consistency check after recombination, guaranteeing that no error occurred during the computations of S_p or S_q and during the recombination.

Coron et al. presented two weaknesses of Vigilant's algorithm and some modifications [103]. The weakness comes from the fact that the integrity of the modulus computation, the $p - 1$ computation, and the $q - 1$ computation are not checked.

8.4.5 Summary of Shamir's Method and Variants

Although many countermeasures based on Shamir's method have been proposed, almost all of them were shown not to be secure enough. For example, the methods of [18, 53, 91] have been cryptanalyzed in [38, 410, 432] respectively.

All methods based on Shamir's cannot guarantee detecting all errors with 100 % certainty, because there always exists a probability that $S^* \equiv Z \bmod r$ in the presence of errors. However, with proper selection of the size of r (for example, 16, 32, or 64 bits), this probability becomes negligible. These methods impact the performance in terms of both running time and memory requirement. In certain variants, extra personalization process is also required.

Infective computation can be used to avoid conditional checks. Therefore, it can be combined with other techniques. For example, Giraud's method and variants in the next section (such as Shamir's method and variants).

8.5 Giraud's Method and Variants

In 2005, Giraud [161] proposed a SPA-(Simple Power Analysis-) and FA-(Fault Attack-)resistant modular exponentiation based on the Montgomery ladder [205, 294]. Then he presented a variant of it as shown in Algorithm 8.9, where k is a 32-bit random number [162]. He used the fact that temporary variables ($S[0]$, $S[1]$)) are of the form (M^{α}, $M^{\alpha+1}$) at each step of the algorithm as shown in Algorithm 8.8. Both values are used to compute the next temporary couple. So if an error occurs on a temporary result $S[0]$ or $S[1]$, or even if a multiplication or a squaring operation is disturbed at any moment of the computation, then the coherence between $S[0]$ and $S[1]$ is lost. Therefore, the last computed couple ($S[0]$, $S[1]$) is not of the form of (M^d, M^{d+1}). The error can thus be detected by checking the coherence between the last two temporary results, $S[0]$ and $S[1]$, i.e., by testing if $M \cdot S[0] \equiv$

Algorithm 8.7: Vigilant's algorithm

Input: $M \neq 0, p, q, d_p, d_q, i_q$
Output: M^d mod N or *error*

begin
  Choose a 32-bit random integer r, four 64-bit random integers R_1, R_2, R_3 and R_4;
  $p^* = pr^2$, $M_p = M \bmod p^*$;
  $i_{pr} = p^{-1} \bmod r^2$, $\beta_p = p \cdot i_{pr}$ and $\alpha_p = 1 - \beta_p \bmod p^*$;
  $\widehat{M}_p = \alpha_p M_p + \beta_p(1+r) \bmod p^*$;
  if $\widehat{M}_p \neq M \bmod p$ **then**
    Return error
  end
  $d_p^* = d_p + R_1(p-1)$;
  $S_{pr} = \widehat{M}_p^{\,d_p^*} \bmod p^*$;
  if $d_p^* \neq d_p \bmod p - 1$ **then**
    Return error
  end
  if $\beta_p S_{pr} \neq \beta_p(1 + d_p^* r) \bmod p^*$ **then**
    Return error
  end
  $S_p^* = S_{pr} - \beta_p(1 + d_p^* r - R_3)$;
  $q^* = qr^2$, $M_q = M \bmod q^*$;
  $i_{qr} = q^{-1} \bmod r^2$, $\beta_q = q \cdot i_{qr}$ and $\alpha_q = 1 - \beta_q \bmod q^*$;
  $\widehat{M}_q = \alpha_q M_q + \beta_q(1+r) \bmod q^*$;
  if $\widehat{M}_q \neq M \bmod q$ **then**
    Return error
  end
  if $M_p \bmod r^2 \neq M_q \bmod r^2$ **then**
    Return error
  end
  $d_q^* = d_q + R_2(q-1)$;
  $S_{qr} = \widehat{M}_q^{\,d_q^*} \bmod q^*$;
  if $d_q^* \neq d_q \bmod q - 1$ **then**
    Return error
  end
  if $\beta_q S_{qr} \neq \beta_q(1 + d_q^* r) \bmod q^*$ **then**
    Return error
  end
  $S_q^* = S_{qr} - \beta_q(1 + d_q^* r - R_4)$;
  $S = S_q^* + q(i_q(S_p^* - S_q^*) \bmod p^*)$;
  $N = pq$;
  if $N[S - R_4 - q \cdot i_q(R_3 - R_4)] \neq \bmod Nr^2$ **then**
    Return error
  end
  if $q \cdot i_q \neq 1 \bmod p$ **then**
    Return error
  end
  Return $S \bmod N$
end

$S[1] \bmod N$ (see Algorithm 8.9). If this test fails, the signature is not returned.[1] Infective computation can also be used here as well to avoid an explicit check.

Algorithm 8.8: SPA-SE-resistant modular exponentiation

Input: $M \neq 0, d = (d_{n-1}, \ldots, d_0)_2 odd, k,\ N$
Output: $M^{d-1} \bmod k \cdot N$, $M^d \bmod k \cdot N$

begin
  $S[0] \longleftarrow M$;
  $S[1] \longleftarrow S[0]^2 \bmod k \cdot N$;
  for *i from* $n-2$ *to* 1 **do**
    $S[\overline{d_i}] \leftarrow S[\overline{d_i}] \cdot S[d_i] \bmod k \cdot N$;
    $S[d_i] \leftarrow S[d_i]^2 \bmod k \cdot N$;
  end
  $S[1] \leftarrow S[1] \cdot S[0] \bmod k \cdot N$;
  $S[0] \leftarrow S[0]^2 \bmod k \cdot N$;
  if *(Loop Counter i not disturbed) & (Exponent d not disturbed)* **then**
    Return ($S[0]$, $S[1]$)
  else
    Return error
  end
end

Algorithm 8.9: Giraud's SPA and FA-resistant CRT-RSA

Input: $M \neq 0,\ p, q, d_p, d_q, i_q,\ N$
Output: $M^d \bmod N$

begin
  Pick a 32-bit random number k;
  $(S_p^*, S_p) \longleftarrow$ Algorithm 8.8 with M, d_p, p, k as inputs;
  $(S_q^*, S_q) \longleftarrow$ Algorithm 8.8 with M, d_q, q, k as inputs;
  $S^* \leftarrow \mathrm{CRT}_{blinded}(S_p^*, S_q^*)$;
  $S \leftarrow \mathrm{CRT}_{blinded}(S_p, S_q)$;
  $S^* \leftarrow M \cdot S^* \bmod (p \cdot q)$;
  if *$(S^* = S)$ & (Parameters p, q, i_q not modified)* **then**
    Return (S)
  else
    Return error
  end
end

[1] The original version of [161] does not have a blinded modulus. That is, every modular computation is done with modulus N instead of $k \cdot N$. Therefore the original version is vulnerable to a relative doubling attack [431]. The CRT recombination with blinded moduli is also used in the modified version to counter other specific SPA attacks (cf. [311]).

Fumaroli and Vigilant proposed a variant based on the Montgomery ladder [149]. However, it was shown that this method could not be used to build a fault-resistant implementation of CRT-RSA [228]. Giraud's idea was extended to a right-to-left exponentiation algorithm by Boscher et al. [61].

Rivain proposed a double exponentiation algorithm [346]. He devised an algorithm taking as input two exponents a and b and returning $(M^a \bmod N, M^b \bmod N)$. Using this algorithm one can detect faults by setting $a = d$ and $b = \varphi(N) - d$ and checking that $M^d \cdot M^{\varphi(N)-d} \equiv 1 \bmod N$.

In conclusion, Giraud's method and variants exploit some redundancy already present in the exponentiation algorithm. This can be a disadvantage they impose the exponentiation algorithm.

8.6 Embedding Method

As we have discussed in Sect. 8.3, we can verify the signature with the public exponent e. And e is in nearly all cases chosen as a small value (a typical value of e is $2^{16} + 1$). However, in some applications (e.g. Java card), it is not allowed to access the public exponent during signature generation.

In 2009, Joye suggested embedding the value of e in the RSA key object [198]. Such a key object is obtained from (p, q, d_p, d_q, i_q) in CRT mode and from (N, d) in standard mode. When computing an RSA signature $S = M^d \bmod N$, it is checked whether the public exponent e is embedded in the representation of the RSA key. If this is the case, the public exponent e is recovered. Then it is verified if $S^e \equiv M \pmod N$ in standard mode or if $S^e \equiv M \pmod{pq}$ in CRT mode.

The public exponent e is embedded in RSA modulus N, which is shared between the standard and CRT RSA key objects (notice that N can be obtained as $p \cdot q$ in CRT mode). This can be done by using the method to generate RSA modulus N with a predetermined portion [197].

8.7 Second-Order Fault Attacks

Until recently, most of the theoretical fault attacks and countermeasures used a fault model that assumed that the attacker was able to disturb the execution of a cryptographic algorithm only once. However, this approach seems too restrictive since the publication in 2007 of the successful experiment of an attack based on the injection of two faults, namely a second-order fault attack, by Kim and Quisquater [226].

Kim and Quisquater introduced a second-order fault model where they were able to practically break the (first-order) countermeasures of [91, 161]. In their model, one fault is dedicated to the corruption of the RSA computation in order to produce an exploitable faulty signature. The other fault is then used to render the countermeasure ineffective.

Kim and Quisquater's countermeasures against second-order fault attacks proposed in [226, 228] were analyzed by Dottax et al. from an implementation standpoint, revealing a potential vulnerability [125]. Trichina and Korkikyan have also showed experimental results on the second-order fault attacks on protected CRT-RSA implementations running on an advanced 32-bit ARM Cortex M3 core [399].

Chapter 9
Fault Attacks on Elliptic Curve Cryptosystems

Abdulaziz Alkhoraidly, Agustín Domínguez-Oviedo and M. Anwar Hasan

Abstract While there is no known subexponential algorithm for the elliptic curve discrete logarithm problem, elliptic curve cryptosystems have been shown to be vulnerable to a wide range of attacks that target their implementation rather than their mathematical foundation. Fault analysis attacks exploit faults that can occur in the implementation of an elliptic curve cryptosystem to discover the secret information partially or fully. Faults can be injected in a variety of ways and almost all parts of the system can be targeted, e.g., the base point, system parameters, intermediate results, dummy operations and validation tests. In this chapter, we review a collection of the known fault analysis attacks on elliptic curve cryptosystems. We also briefly discuss the known countermeasures to various attacks and comment on their effectiveness.

9.1 Introduction

Elliptic curves were first employed in cryptography by Miller and Koblitz as an alternative to many of the dominant public-key cryptosystems [233, 288]. The group of points on an elliptic curve defined over a finite field has many interesting properties that make it suitable for cryptographic applications. Most importantly, the discrete logarithm problem on that group appears to be hard to solve, more so than similar problems like integer factoring and finding discrete logarithms over finite fields. As such, a comparable security level can be achieved using significantly smaller system parameters, which gives elliptic curve cryptography (ECC) an advantage in terms of efficiency for both software and hardware implementations.

A. Alkhoraidly (✉) · M. A. Hasan
University of Waterloo, Waterloo, ON, Canada
e-mail: amalkhor@uwaterloo.ca

A. Domínguez-Oviedo
Tecnológico de Monterrey, Campus Queretaro, Queretaro, Mexico

M. Joye and M. Tunstall (eds.), *Fault Analysis in Cryptography*,
Information Security and Cryptography, DOI: 10.1007/978-3-642-29656-7_9,

However, despite their theoretical security, elliptic curve cryptosystems are vulnerable to a variety of side-channel attacks (SCAs) that target the implementation of the cryptosystem rather than its mathematical weaknesses and exploit the information leaking during the proper or improper use of the cryptosystem. SCAs are generally passive, i.e., the attacker observes a working cryptosystem without influencing its operations. Examples of side-channel attacks include timing attacks, originally presented in [239], and power analysis attacks, introduced in [240].

Fault analysis attacks (FAAs), on the other hand, are active attacks that use faults to influence the operation of the system. FAAs range in cost, and in complexity from the simple to the highly sophisticated. In essence, fault attacks seek to expose the secret information partially or fully using invalid outputs that result from natural or deliberate faults. Most fault attacks on ECC attempt to move the computation from the secure curve to another, probably weaker, curve. This can be achieved by injecting faults into the curve parameters, or the base point, or during the scalar multiplication. Examples of this class of fault attacks include those presented in [44, 90, 122, 143]. On the other hand, the attack presented in [54] targets the sign of an intermediate point in the scalar multiplication, and results in a faulty output point that still belongs to the original curve. Moreover, some attacks like the safe-error fault attack presented in [427] exploit a countermeasure against simple timing analysis, while [430] presents an attack that exploits validation tests as a single point of failure.

In this chapter, we discuss known fault analysis attacks on elliptic curve cryptosystems. We start by giving some background on elliptic curves and their use in cryptography in Sect. 9.2, along with a brief overview of fault injection methods and general techniques to prevent and detect faults. Section 9.3 follows with a survey of the different classes of fault attacks that attempt to move the computation to an invalid elliptic curve. Section 9.4 discusses the inherently different sign change fault attack, while Sect. 9.5 addresses attacks that target specific parts of the implementation, namely, validation and dummy operations. Section 9.6 gives a summary of known countermeasures.

9.2 Background

This section aims to provide a brief overview of the mathematical concepts often referred to throughout this chapter. Reference books in abstract algebra, like [309], and in cryptography, like [98, 176], can be consulted for a more extensive treatment of these concepts. Moreover, we briefly discuss some of the known methods for fault injection, which are covered in more details in Sects. 16.2 and 17.3.

9.2.1 Preliminaries

A group $(G, \oplus)$, written additively, is a set G associated with a binary operation $\oplus$ defined on G such that it is closed, associative, each element of G has an inverse, and has an identity element. Moreover, the group is called commutative or abelian if $\oplus$ is commutative. For a finite group, the number of elements is referred to as the *group order* or *cardinality*. On the other hand, the order of an element a in a finite group G, denoted by $\mathrm{ord}_G(a)$, is the smallest integer c such that

$$ca = \underbrace{a \oplus a \oplus \cdots \oplus a}_{c \text{ times}} = 0$$

where 0 commonly represents the identity element of $\oplus$. The order of a group is divisible by the order of any of its elements.

A ring $(R, \oplus, \otimes)$ is a set R associated with two binary operations, $\oplus$ and $\otimes$, defined on R such that $(R, \oplus)$ is an commutative group and $\otimes$ is closed, associative and distributive over $\oplus$. Furthermore, the ring is commutative if $\otimes$ is commutative. In a ring R, if the operation $\otimes$ has an identity element the ring is called a ring with identity. The operation $\otimes$ is usually written multiplicatively and its identity, if one exists, is denoted by 1. An example of a commutative ring with identity is the set $\mathbb{Z}_n = \{0, 1, \ldots, n-1\}$ under addition and multiplication modulo n.

In a ring with identity R, if the nonzero elements form a commutative group under $\otimes$, then R is a field. In other words, a field can be seen as a commutative group with respect to two binary operations such that one operation is distributive over the other. While the set of rational numbers $\mathbb{Q}$ is an example of a field, $\mathbb{Z}$ is an example of a ring that is not a field since the only elements that have multiplicative inverses are 1 and -1. A field that has a finite number of elements is called a finite field, and can be either a prime field or an extension field. As the name indicates, a prime field $\mathbb{F}_p$ has a prime cardinality p. An extension field, on the other hand, has a cardinality of p^d, where p is prime and $d > 1$ is an integer. Such a field is created by extending the prime field $\mathbb{F}_p$ where d denotes the extension degree. The characteristic of both $\mathbb{F}_p$ and $\mathbb{F}_{p^d}$, denoted by $\mathrm{char}(\mathbb{F}_p)$, is p. An interesting fact is that all finite fields of the same cardinality are isomorphic, i.e., have the same structure even if they are represented differently. In other words, these fields can be made identical by renaming their elements.

9.2.2 The Elliptic Curve Group

An elliptic curve E over a field K, whose algebraic closure is denoted by $\overline{K}$, is the set of points (x, y), $x, y \in \overline{K}$, that satisfy the Weierstrass equation

$$E : y^2 + a_1xy + a_3y = x^3 + a_2x^2 + a_4x + a_6 \tag{9.1}$$

where the coefficients $a_1, a_2, a_3, a_4, a_6 \in K$ and such that the curve is nonsingular. The set of points $(x, y) \in K \times K$ that satisfy the curve equation, along with the point $\mathcal{O}$ at infinity, is denoted by $E(K)$.

The full Weierstrass equation can be simplified depending on the characteristics of the underlying field, K. For prime fields, and when $\text{char}(K) \neq 2, 3$, (9.1) can be simplified to

$$E : y^2 = x^3 + ax + b \tag{9.2}$$

where $a, b \in K$. In a binary field, i.e., when $\text{char}(K) = 2$, and assuming that $a_1 \neq 0$ and that E is nonsupersingular, (9.1) can be simplified to

$$E : y^2 + xy = x^3 + ax^2 + b \tag{9.3}$$

where $a, b \in K$.

The points on an elliptic curve, together with the point $\mathcal{O}$ at infinity, form an abelian group under the operation of point addition, which has an intuitive geometric interpretation that can be used to derive explicit formulas. When $\text{char}(K) \neq 2, 3$ and $P \neq -Q$, we can find $R = (x_3, y_3) = P + Q$ as follows:

$$\lambda = \begin{cases} \dfrac{y_2 - y_1}{x_2 - x_1}, & \text{if } P \neq Q \\ \dfrac{3x_1^2 + a}{2y_1}, & \text{if } P = Q \end{cases}$$

$$x_3 = \lambda^2 - 2x_1, \quad y_3 = \lambda(x_1 - x_3) - y_1 . \tag{9.4}$$

Each point operation on an affine elliptic curve requires a field inversion, which is significantly more expensive in general than a field multiplication. To address this problem, various projective coordinates were introduced, e.g., [99], and are often employed to reduce the number of field inversions at the cost of more field multiplications and storage space. Different projective coordinates have different costs in terms of field operations. Moreover, it has been shown that adding points represented in different projective coordinates can be more efficient than adding of points in the same coordinate system [98].

The security of ECC is based mainly on the hardness of the elliptic curve discrete logarithm problem (ECDLP), which can be defined as finding the scalar k given $P, kP \in E$. Since all the known solutions for this problem in general elliptic curves have an exponential complexity, parameters should be of appropriate size to make the instance intractable. Moreover, the ECDLP is easier, i.e., there exist sub-exponential solutions, for some special elliptic curves. As such, these curves should be avoided in ECDLP-based cryptography.

9.2.3 Elliptic Curve Scalar Multiplication

Scalar multiplication, the operation of repetitively adding a point to itself on an elliptic curve, is the most important primitive in ECC and its execution time dominates the time required to perform ECC operations. This operation is analogous in many ways to the modular exponentiation operation employed in RSA and similar systems.

A naive way to perform a scalar multiplication is repeated addition, which requires a number of group operations proportional to the group size. A much better solution is the double-and-add algorithm, analogous to the square-and-multiply algorithm for modular exponentiation. Algorithm 9.1 illustrates a basic variant of the double-and-add method. This approach requires a number of group operations proportional to the logarithm of the group size.

Algorithm 9.1: Double-and-add scalar multiplication

Input: $P \in E(K)$, $k = (k_{n-1}, k_{n-2}, \ldots, k_0)$
Output: kP

$Q \leftarrow \mathcal{O}$
for $i \leftarrow n - 1$ **downto** 0 **do**
    $Q \leftarrow 2Q$
    if $k_i = 1$ **then**
        $Q \leftarrow Q + P$
    end
end
return Q

It can be seen from Algorithm 9.1 that a doubling operation is required in every iteration, while an addition operation is required only when the corresponding bit of the scalar k is 1, which is true in half of the iterations on average. It is possible to reduce the cost of the scalar multiplication by reducing the number of point additions, which can be achieved by recoding the scalar k. For example, the nonadjacent form (NAF) representation, with the property that no two nonzero digits are adjacent, can be used to reduce the number of 1s in the scalar. The simplest instance is the 2-NAF, which uses the digits 0, 1, and -1 and ensures that no two consecutive digits are nonzero. The use of the 2-NAF representation reduces the number of additions from $n/2$ to $n/3$ on average [297].

Double-and-add-always variant

It can be easily seen that the iterations of Algorithm 9.1 take different times for different values of k_i, which can be exploited by simple power or timing analysis attacks to determine the secret scalar k. To solve this problem, the computations performed should not depend on the secret data and branching should be avoided. This results in regular iterations as illustrated in Algorithm 9.2, presented in [102]. More recently, other regular scalar multiplication algorithms have been introduced, e.g., in [195].

Algorithm 9.2: Double-and-add-always scalar multiplication [102]

Input: $P \in E(K), k = (1, k_{n-2}, \ldots, k_0)$
Output: kP
$Q[0] \leftarrow P$
for $i \leftarrow n-2$ **downto** 0 **do**
  $Q[0] \leftarrow 2Q[0]$
  $Q[1] \leftarrow Q[0] + P$
  $Q[0] \leftarrow Q[k_i]$
end
return $Q[0]$

Montgomery ladder

It has been shown that, for some elliptic curves, the y-coordinate is not essential in the point addition and doubling operations [294], so the x-coordinate of the resulting point, $P+Q$, can be computed from the x-coordinates of the points P, Q, and $Q-P$. In particular, let $P = (x_1, y_1)$, $Q = (x_2, y_2)$ and $Q - P = (x_3, y_3)$ be points on the elliptic curve $E : y^2 = x^3 + ax + b$; then the x-coordinates of $P + Q = (x_4, y_4)$ and $2Q = (x_5, y_5)$ can be computed by the following equations:

$$x_4 = \frac{2(x_1 + x_2)(x_1 x_2 + a) + 4b}{(x_1 + x_2)^2} - x_3 ,$$

$$x_5 = \frac{\left(x_1^2 - a\right)^2 - 8bx_1}{4\left(x_1^3 + ax_1 + b\right)} .$$

It follows that the result of kP can be found by calculating a sequence of pairs (Q, H), which has the property that $H - Q = P$. Algorithm 9.3 demonstrates the steps of a Montgomery ladder for scalar multiplication. It can be noted from Algorithm 9.3 that the Montgomery ladder is a variant of the double-and-add-always method, and hence can be used as a countermeasure against simple timing attacks [193, 205]. Moreover, in [257], and more recently in [383], it was shown that this algorithm is particularly efficient for curves defined over binary fields.

Algorithm 9.3: Montgomery ladder scalar multiplication [294]

Input: $P \in E(K), k = (1, k_{n-2}, \ldots, k_0)$
Output: The x-coordinate of kP
$Q[0] \leftarrow P, Q[1] \leftarrow 2P$
for $i \leftarrow n-2$ **downto** 0 **do**
  $Q[\overline{k_i}] \leftarrow Q[0] + Q[1]$
  $Q[k_i] \leftarrow 2Q[k_i]$
end
return $Q[0]$

9.2.4 Faults in Digital Systems

In general, it is important for a digital system to be fault-free and consistently give the correct results. Faults can occur for a variety of natural and artificial reasons, and various methods have been proposed to counter their effects on the performance and reliability of the system. Fault detection and tolerance is even more important for cryptosystems, due to the existence of attacks that can exploit faults to discover secret information and threaten the security of the whole system.

Faults can occur in a device either naturally or due to deliberate actions, and can be caused for one of many reasons. In general, variations in standard operation conditions can be used effectively to inject faults into a system. For example, the variation in the supply voltage or the clock frequency can disrupt the execution and cause the processor to skip instructions or disrupt input/output operations. Moreover, exposing the device to temperatures outside its operational range can cause random modifications of the memory and inconsistencies in memory access. It is also possible to inject faults more accurately using the photoelectric effects that are inherent in all electric circuits. The exposure to photons induces currents in the circuit that can disrupt normal operation. In effect, targeting and timing can be made more precise using lasers in fault injection. Faults can be injected in packaged circuits without removing the packaging by using X-rays and ion beams [21].

Faults in electronic circuits can be either permanent or transient. Permanent faults are caused by intentional or unintentional defects in the chip. As the name indicates, they have a permanent effect on the behavior of the circuit. On the other hand, a transient fault does not cause a permanent change in the behavior of the circuit. Such faults are caused by local ionization, which induces a current that can be misinterpreted by the circuit as an internal signal. Fault injection is discussed in more detail in Sects. 16.2 and 17.3 of this book.

Several solutions have been devised to avoid or detect faults, or to prevent the attempt to inject them. Other solutions help us recover from the occurrence of faults and produce a correct output in spite of their existence. Some of these methods are implemented in hardware while some are implemented in software. The main countermeasure against faults and errors is the use of redundancy in the design, which makes it possible to detect erroneous results and behavior. It also may permit the recovery from faults. A common form of redundancy is hardware redundancy, which entails replicating some part of the hardware to prevent the existence of a single point of failure. Another form of redundancy is time redundancy, which amounts to repeating the computation or a part of it to confirm the earlier results and detect transient faults. A third form of redundancy is information redundancy, which is commonly employed in data communication through error detecting and correcting codes. The principle behind information redundancy is the use of more bits to represent the data than is actually necessary. This way, some of the representable bit patterns do not correspond to valid data and can be used to detect and correct errors. It is also possible to combine two or more types of redundancy into a single scheme to get the advantages of different types of redundancy. A more elaborate review of the known countermeasures against fault attacks in elliptic curve cryptosystems is the subject Sect. 10.2 of this book.

9.3 Invalid-Curve Fault Attacks

The choice of the elliptic curve and the underlying field is an important one as it significantly influences the security of the system. The general aim of invalid-curve fault attacks is to move the computation from a secure curve to a weaker one, enabling the attacker to use known mathematical attacks against the faulty outputs. This can be achieved by targeting either the system parameters or the running computation. In both cases, there are known countermeasures that can be used to detect the attack and prevent the faulty output.

9.3.1 Targeting the Base Point

The base point is one of the key parameters in a scalar multiplication operation. It is also relatively easy to target as it is commonly presented to the system as an input. Various known fault attacks target the base point. Some of which assume that the attacker knows the faulty value of the base point while others relax this assumption.

9.3.1.1 Known or Chosen Faulty Base Point

In the attacks introduced by [44], the representation of a point P on a strong elliptic curve E is modified as a result of a fault to move the computation to a different, often weaker, curve E'. The resulting faulty output values can be used to deduce partial information about the secret key. Usually, the attack has to be performed repeatedly since in most cases the guessed values are not unique.

Attack description

Let E be a strong elliptic curve defined over a finite field K as

$$E : y^2 + a_1xy + a_3y = x^3 + a_2x^2 + a_4x + a_6$$

and let P and $Q = kP$ be two points on E. To be able to mount this attack, suppose that the device does not check whether P and Q are actually on E.

According to the ANSI X9.63 and IEEE 1363 standards, a_6 is not used in the addition operation. It follows that for a point $P' = (x', y')$ with $x', y' \in K$ and $P' \notin E$, the calculation of $Q' = k P'$ occurs over the curve $E'(a_1, a_2, a_3, a_4, a_6')$ where

$$a_6' = y'^2 + a_1x'y' + a_3y' - x'^3 - a_2x'^2 - a_4x'$$

instead of over the original curve E. If P' is chosen in such a way that E' is a cryptographically weak curve, i.e., P' has a relatively small order over E', then the value of k modulo $\mathrm{ord}(P')$ can be found by solving a DLP over the subgroup generated by P'. The relatively small size of this group enables the use of the known general DLP computation algorithms like Shank's baby-step-giant-step algorithm [374], or Pollard's rho and Pollard's lambda algorithms [326]. It is also possible to select P' such that it has a smooth order, i.e., composite with small factors, over E'. In this case, the Pohlig-Hellman algorithm [325] can be used to break the bigger DLP into smaller problems in the subgroups of prime order, which can be solved independently.

This process can be repeated until sufficient residues of k are collected, and then Chinese remaindering is used to recover k. This attack runs in polynomial time, and has been extended to standardized EC-based cryptographic protocols in [16].

9.3.1.2 Unknown Faulty Base Point

It is also possible to exploit a fault that results in an unknown faulty base point. Biehl et al. [44] discussed this as an extension to their attack presented earlier, where they assume that the faulty base point has an error in one unknown bit position. Ciet and Joye [90] extended this attack to any number of bit errors in the base point.

Attack using a faulty base point with one bit error [44]

Suppose that the device checks whether the base point P lies on E before starting the computation, and assume a fault can be injected into P in an unknown position right between the test and the computation. Again, the modified point P' will lie on a curve E' with a different, yet unknown, a_6' value. Using the incorrect output $Q' = kP'$, the value of a_6' can be computed. For each of the possible values of P' that lie on E', we find k' such that $Q' = k'P'$ by solving a DLP on E'. Then, we proceed to compute k as in the more basic attack discussed earlier.

The elliptic curve digital signature algorithm (ECDSA) is also susceptible to this attack. Since the faulty base point cannot be given as an input as it is fixed by the protocol, the attacker injects a fault in the base point P to replace it by P' at the beginning of the scalar multiplication. Then, the algorithm computes $r' = (kP')_x \bmod p$, where $(\cdot)_x$ is the x-coordinate of the point and p is the prime order of P. It then computes $s' = k^{-1}(H(m) + dr') \bmod p$ where $1 < k < p - 1$ is a random number, $H(m)$ is the hash of the message m, and d is a the secret key. The resulting signature is the pair (r', s'). The attacker can then find the faulty curve E', on which P' lies, and use r' to find a small set of candidates for $(kP')_x$. Then, the corresponding y-coordinates are found using the equation of E'. Assuming the DLP on E' is weak, and given the candidates for kP', the attacker solves the DLP on E' to find candidate values for k, and then uses those to compute candidates for the secret key as $d = r'^{-1}(s'k - H(m)) \bmod p$.

Attack using a faulty base point with multiple-bit error [90]

Let E be an elliptic curve over K and let $P = (x_P, y)$ be the base point. Faults can be assumed to occur in either or both of the coordinates of P. For example, assume that the value stored for the x-coordinate, x', is corrupted in some unknown bit positions and let $P' = (x', y)$. Then, $Q' = kP'$ can be computed and will lie as before on the curve E', which shares all the parameters of E except a_6. The corresponding value for E', a'_6, can be computed from the coordinates of Q' as

$$a'_6 = y_{Q'}^2 + a_1 x_{Q'} y_{Q'} + a_3 y_{Q'} - x_{Q'}^3 - a_2 x_{Q'}^2 - a_4 x_{Q'}.$$

Then, we know that x' is a root in K of

$$x^3 + a_2 x^2 + (a_4 - a_1 y)x + (a'_6 - y^2 - a_3 y)$$

since $P' \in E'$.

Assuming that r, the order of P' in E', is small, the root of the above polynomial with the least Hamming distance from the original x_P can be used as a candidate for x'. Assuming that the DLP on E' is weak, the Pohlig-Hellman reduction can be used to obtain a candidate for k.

Note that this attack applies in a similar fashion when errors are injected into the y-coordinate instead. However, when both coordinates are faulty, it becomes harder to recover the faulty point P' as the attacker only knows that it lies on E'. One way to recover candidate values of P' is to perform the scalar multiplication repeatedly using modified copies of the scalar as illustrated in more detail in [90].

Attacks on the Montgomery ladder

Fouque et al. [143] have proposed a fault attack on the Montgomery ladder algorithm over prime fields. Their work uses the fact that the y-coordinate is not used in the scalar multiplication, and hence the faulty computation can leave the original curve and move to its twist. More specifically, since the y-coordinates are not utilized in the Montgomery ladder elliptic curve scalar multiplication [205], the computation applies also to the set of points (x, y) with $x \in \mathbb{F}_p$ and $y \in \mathbb{F}_{p^2}$ that form a twist of the original elliptic curve E, denoted by $\tilde{E}$.

The order of the group of the original elliptic curve and that of its twist are roughly of the same size. In practice, an elliptic curve is chosen to have a group with prime order, but the group order of the twist is not necessarily prime. In fact, among the five NIST-recommended curves over prime fields only the curve denoted by P-384 has a twist with a prime order. For the remaining curves, the group orders of their twists are composite and are easier to attack.

In [143], two attacks have been presented. The basic attack is when the adversary is able to choose the input point P and the implementation does not use point validation at the end of the scalar multiplication. The second attack assumes that the attacker cannot select P and that the implementation validates the resulting point at the end of the scalar multiplication. In this case, the attacker needs to inject two faults. First, the attacker injects a fault into the x-coordinate of the input point P. Since half of the x's correspond to points that lie on the original curve while the other half correspond to those that lie on the twist, the resulting value will correspond to points on the twist $\tilde{E}$ with probability $\frac{1}{2}$. Second, at the end of the computation, just before the point validation, the attacker needs to inject a fault into the x-coordinate of the result. Note that due to the missing y-coordinate, point validation will accept any x for which $x^3 + ax + b$ is a square, i.e., with probability $\frac{1}{2}$. In effect, the point validation can be bypassed and the faulty output can be obtained with probability $\frac{1}{4}$.

A similar attack that targets the Montgomery ladder over a binary field was proposed independently in [122]. This attack takes advantage of the fact that the parameter a in (9.3) is not used during the scalar multiplication, and the authors adopt the single bit-flip fault model proposed in [56], which has been shown to be practical [379]. It can be shown that for a fixed value of the curve parameter b there are only two isomorphic classes of curves, one for each value of $\mathtt{Tr}(a)$, where $\mathtt{Tr}(\cdot)$ is the trace function. It follows that it is possible to define two elliptic curves, E_0 and E_1, one for each of these isomorphic classes:

$$E_0 : y^2 + xy = x^3 + b,$$
$$E_1 : y^2 + xy = x^3 + x^2 + b.$$

It is known that the orders of E_0 and E_1 are roughly the same. It is not necessary, however, that both curves be strong (in the cryptographic sense). In fact, among the ten NIST-recommended binary curves, there is only one for which the orders of both E_0 and E_1 are almost prime, namely, the curve K-283.

The key idea behind this attack is to produce an incorrect result from the computation being performed in the weaker curve of the pair E_0 and E_1 due to a fault. It is assumed that the degree of extension is odd, i.e., $\mathtt{Tr}(1) = 1$ in the field. The attack starts by injecting a fault into the x-coordinate of the input point $P = (x_P, y_P) \in E(\mathbb{F}_{2^m})$ of a device computing the scalar multiplication. If the resulting finite field pair after the fault injection is known and the result $\widetilde{Q} = k\widetilde{P} = (\widetilde{x}_Q, \widetilde{y}_Q)$ is released it is possible to obtain the full scalar with high probability of success. Moreover, a variant of this attack has been presented where the faulty finite field pair $\widetilde{P}$ is unknown and two computations with the same scalar are obtained. In such a case, the scalar can also be obtained.

9.3.1.3 Countermeasures

It is commonly advisable to check the correctness of the input points, i.e., whether they belong to the original curve. One way to do that is to compute and verify the value of a_6 using the coordinates of P. It is advisable to avoid NIST curves that have cryptographically weak twists. It is also important to ensure that the output points lie on the original curve. Another countermeasure, which has been reported in [122] and extends the approach presented in [162] for RSA, uses the invariant in the Montgomery algorithm (Algorithm 9.3), i.e., $Q[1] - Q[0] = P$, to validate the variables before results are returned.

9.3.2 Targeting the System Parameters

It is also possible to target the parameters of the system, e.g., the field representation, or the parameters of the curve equation. This will generally lead to the computation being performed on a weaker curve or a weaker curve-field combination.

9.3.2.1 Faulty Field Representation

Faults can be injected into the field parameters, either in storage or in transit. This attack, introduced in [90], exploits this fact. Let E be a curve defined over a prime field $\mathbb{F}_p$ of characteristics other than 2 and 3. Assume that a bit error is injected into p to give the almost similar value p' and that all field operations will then be performed modulo p' instead. In particular, the values of P, Q, a and b will be represented modulo p' as P', Q', a' and b', respectively.

Since Q' satisfies the equation of E', it follows that

$$b' \equiv y'^2_Q - x'^3_Q - a' \equiv b \equiv y^2 - x^3 - a \ (\mathrm{mod} p') \,.$$

Hence, $p'|D$ where $D = \left|y'^2_Q - x'^3_Q - a' - (y^2 - x^3 - a)\right|$ and p' can be revealed through factoring D as the product of factors that has the shortest Hamming distance from p. Using these factors, the value of k can be computed by solving a set of small DLPs and then Chinese remaindering.

Moreover, p' can be found more efficiently when p is a (generalized) Mersenne prime. *Generalized Mersenne primes* are primes of the form

$$p = 2^{\omega_0} + \sum_{i=1}^{B} \pm 2^{\omega_i}$$

where B is typically small [381]. These primes are commonly used to enable highly efficient field reduction and to significantly reduce storage requirements for p. For example, when $B \leq 4$, as is the case for all NIST prime curves [141], p can be written as

$$p = 2^{\omega_0} + \sum_{i=1}^{3} \sigma_i 2^{\omega_i} + \sigma_4 \qquad (9.5)$$

where $\sigma_i \in \{-1, 0, 1\}$ and $\sigma_4 \neq 0$. Then, p can be stored efficiently by storing the values of the ω_i's and σ_i's. Since an error injected into p will affect any of these quantities but will not affect the form given by 9.5, the number of possible values of p' that can result from factoring D will significantly decrease as only candidates that satisfy 9.5 need to be considered.

The same principle applies to binary fields [90], where pentanomials (and trinomials, if any exist) are used as they enable efficient reduction and reduced storage requirements. It is known that, for all $d \leq 1{,}000$, there exists an irreducible pentanomial that can be used to represent the elements of $\mathbb{F}_{2^d}$ [187]. Since a pentanomial can be stored efficiently by only storing its exponents, an error will modify one or more of the exponents but will not affect its form. It follows that all candidates for the faulty reduction polynomial that do not have the same form can be discarded, which significantly reduces the search space.

9.3.2.2 Faulty Curve Parameters

Errors injected into one or more of the curve parameters can be exploited in a similar way to enable a more efficient attack [90]. In general, the parameter a_6 in (9.1) is not targeted since it does not affect the final result as it is not used during the point addition operation. Assuming that only one of the other parameters is modified by an error, the valid value of the point P and the faulty resulting point $\tilde{Q}$ can be used to solve for the value of the randomly modified parameter and the associated $\tilde{a_6}$. In effect, E' is known, and if $\mathrm{ord}(P)$ on E' is small or smooth, the DLP can be solved on E' to get k modulo $\mathrm{ord}(P)$.

9.3.2.3 Countermeasures

Checksums can be used to avoid faults in system parameters. For each of the system parameters, the checksum should be computed and checked after reading it from memory and before outputting the resultant point.

9.3.3 Targeting Intermediate Variables

In addition to the attacks discussed earlier, it is also possible to exploit random faults that occur during the scalar multiplication. In particular, faults injected into intermediate variables lead to faulty outputs that can be used, along with the correct result, to guess parts of the scalar. The following attack, introduced in [44], illustrates this.

9.3.3.1 Exploiting Random Faults During the Computation

Assume that E is defined over a field $\mathbb{F}_q$, and that the binary algorithm is used to perform the scalar multiplication, as in Algorithm 9.1. Also, assume that the attacker can repeatedly input a point P and induce a fault during a specific iteration of the computation, and that the correct result $Q = kP$ is known. Let Q_i denote the values of Q at the ith iteration. During the computation, a fault is injected into a random iteration j to modify Q_j to Q'_j and get Q' as an output. Then, the values of Q, Q' and j can be used to determine the intermediate value Q'_j, which can be used to guess the higher $n - j$ bits of k. The process can then be repeated going downwards through the bits of k.

It has been shown in [44] that finding a secret key of length n bits using this attack requires $O(n \log n)$ faulty scalar multiplications and $O(n \log n)$ bit operations for subsequent calculations. It is also possible to decrease the accuracy of random fault injection, i.e., inject faults into blocks of at most m consecutive iterations rather than into a specific iteration. The choice of the block size m presents a trade-off between the number of register faults required and the time needed to analyze the faulty results.

9.3.3.2 Countermeasures

As seen earlier, point validation is essential when fault attacks are considered. In particular, output points should be validated and any point that does not belong to the original curve should be discarded.

9.4 Sign Change Fault Attack

Earlier fault attacks on ECC worked by inducing faults in a way that would move the computation to a different, probably weaker, elliptic curve. This can be achieved by modifying the base point, an intermediate point, or a parameter of the curve. However, this can be easily countered by verifying the correctness of the parameters and that the resulting point belongs to the original curve.

The attack described in [54] does not move the computation to a different curve, but rather results in a faulty point on the original curve. By collecting enough of these faulty results, the secret can be recovered in expected polynomial time. It follows that point validation is not an effective countermeasure against this attack. Actually, it may help the attacker to remove useless faulty points, i.e., points that fall off the curve, and hence it makes a less precise attack more effective.

Attack description

As the name indicates, this attack involves changing a sign of an intermediate variable during the scalar multiplication, which can be achieved by replacing an intermediate point with its inverse or by changing a digit of the NAF-encoded scalar from 1 to -1.

Applying this attack repeatedly enables the attacker to recover the scalar k starting from the least significant bit. A simpler variant of the attack, where it is assumed that the attacker can target a specific iteration, is described here. However, the attack can be extended to allow for uncertainty in fault injection [54].

Suppose that the basic double-and-add algorithm (Algorithm 9.1) is used to compute $Q = k\,P$, and that the attacker has the correct value for Q. It is assumed that the attacker can inject a temporary sign change fault into the doubling step $Q \leftarrow 2Q$ so that it effectively becomes $Q \leftarrow -2Q$. The attack starts by targeting the last iteration of the algorithm, where $i = 0$. This results in the faulty output

$$\tilde{Q} = -\sum_{i=1}^{n-1} 2^i k_i P + k_0 P = -Q + 2\,k_0 P,$$

which lies on the original curve and cannot be detected by point validation. Moreover, depending on whether k_0 is equal to 0 or 1, $\tilde{Q}$ will be equal to either $-Q$ or $-Q+2P$, respectively, which easily reveals k_0 to the attacker. In general, if a sign change fault is injected into iteration i, the resulting faulty point can be written as

$$\tilde{Q} = -Q + 2\sum_{j=0}^{i} k_j 2^j P$$

where only one of the $i + 1$ least significant bits of k is unknown, namely, k_i. As a result, $\tilde{Q}$ will take one of two values,

$$\tilde{Q} = \begin{cases} -Q + 2\sum_{j=0}^{i-1} k_j 2^j P & k_i = 0 \\ -Q + 2P + 2\sum_{j=0}^{i-1} k_j 2^j P & k_i = 1 \end{cases}.$$

The attack can be generalized to cases where the attacker may not know the precise iteration in which the change happened. The idea is to recover the bits of k in blocks of $1 \le r \le m$, where m is chosen to control a trade-off between the required exhaustive search and the number of faulty results required to give a success probability of at least $\frac{1}{2}$.

9.4.1 Countermeasures

The sign change fault attack is not applicable to Montgomery's scalar multiplication algorithm [294] since it does not use the y-coordinate, which prevents the attacker from changing the sign of intermediate points. Moreover, randomizing the scalar, e.g., by splitting, is effective since the same fault at the same stage of the algorithm would generate a different result every time.

Another countermeasure to the sign change attack has been proposed in [54], and extends a countermeasure presented in [372] for RSA. The idea is to employ another curve E_t defined over a smaller field $\mathbb{F}_t$ for some prime t. Moreover, a base point P_t is chosen to have a large order within E_t. A combined curve E_{pt} and a new base point P_{pt} can be constructed using the Chinese remainder theorem, and then the scalar multiplication is performed once on each of E_{pt} and E_t. If the results do not match modulo t, they are rejected.

A similar countermeasure has been proposed in [19]. Most notably, in this countermeasure, the integer t, the curve E_t and the point P_t are selected randomly and t is not necessarily prime. However, it has been shown in [196] that this setup allows a nonnegligible proportion of faults to pass through, and that a nonrandom setup like the one proposed in [54] is significantly more secure.

A countermeasure based on coherency checking is described [124]. This countermeasure extends the coherency-based countermeasure presented in [162] for RSA signature algorithms. As illustrated in Algorithm 9.4, this countermeasure is based on the right-to-left double-and-add-always scalar multiplication algorithm. It includes within it point validation of resulting points, and since it is a double-and-add-always algorithm, it is inherently resistant to simple power and timing analysis attacks. It is worth mentioning, however, that only single-fault attacks are considered in this algorithm. By tracing the iterations in this algorithm, we notice that in an error-free run, Q_1 is expected to have the value kP, while Q_0 will have the value $\overline{k}P$, where $\overline{k} = 2^n - 1 - k$. Since Q_2 has the value $2^n P$, it follows that $Q_0 + Q_1 + P = Q_2$. As shown in [124], a sign change fault in any of the intermediate values can be detected by checking this invariant. This countermeasure incurs significantly less overhead than the ones based on combined curves.

Algorithm 9.4: Right-to-left double-and-add-always scalar multiplication algorithm with point validation and coherency checking

Input: $P \in E(K)$, $k = (k_{n-1}, k_{n-2}, \ldots, k_0)$
Output: kP

$Q_0 \leftarrow \mathcal{O}$, $Q_1 \leftarrow \mathcal{O}$, $Q_2 \leftarrow P$
for $i \leftarrow 0$ **to** $n-1$ **do**
  $Q_{k_i} \leftarrow Q_{k_i} + Q_2$
  $Q_2 \leftarrow 2Q_2$
end
if $Q_0 \in E(K)$ ***and*** $Q_1 \in E(K)$ ***and*** $Q_2 = Q_0 + Q_1 + P$ **then**
  return Q_1
else
  return $\mathcal{O}$
end

9.5 Fault Attacks on Dummy and Validation Operations

9.5.1 Safe-Error Fault Attacks

A common countermeasure against simple power and timing analysis attacks is the use of dummy operations and variables, e.g., a double-and-add-always algorithm like Algorithm 9.2. When a fault is injected into a dummy variable or during a dummy operation, it results in a safe-error, i.e., an error that does not affect the final result. A safe-error fault attack, as presented in [427, 429], works by injecting a fault resulting in a potential safe-error during an iteration of the scalar multiplication and observing the resulting output. A correct output confirms that the targeted operation or variable was a dummy, while a faulty output indicates the opposite. In both cases, the corresponding bit of the scalar is exposed. Note that input or output validation does not help in countering this attack.

9.5.1.1 Countermeasures

To counter a safe-error fault attack, it is essential to eliminate the potential of safe-errors. In particular, it should be possible to detect errors resulting from any injected fault even when they do not affect the final result. As an example, Algorithm 9.4, which employs coherency checking, allows for the detection of safe-error attacks by validating the dummy intermediate variable.

9.5.2 Double-Fault Attacks

Most countermeasures for fault attacks, e.g., point validation, integrity checking and coherency checking, are based on invariants that are assumed to hold in an error-free computation. As such, they are commonly implemented as logical tests that indicate whether the final output is faulty or not. It is possible, however, for an attacker to influence the result of such a test by injecting a fault in the testing hardware or the status register [430]. This leads to these validation tests becoming a single point of failure.

9.5.2.1 Countermeasures

This can be avoided using infective computation, presented originally in [430] and adapted for elliptic curves in [124]. In essence, the validation test is replaced by a computation that allows the correct result to pass through unchanged and masks any faulty results randomly. For example, instead of testing whether $a = b$ and returning a if the equality holds, the algorithm returns $(a - b)r + a$ for a random r, so the equality test becomes implicit and is no longer dependent on the value of a single bit. This way, the attacker cannot use the faulty result since it is no longer correlated with the secret information.

9.6 Summary of Countermeasures Against Fault Attacks

As discussed earlier, fault attacks on elliptic curve cryptosystems can be divided into various classes, and each class has its corresponding countermeasures. Generally, these countermeasures work by preventing the injection of faults, detecting the resulting errors, or masking the faulty result randomly. While some of these techniques are only applicable to ECC or to specific classes of faults, others apply more generally. In this section, we briefly review the known countermeasures for fault attacks and comment on their effectiveness and limitations. A more elaborate discussion of these countermeasures is the subject of the next chapter.

The occurrence of some types of faults can be prevented through physical means like sensors and metal shields [21]. Moreover, sign change fault attacks can be prevented using the Montgomery ladder [294] since it does not use the y-coordinate, and hence does not allow sign change.

As for detection, checksums can be used to detect errors in system parameters. Moreover, point validation can be used to detect invalid-curve errors as the representation of an elliptic curve point has some inherent information redundancy [44]. It is also possible to use time and/or hardware redundancy, accompanied by comparison, to detect faulty results [123]. Randomization can also be used in a variety of ways while encoding the scalar, base point or curve parameters. Combined with hardware or time redundancy, it prevents similar errors from generating similar faulty results, thus aiding in the detection of errors [123]. Most notably, some algorithms involve redundant computations that allow for detecting errors by checking the coherency of the results. For example, in Algorithm 9.4, intermediate variables satisfy a set of invariants that can be used to check for coherency. These invariants allow for detection of a wide range of errors, including those resulting from the three classes of FAAs discussed earlier [124].

It is also possible to use some techniques to mask the faulty results. For example, randomization is effective in masking some types of errors, particularly those resulting from sign change faults [54]. Moreover, since a validation test is a logical test, its outcome can be manipulated by flipping a single bit, effectively becoming a single point of failure. Infective computation, as presented in [430], circumvents this problem by masking faulty results randomly.

9.7 Conclusion

The use of elliptic curves in cryptography has gained wide acceptance since it was first proposed and has become a core component of many industry standards. This can be attributed to the rich mathematical structure of elliptic curves that enables novel applications and implementations, and to their apparent resistance to general mathematical attacks.

However, elliptic curve cryptosystems are vulnerable, like other cryptosystems, to side-channel attacks, which target the implementation weaknesses rather than the mathematical structure. Similarly, they are vulnerable to the various classes of fault analysis attacks, which include invalid-curve attacks, sign change attacks and attacks on validation and dummy operations. The feasibility and effectiveness of a specific fault attack depends on the properties of the implementation and the level of access the attacker has. Some fault attacks are generic with relatively relaxed assumptions and simple procedures, while others involve a sophisticated attacker and expensive apparatus.

It is essential for designers to consider and evaluate the risk of fault attacks against their implementations and implement the necessary countermeasures to defeat these and other side-channel attacks. It is also important for them to consider the interactions between countermeasures and attacks since a countermeasure of one attack can enable another one.

Chapter 10
On Countermeasures Against Fault Attacks on Elliptic Curve Cryptography Using Fault Detection

Arash Hariri and Arash Reyhani-Masoleh

Abstract In this chapter, we study fault detection in finite field and elliptic curve arithmetic operations as a countermeasure against fault attacks in elliptic curve cryptography. In this regard, we review parity-based and time redundancy-based approaches described in the literature for finite field operations. For elliptic curve cryptography, we also present some approaches based on input randomization and point validation.

10.1 Introduction

The cryptographic systems are used to ensure the protection of data within an application or organization. To achieve this objective, a cryptographic system should satisfy security requirements such as the ones proposed by the National Institute of Standards and Technology (NIST, FIPS 140-2), which include physical security, mitigation of attacks, authentication, self-tests, and so on. Mitigation of attacks includes different mechanisms to prevent the various attacks that are common to cryptographic systems. Fault induction is one of the attacks which has received considerable attention. In this type of attack, an attacker manipulates the cryptosystem (e.g. through laser, glitch, magnetic attacks) and induces errors in the computation of cryptographic algorithms. To overcome the problems which might be caused by this type of attack, different countermeasures are used.

In this chapter, we study the countermeasure for the implementations of cryptosystems based on Elliptic Curve Cryptography (ECC). These countermeasures are reviewed for ECC and the underlying finite field operations. More precisely, we consider concurrent error detection/correction in finite field arithmetic operations

A. Hariri (✉) · A. Reyhani-Masoleh
Department of Electrical and Computer Engineering,
The University of Western Ontario, ON, Canada
e-mail: arash.hariri@gmail.com

M. Joye and M. Tunstall (eds.), *Fault Analysis in Cryptography*,
Information Security and Cryptography, DOI: 10.1007/978-3-642-29656-7_10,

in Sects. 10.2 and 10.3. Then, we briefly review the countermeasures based on input randomization, point validation, and concurrent error detection for the ECC in Sect. 10.4.

10.2 Parity Code-Based Fault Detection

In this section, we review the fault detection approaches using parity codes in $GF(2^m)$ finite field multipliers. For each design, we briefly provide some background information to clarify the terms and the formulation.

Let $F(z)$ be an irreducible polynomial of degree m defined as

$$F(z) = z^m + f_{m-1}z^{m-1} + \cdots + f_1 z + f_0, \tag{10.1}$$

where $f_i \in GF(2)$ and $0 \le i \le m-1$. The binary extension field $GF(2^m)$ can be constructed using $F(z)$. Assuming x is a root of the irreducible polynomial $F(z)$, i.e. $F(x) = 0$, the polynomial basis (PB) is defined as the set $\{1, x, x^2, \ldots, x^{m-1}\}$. Now, $A, B \in GF(2^m)$ can be defined as

$$A = \sum_{i=0}^{m-1} a_i x^i, \; B = \sum_{i=0}^{m-1} b_i x^i, \tag{10.2}$$

where $a_i, b_i \in GF(2)$.

10.2.1 Single-Bit Parity-Based Approaches

In [139], a single-bit parity code-based approach was proposed to implement online error detection for all one polynomial (AOP) multipliers. The AOP is defined as $F(z) = z^m + z^{m-1} + \cdots + z + 1$ and the multiplication is done modulo $F(x)$, i.e. $C = A \cdot B \bmod F(x)$, where x is a root of $F(z)$. This multiplication is carried out by the following matrix formulation using the extended PB $\{1, x, \ldots, x^{m-1}, x^m\}$:

$$\begin{bmatrix} c_m \\ c_{m-1} \\ \vdots \\ c_1 \\ c_0 \end{bmatrix} = \begin{bmatrix} a_0 & a_1 & \cdots & a_m \\ a_m & a_0 & \cdots & a_{m-1} \\ \vdots & \vdots & \vdots & \vdots \\ a_2 & a_3 & \cdots & a_1 \\ a_1 & a_2 & \cdots & a_0 \end{bmatrix} \begin{bmatrix} b_m \\ b_{m-1} \\ \vdots \\ b_1 \\ b_0 \end{bmatrix}, \tag{10.3}$$

where $A = \sum_{i=0}^{m} a_i x^i$, $B = \sum_{i=0}^{m} b_i x^i$, and $C = \sum_{i=0}^{m} c_i x^i$.

Now, let $\hat{A}'$ and $\hat{B}'$ be the parities of the operands A and B using the redundant representation, respectively. It is easy to write the following for the parity of C, i.e. $\hat{C}'$, based on (10.3)

$$\hat{C}' = \hat{A}' \cdot \hat{B}'.$$

Another multiplier has also been described in [139], the modified AOP multiplier. This is achieved by forcing $a_m = 0$ and $b_m = 0$ in (10.3), which results in the following equation:

$$\begin{bmatrix} c_m \\ c_{m-1} \\ \vdots \\ c_1 \\ c_0 \end{bmatrix} = \begin{bmatrix} a_1 & a_2 & \cdots & a_{m-2} & a_{m-1} & 0 \\ a_0 & & \cdots & a_{m-3} & a_{m-2} & a_{m-1} \\ 0 & & \cdots & a_{m-4} & a_{m-3} & a_{m-2} \\ \vdots & \vdots & \vdots & \vdots & \vdots & \vdots \\ a_3 & a_4 & \cdots & 0 & a_0 & a_1 \\ a_2 & a_3 & \cdots & a_{m-1} & 0 & a_0 \end{bmatrix} \begin{bmatrix} b_{m-1} \\ \vdots \\ b_1 \\ b_0 \end{bmatrix}. \tag{10.4}$$

Using (10.4), one can write the following on parity prediction in this multiplier:

$$\hat{C} = \hat{A}\,\hat{B} + \sum_{i=1}^{m-1} a_i b_{m-i},$$

where $\hat{A}$, $\hat{B}$, and $\hat{C}$ are the parities of A, B, and C using the standard polynomial basis, respectively.

The third multiplier considered in [139] is a bit-serial Massey-Omura multiplier. This multiplier is defined using the normal basis $\{\beta, \beta^2, \ldots, \beta^{2^{m-1}}\}$, with $\beta \in GF(2^m)$. The normal basis multiplication is defined as $C = A \cdot B$, where $A = \sum_{i=0}^{m-1} a_i \beta^{2^i}$, $B = \sum_{i=0}^{m-1} b_i \beta^{2^i}$, $C = \sum_{i=0}^{m-1} c_i \beta^{2^i}$, and $a_i, b_i, c_i \in GF(2)$. The Massey-Omura multiplier is constructed using the function $f(a_0, a_1, \ldots, a_{m-1}, b_0, b_1, \ldots, b_{m-1})$. To generate the coordinates of C, the inputs of the function f are shifted cyclically [273]. The following lemma is described in [139] to derive the parity prediction formulation for the Massey-Omura multiplier.

Lemma 10.1 *Let $A = \sum_{i=0}^{m-1} a_i \beta^{2^i}$, $B = \sum_{i=0}^{m-1} b_i \beta^{2^i}$, and $C = \sum_{i=0}^{m-1} c_i \beta^{2^i}$ be field elements of $GF(2^m)$, where $C = A \cdot B$. Also, let $\hat{A}$, $\hat{B}$, and $\hat{C}$ be the parity bits of A, B, and C, respectively. Now, using a cyclic function $H^{(j)} = H(a_{0+j}, a_{1+j}, \ldots, a_{m-1+j})$, the parity of the multiplication product, $\hat{C}$, can be obtained by [139]*

$$\hat{C} = \hat{A}\hat{B} + \sum_{i=0}^{m-1} b_i H^{(i)},$$

where the index addition in H is modulo m.

The generation of $\hat{C}$ using the function H resembles the generation of the coordinates of C using the function F by shifting the coordinates of A.

In [343], a single-bit parity code is used for fault detection in bit-serial and bit-parallel PB multipliers. To explain this work, we recall that $F(x) = 0$, and as a result one can write (10.1) as

$$x^m = f_{m-1}x^{m-1} + \cdots + f_1 x + f_0. \tag{10.5}$$

First, we present the following note to show a property of the irreducible polynomial $F(z)$.

Note 10.1 Since $F(z)$ is an irreducible polynomial, it is not divisible by $(x+1)$ and consequently $F(1) = 1$. We can also state that $\sum_{i=0}^{m-1} f_i = 0$ [343].

We start with the least significant bit (LSB) first bit-serial PB multiplier which is shown in Fig. 10.1 using white blocks. This multiplier includes two m-bit registers (X and Y), the x module, which computes a multiplication by x, followed by a reduction modulo $F(x)$. Also, the AND and XOR blocks perform logical AND and XOR operations, respectively. To study the fault detection circuit for this multiplier, the following lemma is presented.

Lemma 10.2 *Let $Q = A \cdot x \bmod F(x)$ and $\hat{A} = \sum_{i=0}^{m-1} a_i$ be the parity bit of A. Then, the parity bit of Q, i.e. $\hat{Q}$, is obtained as follows [343]:*

$$\hat{Q} = \hat{A} + a_{m-1},$$

where a_{m-1} is the most significant coordinate of A defined in (10.2).

The other two operations in the LSB-first PB multiplier are field addition and multiplication by one bit. For these operations, one can define the following properties.

Property 10.1 Let A and B be two field elements of $GF(2^m)$, and $\hat{A}$ and $\hat{B}$ be their parity bits, respectively. The parity bit of $Q = A + B$ can be obtained as $\hat{Q} = \hat{A} + \hat{B}$.

Property 10.2 Let A be a field element of $GF(2^m)$ and $\hat{A}$ be its parity bit. The parity bit of $Q = b \cdot A$ can be obtained as $\hat{Q} = b \cdot \hat{A}$, where $b \in GF(2)$.

The fault detection scheme for LSB-first bit-serial PB multiplier using a single-bit parity code can be obtained using Lemma 10.2 and Properties 10.1 and 10.2. The fault detection in the x module is based on Lemma 10.2. The fault detection in the XOR and AND blocks are based on Properties 10.1 and 10.2, respectively. The final fault detection circuit for the LSB-first bit-serial PB multiplier is shown in Fig. 10.1a using grey blocks.

The other bit-serial PB multiplier is based on a most significant bit (MSB) first multiplication algorithm and is shown in Fig. 10.1b. The building blocks of this multiplier are similar to the ones of the LSB-first multiplier. The fault detection circuit of this multiplier is shown in Fig. 10.1b using gray blocks.

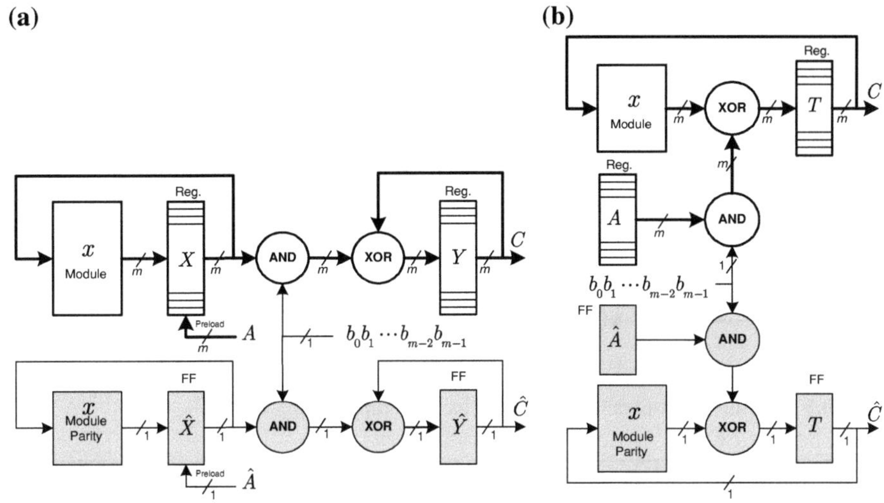

Fig. 10.1 **a** The LSB-first bit-serial PB multiplier with fault detection, **b** The MSB-first bit-serial PB multiplier with fault detection

Traditionally, the bit-parallel PB multiplication can be based on the following formulation:

$$C = A \cdot B \mod F(x) = \sum_{i=0}^{m-1} b_i \cdot \left((Ax^i) \mod F(x) \right). \tag{10.6}$$

Assuming $A^{(i)} = x \cdot A^{(i-1)} \bmod F(x)$ for $1 \le i \le m-1$, and $A^{(0)} = A$, (10.6) can be written as

$$C = \sum_{i=0}^{m-1} b_i \cdot A^{(i)}. \tag{10.7}$$

The structure of the traditional bit-parallel PB multiplier is shown in Fig. 10.2, based on (10.7). The main components of this multiplier are the same as the ones discussed for bit-serial PB multipliers, and consequently the fault detection circuits for this multiplier can be designed similarly.

A single-bit parity-based approach has been used in [177] for fault detection in the bit-serial Montgomery multiplier of [235]. Properties 10.1 and 10.2 mentioned for the single-bit parity code in PB are also applicable in this approach. Assuming A, B, and $C \in GF(2^m)$, the bit-serial Montgomery multiplication algorithm proposed in [235] computes $C = A \cdot B \cdot x^{-m} \bmod F(x)$ and is shown in Algorithm 10.1.

The following lemma addresses parity prediction in the bit-serial Montgomery multiplication algorithm.

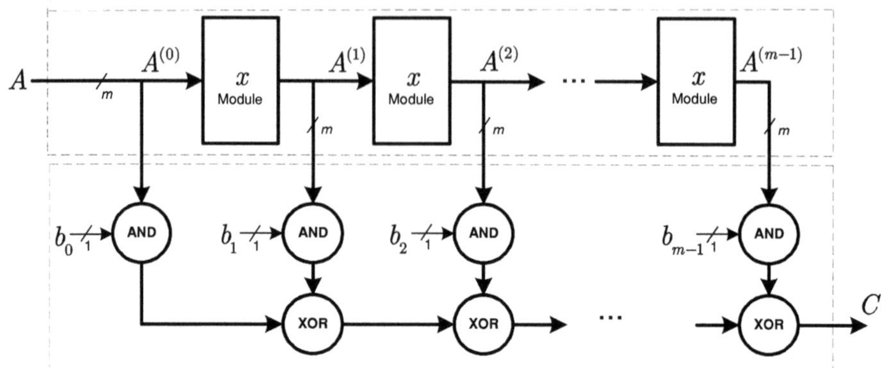

Fig. 10.2 Traditional bit-parallel polynomial basis multiplier

---

Algorithm 10.1: Bit-serial Montgomery multiplication over $GF(2^m)$ [235]

Input: A, B, $F(x)$
Output: $C = A \cdot B \cdot x^{-m} \bmod F(x)$

$T \leftarrow 0$;
for $i = 0$ ***to*** $m - 1$ **do**
  $T' \leftarrow T + b_i A$;
  $T'' \leftarrow T' + t'_0 F(x)$;
  $T \leftarrow T''/x$;
end
return T

Lemma 10.3 *Let $T^{(i)}$ and $T^{(i-1)}$ be the value of T at the ith and $(i-1)$th iterations of the Montgomery multiplication algorithm, respectively. Also, let $\hat{T}^{(i)}$ and $\hat{T}^{(i-1)}$ be their corresponding parity bits. Now, $\hat{T}^{(i)}$ can be obtained as [177]*

$$\hat{T}^{(i)} = \hat{T}^{(i-1)} + b_i \cdot \hat{A} + t_0^{(i-1)} + b_0 \cdot a_0,$$

where $\hat{A}$ and $t_0^{(i-1)}$ are the parity bit of A and the LSB of $T^{(i-1)}$, respectively.

The fault detection circuit for the bit-serial Montgomery multiplication algorithm can be constructed from Lemma 10.3 directly.

10.2.2 Multiple-Bit Parity-Based Approaches

Another approach to fault detection in bit-parallel PB multiplication was proposed in [83]. This approach is based on using multiple interlacing parity bits. Using this method, the parity bits for the operand A can be defined as follows:

$$\hat{A}_i = \begin{cases} \sum_{k=0}^{s-1} a_{kr+i} & m - sr \leq i < r \\ \sum_{k=0}^{s} a_{kr+i} & 0 \leq i < m - sr \end{cases},$$

where r is the number of parity bits and $s = \lfloor \frac{m}{r} \rfloor$ is the minimum number of the coordinates of A represented by a parity bit. Note that if $m = sr$, each parity bit covers s coordinates of A. Otherwise (which is the common case in ECC), some parity bits cover s bits and the rest cover $s + 1$ bits.

Predicting the parity of the final multiplication product C is done in two steps. First, the parity prediction is performed for the first network shown in Fig. 10.2, which includes the x modules. The following lemma addresses the parity prediction in the x modules.

Lemma 10.4 *Assuming $A^{(i-1)}$ and $\hat{A}^{(i-1)}$ are the input of the x module and its parity, respectively, the parity of the x module's output $A^{(i)}$ can be obtained as follows [83]:*

$$\hat{A}_j^{(i)} = \begin{cases} a_{m-1}^{(i-1)}(\hat{F}_j + 1) + \hat{A}_{(j-1) \bmod r}^{(i-1)} & j = m - sr \\ a_{m-1}^{(i-1)}\hat{F}_j + \hat{A}_{(j-1) \bmod r}^{(i-1)} & 0 \leq j < r, j \neq m - sr \end{cases},$$

where $\hat{F}_j$ is the jth parity bit of $F(z)$.

The formulation above can be used in each x module to predict the parity. The second network of the bit-parallel PB multiplier shown in Fig. 10.2 is based on (10.7) and consists of AND operations with the coordinates of B and the final sum using XOR gates. The parity prediction formulation for this network is

$$\hat{C}_j = \sum_{i=0}^{m-1} b_i \hat{A}_j^{(i)}, \quad 0 \leq j < r.$$

The parity based approach used in [343] to design fault detection circuits for the PB multipliers was extended in [355] using multiple-bit parity codes. The multipliers considered in [355] include the LSB-first bit-serial PB multiplier (shown in Fig. 10.1a using white blocks) and the traditional bit-parallel PB multiplier (shown in Fig. 10.2).

Assuming $A = \sum_{i=0}^{m-1} a_i x^i$ is a field element of $GF(2^m)$ and $a_i \in GF(2)$, the multiple-bit parity of A is defined by dividing A into k parts. Assuming m is divisible by k (for simplicity), the jth part of A is defined as

$$A_j = x^{jk} \sum_{i=0}^{l-1} a_{jk+i} x^i,$$

where $0 \leq j < k$ and $l = m/k$. Now, the parity bit is computed for each part, i.e. A_j, and denoted by $\hat{A}_j$. Also, the parity is defined for the irreducible polynomial $F(x)$ by excluding the term x^m, and the jth parity bit is denoted by $\hat{F}_j$, $0 \leq j < k$.

The x module is the main building block in both the bit-serial and bit-parallel PB multipliers. The following lemma can be presented to obtain the parity prediction formulation for this module using the multiple-bit parity code.

Lemma 10.5 *Let A and $\hat{A}$ be the input of the x module and its k-bit parity, respectively. The parity of the output of the x module, i.e., $A' = A \cdot x \bmod F(x)$, is obtained by [355]:*

$$\hat{A}'_j = a_{jl-1} + \hat{A}_j + a_{(j+1)l-1} + a_{m-1}\hat{F}_j.$$

The remaining blocks of the bit-serial and bit-parallel PB multipliers are the XOR and AND blocks. As in, the single-bit parity approach, the parity prediction in these blocks are performed using Properties 10.1 and 10.2, respectively.

An alternative approach proposed in [355] is to partition A and $F(x)$, which in fact results in an interlacing parity code and is similar to the approach proposed in [83].

In [354], a modified multiple-bit parity code-based approach has been proposed which uses the parities of both operands A and B. In this approach, the partitioning of the operands is similar to the one explained in [355], i.e., the operands are divided into k l-bit slices. The parity prediction in the x module is the same as the one outlined in Lemma 10.5. However, the parity prediction in the XOR and AND blocks should be modified to incorporate the parity of the operand B. In [354], it has been shown that using multiple-bit parity for both operands increases the fault detection capability in comparison to the approach used in [355]. However, this approach also has greater area overhead in comparison to the one proposed in [355].

In [356], a fault detection approach has been presented for bit-serial (shown in Fig. 10.1a using white blocks) and bit-parallel (shown in Fig. 10.2) PB multipliers using (n, m) linear codes. The codeword is defined as $V = (v_0, v_1, \ldots, v_{n-1})$ and the code polynomial is a polynomial whose coefficients are the components of V. A polynomial of degree $n - m$ is used to generate the code polynomials of degree $n - 1$ or less and is defined as follows:

$$G(x) = x^{n-m} + g_{n-m-1}x^{n-m-1} + \cdots + g_2x^2 + g_1x + 1,$$

where $G(x)$ is known as the generator polynomial and $g_i \in GF(2)$ for $1 \leq i < n-m$. Let A, B, and Q be the field elements of $GF(2^m)$, and $b \in GF(2)$. Also, let $\hat{A}$, $\hat{B}$, and $\hat{Q}$ be the encoding of A, B, and Q, respectively. Now, the following properties can be defined [356].

Property 10.3 If $Q = b \cdot A$, then $\hat{Q} = b \cdot \hat{A}$ since $\hat{Q} = Q \cdot G(x) = b \cdot A.G(x) = b \cdot \hat{A}$.

Property 10.4 If $Q = A + B$, then $\hat{Q} = \hat{A} + \hat{B}$ since $\hat{Q} = Q \cdot G(x) = (A + B) \cdot G(x) = \hat{A} + \hat{B}$.

The fault detection in the AND and XOR blocks of the bit-serial and bit-parallel PB multipliers can be implemented using Properties 10.3 and 10.4, respectively. Now, the main remaining block in these multipliers is the x module. The following lemma can be presented to implement the fault detection circuit in this module [356].

Lemma 10.6 *Let $A = \sum_{i=0}^{m-1} a_i x^i$ and $\hat{A} = \sum_{i=0}^{n-1} \hat{a}_i x^i$ be the input of the x module and its encoding, respectively. The encoded output of the x module, i.e., $\hat{A}'$, can be obtained as follows:*

$$\hat{A}' = x\hat{A} + \hat{a}_{n-1}\hat{F}(x),$$

where $A' = A \cdot x \bmod F(x)$ and $\hat{F}(x)$ is the encoding of $F(x)$ [356].

Lemma 10.6 and Properties 10.3 and 10.4 are enough to implement the fault detection circuit for the bit-serial and bit-parallel PB multipliers using linear codes.

10.3 Time Redundancy-Based Fault Detection

The second common approach for fault detection in finite field arithmetic operations is based on time redundancy and has been used in many works such as [28, 87, 86, 248, 249]. This approach is efficient for pipelined multipliers, including systolic and semi-systolic array multipliers. The time redundancy-based approach takes advantage of some well-known techniques such as recomputing with shifted operands (RESO) [320, 321] or alternate-data retry [375]. In this section, we describe some of the existing works for fault detection in finite field arithmetic using time redundancy.

In [249], an RESO-based approach has been proposed for fault detection in a semi-systolic implementation of the PB multiplication. In this approach first the multiplication is performed using the main inputs $A, B \in GF(2^m)$, which results in $C = A \cdot B \bmod F(x)$. In the second round, A and B are represented using the basis $\{1, x, x^2, \ldots, x^{m-1}, x^m\}$ and the final result is converted to the main basis, i.e., $\{1, x, x^2, \ldots, x^{m-1}\}$, and compared with C.

Another time redundancy-based approach is proposed in [87] for semi-systolic implementation of the Montgomery multiplication. The first set of the inputs used in this approach comprises A and $B \in GF(2^m)$ and the multiplication result is $C = A \cdot B \bmod F(x)$. The second set of the inputs comprise the Montgomery residues, defined as $\overline{A} = A \cdot x^m \bmod F(x)$ and $\overline{B} = B \cdot x^m \bmod F(x)$. The output of this operation is $\overline{C} = \overline{A} \cdot \overline{B} \cdot x^{-m} \bmod F(x)$. After completing both multiplications, one can compute $C \cdot x^m \bmod F(x)$ and compare it with $\overline{C}$ to detect the errors. To increase the fault detection capability, the operation is repeated by shifting the operands A and $\overline{A}$ as well.

The approach proposed in [87] has been modified in [177] to reduce the overhead. The difference between [87] and [177] is that the latter uses a multiplication by x^{-1}

modulo $F(x)$ for the inputs and a multiplication by x modulo $F(x)$ for the output, which results in lower time and area overheads.

In [28], time redundancy-based techniques are proposed for polynomial basis, dual basis, and normal basis finite field arithmetic units. Here, we briefly study the techniques without considering the architecture of the arithmetic unit, which is a pipelined architecture.

In the case of polynomial basis multiplication, four finite field arithmetic operations have been considered in [28] and the fault detection techniques can be summarized as follows.

- Addition: The main inputs are $A, B \in GF(2^m)$ and the main output is obtained as $C = A + B$. The second set of inputs comprises $A' = A \cdot x \bmod F(x)$ and $B' = B \cdot x \bmod F(x) \in GF(2^m)$, and therefore the second output is obtained as $C' = A' + B'$. Then, one can compute $C' \cdot x^{-1} \bmod F(x)$ and compare it with C.
- Multiplication: The main inputs are $A, B \in GF(2^m)$ and the main output is obtained as $C = A \cdot B \bmod F(x)$. The second set of inputs comprises $A' = A \cdot x \bmod F(x)$ and $B' = B \cdot x \bmod F(x) \in GF(2^m)$, and therefore the second output is obtained as $C' = A' \cdot B' \bmod F(x)$. Then, one can compute $C' \cdot x^{-2} \bmod F(x)$ and compare it with C.
- Inversion: The main input is $A \in GF(2^m)$ and the main output is obtained as $A^{-1} = 1/A \bmod F(x)$. The second input is $A' = A \cdot x \bmod F(x)$ and the second output is $A'^{-1} = 1/A' \bmod F(x)$. Then, one can compute $A'^{-1} \cdot x \bmod F(x)$ and compare it with A^{-1}.
- Division: The main inputs are $A, B \in GF(2^m)$ and the main output is obtained as $C = A/B$. The second set of inputs comprises $A' = A \cdot x \bmod F(x)$ and $B' = B \cdot x^{-1} \bmod F(x) \in GF(2^m)$ and the corresponding output is $C' = A'/B'$. Then, one can compute $C' \cdot x^{-2}$ and compare it with C.

Fault detection in the dual basis arithmetic operations presented in [28] is similar to the one explained above for the polynomial basis operations.

In the case of normal basis, it is well known that the square and the square root operations are performed by a circular left shift and a circular right shift, respectively, at no cost. Therefore, the following techniques are used for fault detection in normal basis arithmetic units [28].

- Addition: The main inputs are $A, B \in GF(2^m)$ and the main output is obtained as $C = A + B$. The second set of inputs are $A' = A^2$, $B' = B^2$ and the output is $C' = A' + B'$. Then, one can take the square root of C' and compare it with C.
- Multiplication: The main inputs are $A, B \in GF(2^m)$ and the main output is obtained as $C = A \cdot B$. The second set of inputs comprises $A' = A^2$ and $B' = B^2$ and the output is $C' = A' \cdot B'$. Then, one can take the square root of C' and compare it with C.

Inversion: The main input and output are $A \in GF(2^m)$ and $A^{-1} = 1/A$, respectively. The second input is $A' = 1/A^2$ and the corresponding output is $C' = A'^{-1}$. Then, one can take the square root of C' and compare it with C.

Division: The main inputs are $A, B \in GF(2^m)$ and the main output is obtained as $C = A/B$. The second inputs are $A' = A^2$, $B' = B^2$ and the output is $C' = A'/B'$. Then, one can take the square root of C' and compare it with C.

10.4 Fault Detection in Elliptic Curve Scalar Multiplication

The fault detection approaches explained in the previous sections can be used to detect the faults in the ECC scalar multiplication. However, there are other approaches in the literature which focus on the scalar multiplication. In this section, we explain the point validation approach proposed in [16], the time redundancy-based approach proposed in [387], and the input randomization approach proposed in [123].

In [387], fault detection circuits have been proposed for scalar multiplication in ECC. This approach is based on time redundancy and resembles the approaches explained in the previous section. Assuming P is a point on the elliptic curve E and k is a positive integer, the operation to compute $kP = P + P + \cdots + P$ (k times) is called the scalar multiplication in elliptic curve cryptography. The scalar operation is performed using point addition and point doubling operations and these operation are based on finite field arithmetic operations such as addition, multiplication, squaring, and inversion. The following techniques are used in [387] for fault detection.

Addition: The addition is done with inputs A and B in $GF(2^m)$ and the result is $C = A + B$. Now, there are two scenarios. In the first one, the output C is added to B again and compared with A to detect errors. The other scenario divides the inputs into two halves and the XOR network is also divided into two sub-networks. In the first round, the left sub-network processes the left halves of the inputs and the right sub-network processes the right halves. In the second round, the inputs to the sub-network are interchanged. Finally the results are compared to detect any possible errors.

Inversion: Let A^{-1} be the inverse of A modulo $F(x)$. It is known that $(A^{-1})^{-1} = A$. In the first round, A is the input of the inverter and A^{-1} is the expected output. Now, A is stored in a register and the inversion result becomes the input of the inverter in the second round. The output of the inverter in the second round is expected to be A. A comparison is performed to compare the second output with A to detect possible errors.

Multiplication: In this operation, A and B are the inputs and $C = A \cdot B \bmod F(x)$ is the output of the multiplier in the first round. In parallel to this operation B^{-1} is computed. In the second round, B^{-1} and the computed C are the inputs of the multiplier and the result is expected to be A. A comparison between the output of the second round and A can detect the possible errors. The technique for the squaring operation is very similar and the only difference is that both inputs of the multiplier in the first round are connected to A to compute A^2.

The overall architecture presented in [387] uses the inversion block, which is free most of the time during the scalar multiplication, to provide the inputs for the error detection algorithm. It should be noted that this approach requires a separate inverter since it is very common to implement inversion in ECC using multiplication and squaring without having a separate module for inversion.

In [16], a method has been proposed against fault attacks on ECC which is known as point validation. In this approach, the point is verified to be on the elliptic curve assuming that the attack might change the results so that the resulting point is no longer on the curve. This approach fails if the attack is a Sign Change Fault attack [54], because the points will not leave the elliptic curve.

Different fault detection approaches have been proposed in [123] for the scalar multiplication using input randomization to also provide countermeasures against sign change fault attacks. Two properties, introduced in [102] as countermeasures against differential power analysis, are used in [123] to construct a randomized input.

Property 10.5 Let (X, Y, Z) be the projective representation of P. It can be verified that $(\gamma^c X, \gamma^d Y, \gamma Z)$ also represents P, where $\gamma \in GF(2^m)$, $c = 1$, and $d = 2$ for the López and Dahab representation, and $\gamma \in GF(p)$, $c = 2$, and $d = 3$ for the Jacobian representation [102].

Property 10.6 Let k be a positive integer, j be a at least 20-bit random integer, and $\#E$ be the order of the elliptic curve E. Now, one can define $k'' = k + j(\#E)$ and use it to compute $k''P$. This can be written as $k''P = (k + j(\#E))P$, and since $(\#E)P = \infty$, it is concluded that $kP = k''P$ [102].

One of the approaches proposed in [123] is based on time redundancy and recomputation. In this approach, one scalar multiplication module is used; however, the scalar multiplication is performed using the parameters γ_0 and j_0 in the first round, and γ_1 and j_1 in the second round, to obtain different inputs based on Properties 10.5 and 10.6. The output of the first round is stored in a register and compared with the output of the second round.

The drawback of this recomputation is that the scalar multiplication is done twice, and therefore this approach has a large time overhead. To reduce this overhead; a modified approach has been proposed in [123]. In the modified approach (partial recomputation), the point P is encoded twice, similarly to the original approach; however, the scalar k is encoded once and used in the first round. In the second round, l least significant bits of the encoded k is used and the output of the second round is compared with the output of the first round after l iterations. This results in reducing the time overhead as the second round is not a complete scalar multiplication.

Another architecture is presented in [123] and uses the combination of point validation and parallel computation. In this approach, two scalar multiplication modules are used with different encodings of the point P and the scalar k using Properties 10.5 and 10.6. The outputs of the modules are compared for fault detection purposes. Also, point verification modules are used to validate the outputs of the scalar multiplication modules. This approach has higher area overhead in comparison to the recomputation approach.

Parallel computation and recomputing approaches are also merged in [123] to implement another fault detection circuit. In this architecture, two scalar multiplication modules are used (i.e., parallel computation) and each module uses two different encodings of P and k based on Properties 10.5 and 10.6 (i.e., recomputation) to increase the fault detection capability.

In [123], it has been concluded that the recomputation, the partial recomputation ($l = 16$), and the parallel computation approaches have area overheads of 42.6, 42.6, and 137.7 %, respectively, when implemented using a NIST-recommended curve over $GF(2^{163})$ on a Xilinx Virtex 2000E FPGA. Also, the parallel computation with point validation and the parallel computation with recomputing approaches have the area overheads of 157.8 and 162.2 %, respectively.

The time overheads of these approaches, based on the number of clock cycles, are as follows: recomputation 108 %, partial recomputation 18.4 %, parallel computation 13.35 %, parallel computation with point validation 13.35 %, and parallel computation with recomputation 13.35 % if no error is detected in the first round and 123.5 % otherwise. For more information, one can refer to [123].

Chapter 11
Design of Cryptographic Devices Resilient to Fault Injection Attacks Using Nonlinear Robust Codes

Kahraman D. Akdemir, Zhen Wang, Mark Karpovsky and Berk Sunar

Abstract This chapter mainly discusses robust and partially robust codes and their application to various cryptographic primitives. Initially, robust nonlinear codes are described in detail and their error detection capabilities are measured theoretically. Next, various nonlinear constructions are provided and their potential applications are described. More specifically, we discuss the protection of the AES data path, finite state machines (FSMs), and elliptic curve cryptosystems (ECCs). The main advantage of robust codes is that they are nonlinear and hence the success of an injected fault is data-dependent. As a result, error detection using nonlinear robust codes is one of the most effective solutions to active fault injection attacks.

11.1 Introduction

Active fault injection attacks pose a serious threat to many cryptographic applications, such as smart cards. Various countermeasures have been proposed to provide security against these attacks.

In [263, 264], a solution based on time redundancy by means of a double-data-rate (DDR) computation template was presented. Each computation is conducted twice and the results are compared to detect injected faults. Both clock edges were exploited to control the computation flow for the purpose of improving the throughput of the system. In [244, 295], the authors investigated the usage of dual rail encoding for the protection of cryptographic devices from different types of side-channel attacks in asynchronous circuits.

K. D. Akdemir (✉) · B. Sunar
Department of Electrical and Computer Engineering, Worcester Polytechnic Institute, Worcester MA, USA
e-mail: kahraman.akdemir@gmail.com

Z. Wang · M. Karpovsky
Reliable Computing Laboratory, Boston University, Boston MA, USA

M. Joye and M. Tunstall (eds.), *Fault Analysis in Cryptography*, Information Security and Cryptography, DOI: 10.1007/978-3-642-29656-7_11,

The most commonly used fault detection technique is concurrent error detection (CED), which employs circuit-level coding techniques, e.g. parity schemes, modular redundancy, etc., to produce and verify check digits after each computation. In [34], a secure AES architecture based on linear parity codes was proposed. The method can detect all errors of odd multiplicities with reasonable hardware overhead. In [155], an approach to fault-tolerant public key cryptography based on redundant arithmetic in finite rings was presented. The method is closely related to cyclic binary and arithmetic codes. In [218], the authors proposed a CED technique that exploits the inverse relationships existing between encryption and decryption at various levels. A decryption is immediately conducted to verify the correctness of the encryption operation. A lightweight concurrent fault detection scheme for the S-Boxof AES was proposed in [221]. The structure of the S-Boxis divided into blocks and the predicted parities for these blocks are obtained and used for the fault detection. Various fault attack countermeasures were compared in terms of the hardware overhead and the fault detection capabilities in [266].

Error detecting codes [260] are often used in cryptographic devices to detect errors caused by injected faults and prevent the leakage of useful information to attackers. Most of the proposed error detecting codes are linear codes like parity codes, Hamming codes and AN codes [335]. Protection architectures based on linear codes concentrate their error detecting abilities on errors with small multiplicities or errors of particular types, e.g. errors with odd multiplicities or byte errors. However, in the presence of unanticipated types of errors, linear codes can provide little protection. Linear parity codes, for example, can detect no errors with even multiplicities.

In [63], the author compared several concurrent fault detection schemes for the Advanced Encryption Standard based on linear codes. As expected, the simulation results showed that the error detecting capabilities of systems protected by linear codes largely depend on the error profiles at the output of the device due to the injected faults. The spectrum of available fault injection methods and the adaptive nature of an attacker suggest that it would be possible to bypass such protection by injecting a class of faults or errors which the cryptographic device does not anticipate. Considering even only inexpensive noninvasive or semi-invasive fault attacks, there is a wide spectrum of the types of faults and injection methods an attacker has at his disposal [21].

Robust codes have been proposed as a solution to the limitation of linear error detecting codes for detection of fault injection attacks [212]. These nonlinear codes are designed to provide equal protection against all errors, thereby eliminating possible weak areas in the protection that can be exploited by an attacker. Several variants of robust codes have been used to protect both private and public cryptographic algorithms. These variants allow several trade offs between robustness and hardware overhead for many architectures. Robust and partially robust codes have been used for the protection of both private [211, 212] and public key cryptosystems [156].

In this chapter we will present the basic constructions of robust codes and their application to the design of secure cryptographic devices. As case studies, we discuss secure AES, secure Elliptic Curve Cryptography (ECC) and secure Finite State Machine (FSM) architectures based on robust codes, which are resilient to strong fault injection attacks. More details related to robust codes and their application to the design of cryptographic devices can be found in [9, 10, 155–157, 211, 212, 216, 245].

11.2 Adversarial Fault Model

In order to model the effects of an active fault attack, we need to reflect the effect of this fault in the circuit parameters. As a result, we model a fault injected into the circuit as an erroneous result at the output of the device, i.e. the erroneous output $\tilde{x} = x + e$, where x is the expected output.

Fault injection attacks on hardware implementing different cryptographic algorithms are discussed in Sect. 16.2. In this chapter, we make the following advanced attacker assumptions as part of our fault model.

1. The structure/function of the device is public and hence known by the attacker.
2. The attacker does not have to use a particular fault injection methodology. He can inject faults using any method he prefers.
3. The advanced attacker we assume has high temporal and spatial fault injection capability.
4. The attacker cannot overwrite the output values of the device by injecting faults. The errors observed at the output of the device are always additive in nature.
5. The attacker cannot observe any existing data on the circuit at the time of fault injection i.e. in the same clock cycle. This means that the attacker will not be able to adaptively attack the circuit by first reading the existing data and then choosing the appropriate error vector.
6. The attacker can reflect any specific error vector he desires at the output, i.e. he can pick the value of e that will be observed at the output of the device.
7. Every error vector (all multiplicities) can be observed at the output of the device.

We assume that the device is disabled or resets the secret information after an injected fault is detected. Hence, in our detection model, the attacker will not be able to try many different error vectors until he breaks the device. He has only one chance to successfully inject a fault.

Obviously, under such an adversarial fault model, protections of cryptographic devices based on linear error detecting codes will not be sufficient and can be easily bypassed. For instance, the attacker can break systems protected using double module redundancy (DMR), which is basically based on linear duplication codes, by reflecting the same error vectors at the output of both the original and the redundant devices.

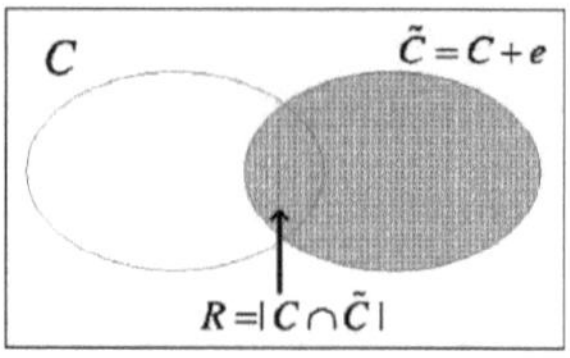

Fig. 11.1 Definition of robustness

11.3 Definition and Basic Properties of Robust Codes

We present most of the definitions in terms of binary codes. These results can be easily generalized for codes over nonbinary fields.

Definition 11.1 (*R-Robust Code*) A code $C \subseteq GF(2^n)$ is R-robust if the size of the intersection of the code C and any of its translations $\tilde{C} = \{\tilde{x} \mid \tilde{x} = x + e, x \in C, e \in GF(2^n), e \neq 0\}$ is upper bounded by R:

$$R = \max_{0 \neq e \in GF(2^n)} |\{x \mid x \in C, x + e \in C\}|.$$

where $+$ is the component-wise addition modulo 2.

A graphic depiction of the definition of a robust code is shown in Fig. 11.1. Let $C \subseteq GF(2^n)$, and $\tilde{C}_e$ be the set of all codewords of C shifted by an element $e \in GF(2^n)$. The code C is R-robust if for any nonzero $e \in GF(2^n)$, the size of the intersection of C and $\tilde{C}_e$ is upper bounded by R.

The above-defined robust codes have beneficial properties when the worst case error masking probability of the codes is considered. Let $M = |C|$ be the number of codewords in a code C. By the definition of an R-robust code there are at most R codewords which can mask any fixed error e. The error masking probability $Q(e)$ can be thus defined as

$$Q(e) = \frac{|\{x \mid x \in C, x + e \in C\}|}{M}. \tag{11.1}$$

Robust codes have no undetectable errors. For an R-robust code, the worst case probability of masking an error is at most R/M for any error when the codewords of the robust code are assumed equiprobable. Clearly, robust codes which have a minimum R for a given M will also have the lowest probability of error masking and hence predictable behavior in the presence of unpredictable error distributions, since the worst case probability of masking any error is bounded. In the following sections we investigate the construction and optimality of the codes followed by some examples of applications.

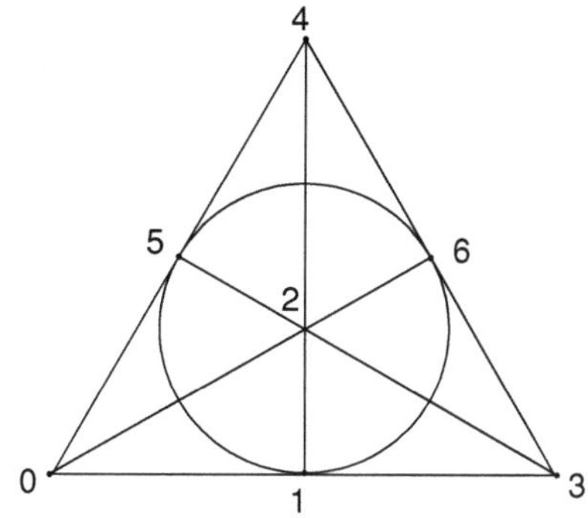

Fig. 11.2 Projective plane of order 2

11.4 Bounds, Optimality, and Perfect Robust Codes

Based on the above definitions of robust codes it is possible to derive the following main property for an R-robust code.

Property 11.1 *If the code C is R-robust, then in the multiset $S_C = \{x_j + x_i \mid x_i, x_j \in C, x_i \neq x_j\}$, any element appears at most R times.*

Robust codes are optimal if they have the maximum number of codewords M for a given R and length n. From Property 11.1, a relation on R, n and M of the code can be established:

$$M^2 - M \leq R(2^n - 1). \qquad (11.2)$$

Definition 11.2 (*Perfect Robust Code*) An R-robust code with n bits and M codewords satisfying $M^2 - M = R(2^n - 1)$ is **perfect**.

Perfect robust codes correspond to extensively studied combinatorial structures known as difference sets and symmetric designs [206].

Definition 11.3 (*Difference Set*) Let G be a group of order v, and C an M-subset of G. Then C is called a (v, M, R)-difference set if the list of differences $S = (\alpha - \beta : \alpha, \beta \in C, \alpha \neq \beta)$ contains each nonzero element of G exactly R times.

Clearly, (q^n, M, R) difference sets in $GF(q^n)$ are exactly perfect q-ary R-robust codes of length n and M codewords.

Despite extensive research on the combinatorial structures it is still not known in the general case for which parameters such difference sets and hence perfect robust codes exist, and we note that the perfect robust codes based on the known difference sets have high complexity of decoding (detecting for a given x whether $x \in C$ or $x \notin C$). A good summary of existing difference sets can be found in [42].

Example 11.1 The projective geometry $PG(m, q)$ [110] is an example of a classic combinatorial balanced-incomplete design [42] that generates a difference set with parameters ($v = \frac{q^{m+1}-1}{q-1}$, $M = \frac{q^m-1}{q-1}$, $R = \frac{q^{m-1}-1}{q-1}$), where q is a prime power and $m > 1$ is an integer. These are known as Singer difference sets.

$PG(2, 2)$ is shown in Fig. 11.2. The set of any collinear points forms a perfect robust code defined over $GF(7)$ with $n = 1$, $M = 3$ and $R = 1$. The differences of the three codewords cover all nonzero elements of $GF(7)$ exactly once.

It has been shown that all difference sets over binary fields, and thus binary robust codes which are the most important and practical for hardware design, exist only for even dimensions and have the following parameters: $n = 2s$, $M = 2^{2s-1} \pm 2^{s-1}$, $R = 2^{2s-2} \pm 2^{s-1}$ [42]. These codes can be constructed using the method presented in [215].

Systematic codes, which are often more practical for many applications due to their separation of data and check bits, cannot be perfect.

Theorem 11.1 *[214] For any systematic R-robust code with length n and k information bits, there are at least 2^{n-k} elements in $GF(2^n)$ which cannot be expressed as differences of two codewords.*

Proof For any systematic codeword $x = (x_1, x_2 = f(x_1))$ an error $e = (e_1, e_2)$ is masked iff $f(x_1 + e_1) = x_2 + e_2$. An error $e = (e_1 = 0, e_2 \neq 0)$ is never masked since $f(x_1) = x_2 + e_2$ iff $e_2 = 0$. An error that is never masked cannot be expressed as a difference of two codewords. Hence elements from $GF(2^n)$ of the form $x(0, x_2 \in GF(2^r))$, where $r = n - k$, cannot be expressed as a difference of two codewords.

Corollary 11.1 *There are no perfect systematic robust codes.*

When perfect robust codes are not available, the best possible codes which maximize M for a given n and R are referred to as optimum robust codes.

Definition 11.4 (*Optimum Robust Code*) Robust codes which have the maximum possible number of codewords M for a given length n and robustness R with respect to (11.2) are called optimum. For optimum codes, adding any additional codewords would violate bound (11.2) and

$$M^2 - M \leq R(2^n - 1) < M^2 + M\,. \tag{11.3}$$

Example 11.2 Consider the binary code $C = \{000, 001, 010, 100\}$. It is easy to verify that for any nonzero element $e \in GF(2^3)$, there are at most two pairs of c_1, c_2 satisfying $e = c_1 + c_2$, where $+$ is the XOR operation. Hence the code is 2-robust.

The code is not perfect since equality does not hold for (11.2). The code, however is an optimum robust code with $n = 3$, $M = 4$, $R = 2$. No other code can exist with the same n and R that has more codewords since five codewords would violate condition (11.2).

Several constructions of optimum systematic robust codes will be presented in the next section.

11.5 Constructions of Optimal Systematic Robust Codes

There is a strong relationship between robust codes, nonlinearity, and nonlinear functions since all robust codes are nonlinear. The parameters of systematic robust codes depend on the nonlinearity of the encoding function of the codes.

We first review some basic definitions and properties of nonlinearity; a good survey of nonlinear functions can be found in [80].

Let f be a function that maps elements from $GF(2^k)$ to $GF(2^r)$.

$$f : GF(2^k) \to GF(2^r) : a \to b = f(a) .$$

The nonlinearity of the function can be measured by using derivatives $D_a f(x) = f(x+a) + f(x)$. Let

$$P_f = \max_{0 \neq a \in GF(2^k)} \max_{b \in GF(2^r)} Pr(D_a f(x) = b) ,$$

where $Pr(E)$ denotes the fraction of cases when E occurred. The smaller the value of P_f, the higher the corresponding nonlinearity of f. For linear functions, $P_f = 1$.

Definition 11.5 A binary *function* $f : GF(2^k) \to GF(2^r)$ has perfect nonlinearity iff $P_f = \frac{1}{2^r}$.

Theorem 11.2 *[213] Let f be a nonlinear function that maps $GF(2^k)$ to $GF(2^r)$ where $k \geq r$; the set of vectors resulting from the concatenation of*

$$x_1, x_2 : (x_1, x_2 = f(x_1))$$

where $x_1 \in GF(2^k)$ and $x_2 \in GF(2^r)$, forms a robust systematic code with $R = 2^k P_f$, $n = k + r$ and $M = 2^k$.

Proof The error $e = (y_1, y_2)$ with $(y_1 \in GF(2^k), y_2 \in GF(2^r))$ will be masked iff $f(x_1 + y_1) + f(x_1) = y_2, x_1 \in GF(2^k)$, which is exactly when $D_{y_1} f(x_1) = y_2$.

From Theorem 11.1, there are at least 2^{n-k} errors which will be detected with probability 1 by any systematic code with length n and k information bits. Thereby a more strict bound can be derived for systematic codes. In this case we have

$$M^2 - M \leq R(2^n - 2^{n-k}). \tag{11.4}$$

Corollary 11.2 *A systematic robust code*

$$C = \{(x_1, x_2 = f(x_1)) \mid x_1 \in GF(2^k), x_2 \in GF(2^r)\}$$

is optimum if the encoding function f is a perfect nonlinear function.

Remark 11.1 The nonlinearity of the encoding function f for systematic codes corresponds to the worst case error masking probability of the codes [213, 214]. We have

$$P_f = \max_{e=(e_1,e_2),e_1\neq 0} Q(e) = \max_{e\in GF(2^{k+r})} Q(e).$$

where $e_1 \in GF(2^k)$, $e_2 \in GF(2^r)$.

The following two constructions are examples of optimum robust codes based on perfect nonlinear functions.

Construction 11.1 (Quadratic Systematic Code [213]) *Let*

$$x = (x_1, x_2, \ldots, x_{2s}, x_{2s+1}),$$

$x_i \in GF(2^r)$, $s \geq 1$. *A vector* $x \in GF(2^{(2s+1)r})$ *belongs to the code iff*

$$x_1 \bullet x_2 + x_3 \bullet x_4 + \cdots + x_{2s-1} \bullet x_{2s} = x_{2s+1}, \tag{11.5}$$

where $\bullet$ *is the multiplication in* $GF(2^r)$ *and* $\sum_{i=1}^{s} x_{2i-1} \bullet x_{2i}$ *is a perfect nonlinear function from* $GF(2^{2sr})$ *to* $GF(2^r)$. *The resulting code is a robust code with* $R = 2^{(2s-1)r}$, $n = (2s+1)r$ *and* $M = 2^{2sr}$. *The code is optimum with respect to Definition 11.4.*

Applications of binary quadratic systematic codes for designing of memories with self-detection, for data transmission in noisy channels and for data verification can be found in [213].

Example 11.3 (Robust Parity) s Methods based on linear parity check codes are often used for online error detection in combinational circuits [247]. The linear 1-dim parity codes can detect all errors of odd multiplicities but offer no protection for errors of even multiplicities.

As an alternative to the 1-dim parity codes, the quadratic systematic robust codes defined in Construction 11.1 can be used. When $r = 1$, the function defined in (11.5) is known as the bent function. The resulting systematic robust code has the same redundancy as the linear parity code. Unlike the linear parity code, the robust code will mask an error with a probability of at most $\frac{1}{2}$ regardless of the error multiplicity, providing predictable error detection regardless of the error distribution.

Perfect nonlinear functions from $GF(2^k)$ to $GF(2^k)$ do not exist [80]. Functions with optimum nonlinearity in this case have $P_f = 2^{-k+1}$ and are called almost perfect nonlinear (APN) functions [276]. When the f are APN functions, the robust codes constructed as in Theorem 11.2 have $R = 2$. These codes are not optimum.

Construction 11.2 (Robust Duplication Code) *Let* $x = (x_1, x_2)$, $x_1, x_2 \in GF(2^r)$ $(k = r)$. *The robust duplication code* C *contains all vectors* $x \in GF(2^{2r})$ *which*

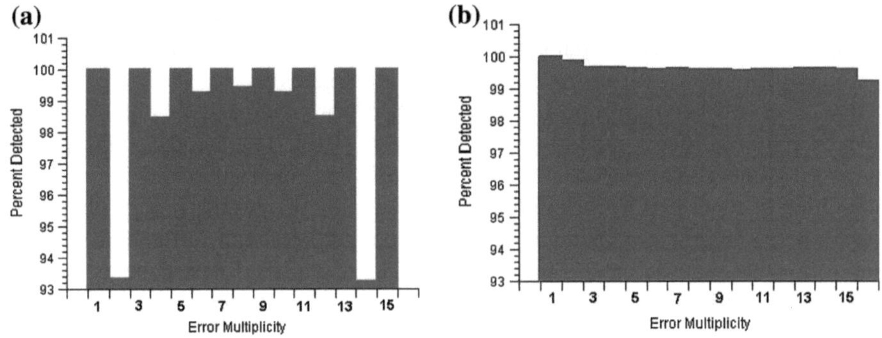

Fig. 11.3 Percentages of errors detected versus error multiplicities for **a** 8-bit linear duplication **b** 8-bit robust duplication

satisfy $x_1^3 = x_2$, *where all the computations are in* $GF(2^r)$. *The code is a* 2-*robust code with* $n = 2r$ *and* $M = 2^r$.

As an example, Fig. 11.3 shows the percent of detectable errors as a function of error multiplicity (number of distorted bits) for eight-bit linear and robust duplication codes ($k = r = 8$). The detection capability of linear duplication codes depends largely on the multiplicity and type of the error. The scheme offers relatively poor protection for errors of even multiplicities, which can be exploited by the attacker to increase his chance of implementing a successful fault injection attack. In contrast the robust duplication code has almost completely uniform error detection. This robust code has $R = 2$. Any error can be masked for at most two messages. Unlike in linear codes, regardless of which subset of errors is chosen for this robust code the error masking probability is upper-bounded by 2^{-7}.

Robust duplication codes can be a viable alternative to standard duplication techniques. The application of binary robust duplication codes to memories with self-error detection and to the comparison between standard and robust duplication techniques can be found in [213].

The detection of errors for robust codes are message-dependent. If the same error stays for more than one clock cycle, even if the injected fault manifests as an error that cannot be detected at the current clock cycle, it is still possible that the error will be detected at the next clock cycle when a new message arrives. Therefore, the advantage of robust codes will be more significant for lazy channels where errors have high probabilities of repeating themselves over several clock cycles. Applications of robust codes for the protection of lazy channels can be found in [213].

11.5.1 Partially Robust Codes

Robust codes generally have higher complexity of encoding and decoding than classical linear codes. The quadratic systematic codes from Construction 11.1 require that sr-bit multipliers and $s - 1r$-bit component-wise additions. Assuming an r-bit multiplier requires r^2 two-input gates, the encoder for the systematic quadratic code can be implemented with $sr^2 + r(s-1)$ two-input gates.

As a trade off between robustness and the hardware overhead for computational devices, partially robust codes were introduced in [216]. These codes combine linear and nonlinear mappings to decrease the hardware overhead associated with the generation of check bits by the predictor. The encoding of systematic partially robust code is performed first by using a linear function to compute the redundant r check bits, followed by a nonlinear transformation. The use of the linear code as the first step in the encoding process typically results in hardware savings in the encode or predictor since the nonlinear function only needs to be computed based on the r-bit output of the linear block. The application of the nonlinear transformation reduces the number of undetectable errors, thus increasing the robustness of the linear codes.

Construction 11.3 (Partially Robust Codes [245]) *Let $f : GF(2^r) \rightarrow GF(2^r)$ be a nonlinear function with $P_f < 1$ and let $P : GF(2^k) \rightarrow GF(2^r)$, $r \leq k$, be a linear onto function. The set of words in the form $(x, f(P(x))$ form a code with 2^{k-r} undetectable errors.*

For partially robust codes described in Construction 11.3, the number of undetectable errors is reduced from 2^k to 2^{k-r} compared to the linear code with the same redundancy. Partially robust codes with $k = 128$ and $r = 32$ have been used in [211] for the design of private key security devices on AES resistant to fault injection attacks. Implementation of this approach resulted in about 80 % hardware overhead.

11.5.2 Variants of Robust and Partially Robust Codes

A possible variant of traditional robust codes is from including a minimum distance into the design criteria. Let $\|e\|$ denote the multiplicity of an error e (the number of 1s in e). A robust code where the $Q(e) = 0$ for all errors with $\|e\| < d$ is a d-minimum distance robust code.

Minimum distance robust codes are fully robust codes but also have minimum distance larger than 1. Since these codes are robust they have no undetectable errors and the worst case error masking probability is upper-bounded and predictable. Moreover, they also provide a guaranteed 100 % probability of detection for errors with small multiplicities. Such codes can be useful for providing the highest protection against the most likely or most dangerous threat while maintaining a detection guarantee in the case of an unexpected behavior or attack. Moreover, minimum distance robust or partially robust codes with Hamming distance at least 3 can be used for not only error detection but also for error correction [241]. For constructions of minimum distance robust and partially robust codes and for applications of these codes to the design of secure cryptographic devices and reliable memory architectures, please refer to [245, 414, 415].

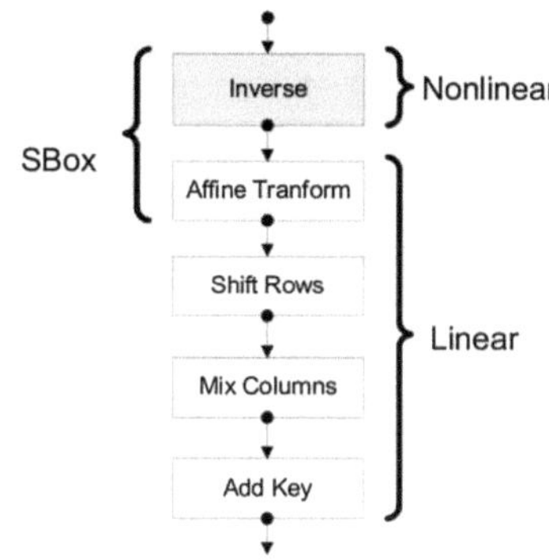

Fig. 11.4 Transformations involved in one typical round of encryption of AES

Another variant of robust codes is the multilinear code. Multilinear codes are based on randomly selecting a linear code from a set of linear codes for each encoding and the corresponding decoding operation. The proposed method can achieve as small a number of undetectable errors as classical robust codes while requiring much less hardware overhead. Constructions of algebraic and arithmetic multilinear codes and the applications of these codes for the protection of multipliers, lazy channels and finite state machines can be found in [412, 413, 416].

11.6 Secure AES Architectures Based on Nonlinear Codes

In this section, we discuss the protection of the data path of AES devices based on nonlinear codes. The key expansion block can be protected in a similar way. For a discussion on the protection of the control circuit of the system (e.g. Finite Stage Machines, State Registers), please refer to Sect. 11.7.

Encryption in AES-128 (AES with a 128-bit key) involves performing ten rounds of transformations on a block of 128 bits with the last round having one less transformation and with the first round being preceded by a round key addition. (The complete AES specification can be found in [142].) In each of the nine typical rounds there are four transformations: SubBytes (SBox), ShiftRows, MixColumns, and AddRoundKey. The last round differs from the rest in that it does not contain the MixColumns transformation. The SBox transformation actually involves two operations: inversion in $GF(2^8)$ followed by an affine transform which involves a matrix multiplication over $GF(2)$, followed by the addition of a constant vector. With the exception of inversion, all other transformations and operations are linear (Fig. 11.4), i.e. they can be implemented using XOR gates only.

When considering only one round, the 128-bit data path can be divided into four identical independent 32-bit sections. Furthermore, in each of the four partitions the nonlinear inversion is performed on an eight-bit data block. Thus, the nonlinear

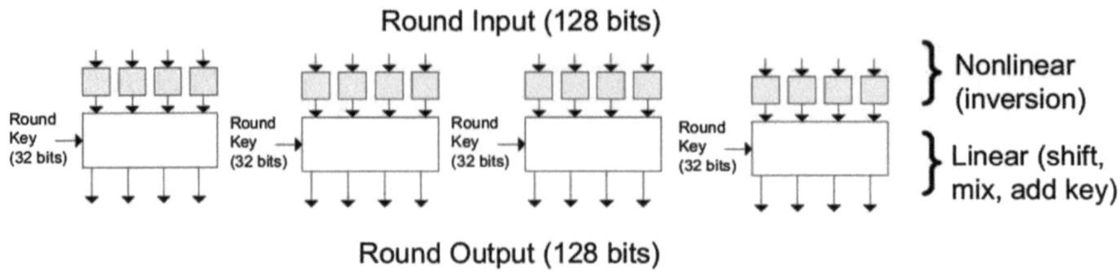

Fig. 11.5 The nonlinear portion of one round can be separated into 16 identical independent blocks. The linear portion can be separated into 4 identical independent blocks

section is composed of 16 disjoint blocks and the linear portion is composed of four identical disjoint blocks (Fig. 11.5).

Based on this partitioning, redundant protection hardware can be designed for each of the two types of blocks. The details of each block as a method of protection are discussed in the next section.

11.6.1 Protection of the Nonlinear Block of AES

The nonlinear block performs an inversion in $GF(2^8)$. Since 0 does not have an inverse, it is defined that the result of the inverse operation on 0 is 0.

Let x be the input to the nonlinear block. The fault detection circuitry for the nonlinear block can be based on multiplication in $GF(2^8)$ of input and output vectors to verify the condition (Fig. 11.6)

$$x \bullet x^{-1} = \begin{cases} 00000001 & \text{if } x \neq 0, \\ 00000000 & \text{if } x = 0. \end{cases}$$

Remark 11.2 x^{-1} is an APN function over $GF(2^8)$ [276]. Hence the code C defined by $\{(x, x^{-1})\}$ is a robust code with no undetectable errors.

To reduce the hardware overhead, we can compute the least $r < 8$ significant bits of the product instead of the whole eight-bit product. The probability that an error in the inverter will be missed is equal to the probability that two eight-bit vectors multiplied together will produce the expected r-bit constant I_r. When $x = 0$, $I_r = 0$. Otherwise I_r has 1 for the least significant bit and 0 elsewhere.

Let $e = (e_1, e_2)$ be the error vector, where $e_1 \in GF(2^8)$ is the error at the output of the original inverter and e_2 is the error at the output of the redundant portion. Then e is missed iff

$$((x^{-1} + e_1) \bullet x)_r = I_r + e_2, \tag{11.6}$$

i.e. the least significant r bits of the product in $GF(2^8)$ are equal to $I_r + e_2$ or, equivalently, $x \bullet e_1 = e_2$.

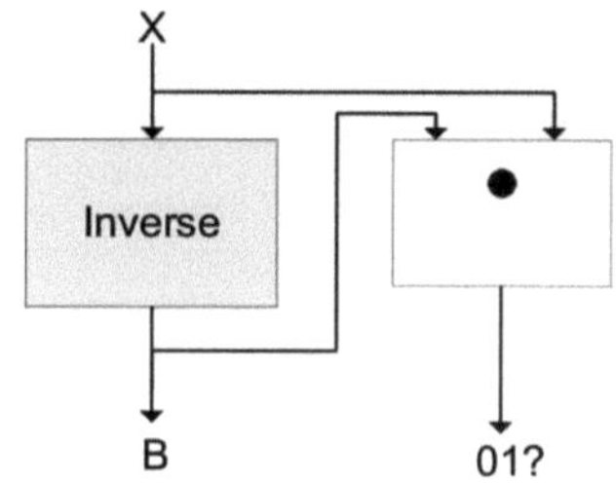

Fig. 11.6 Architecture for protection of nonlinear block. The redundant portion performs partial multiplication in $GF(2^8)$, $(r = 2)$

For every nonzero $e = (e_1, e_2)$, there is at most one solution for x. Therefore all nonzero errors e will be detected with a probability of at least $1 - 2^{-r}$. Moreover, since error detection depends on the input x, the probability that e will be missed after m random inputs is 2^{-rm}.

In one round of encryption of AES there are $T = 16$ disjoint inverters, each with its own independent error detection. While for a single inverter the probability of missing an error is constant for all fault multiplicities, this is not the case when multiple inverters are considered together. The probability that a fault will not be detected if it affects t inverters is q^t where q is the probability of missing a fault in one inverter (For the proposed architecture, $q \leq 2^{-r}$).

Assuming that the distribution of faults is uniform, the probability that a fault of multiplicity l will affect t out of T inverters can be determined as $P_T(t, l) = \frac{N_T(t,l)}{2^l}$, where $N_T(t, l) = \binom{T}{t}(t^l - \sum_{j=1}^{t-1} N_t(j, l))$. Thus, for AES and its $T = 16$ inverters the probability of missing a fault of multiplicity l in the whole nonlinear portion of the encryption of one round is $Q_T(l) = \sum_{i=1}^{\min\{T,l\}} q^i P_T(i, l)$.

11.6.2 Protection of the Linear Block of AES

The general architecture used for protection with linear codes is presented in Fig. 11.7. The architecture is composed of three major hardware components: original hardware, redundant hardware for predicting the r-bit signature v (which is a linear combination of components of the output x of the original device), and an error detection network (EDN).

The signature predictor contains most of the redundant hardware. The k bits of output of the original hardware and the r redundant output bits of the signature predictor form the $n = k + r$ extended output of the device. The extended output forms a codeword of the systematic error-detecting code which can be used to detect errors in the original hardware or in the Predictor. It is the EDN that verifies that the extended output of the device belongs to the corresponding code; if it does not, the EDN raises an error signal. In a linear protection scheme the predicted r-bit signature v of the predictor is a linear combination of the k-bit output of the original device ($v = Px$, where P is an $r \times k$ check matrix for the linear code used for protection).

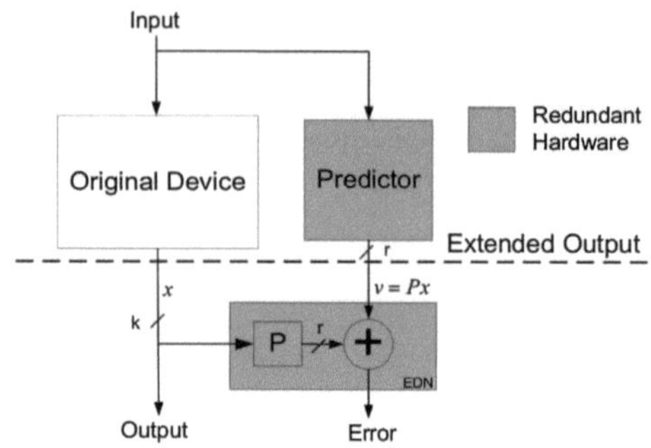

Fig. 11.7 General architecture for protection of hardware with linear error detecting codes

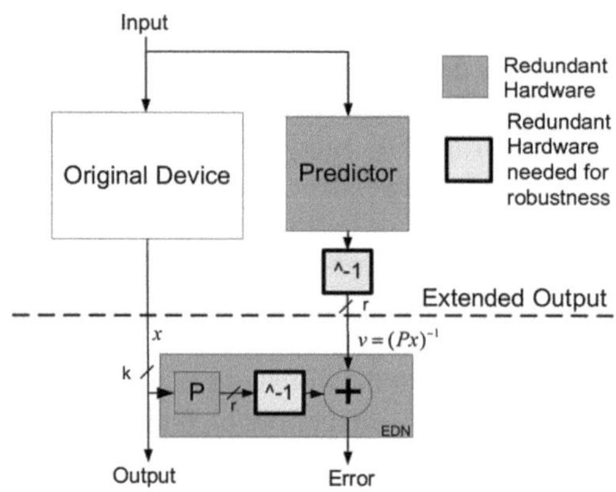

Fig. 11.8 Architecture for protection of hardware with nonlinear partially robust error detecting codes

With only slight modification, the same architecture used for protection with linear codes can be used to provide protection based on the systematic nonlinear partially robust codes $\{x, (Px)^{-1}\}$. The transformation only requires an addition of two copies of one extra component for the multiplicative inverse in $GF(2^r)$. The modified architecture is shown in Fig. 11.8. The extended output of the device is now protected with the partially robust nonlinear code with the properties outlined above. An additional (and identical) multiplicative inverse is also needed in the EDN to verify the nonlinear signature. This transformation can be applied to any linear protection method regardless of the algorithm it is protecting.

As Table 11.1 shows, one very desirable consequence of the addition of inversion to create a robust code is the reduction in the number of undetectable errors. The number of undetectable errors is reduced from 2^k to 2^{k-r}. When $k = r$, the code is robust and all nonzero errors are detectable.

Since large r may be necessary to provide for a sufficiently high error-detecting probability, the use of one large device which takes the multiplicative inverse of all of the r-redundant bits might not be practical. Transforming an implementation protected by a linear code with $r = 32$ into one protected by a robust systematic code would require several thousands additional two-input gates.

Table 11.1 Error detection capabilities of linear and nonlinear codes

Error detection probability	Number of errors		
	Linear	Robust (r is odd)	Robust (r is even)
0	2^k	2^{k-r}	2^{k-r}
1	$2^n - 2^k$	$2^{n-1} + 2^{k-1} - 2^{k-r}$	$2^{n-1} + 2^{k-1} - 2^{k-r} + 2^k - 2^{k-r}$
$1 - 2^{-r+1}$	0	$2^{n-1} - 2^{k-1}$	$2^{n-1} - 2^{k-1} - 2^{k+1} + 2^{k-r+1}$
$1 - 2^{-r+2}$	0	0	$2^k - 2^r$

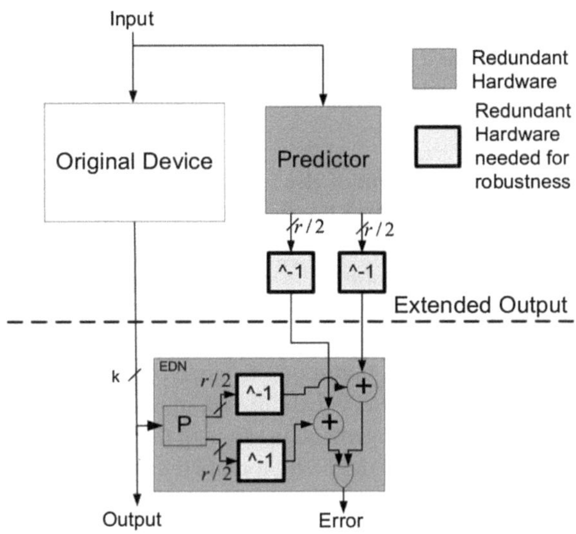

Fig. 11.9 Optimized architecture, the multiplicative inverse is split into $t = 2$ separate modules

It is possible to trade off the level of robustness for the amount of hardware overhead required to transform linear protection to protection based on systematic robust codes. Instead of taking one multiplicative inverse for all r-bit vectors, it is possible to divide the one large inversion into disjoint smaller inversions while retaining many of the robust properties outlined earlier. That is, we can replace multiplicative inverse in $GF(2^r)$ by t s-bit disjoint inverses in $GF(2^s)$ to produce the nonlinear r bit output ($r = t\,s$). Thus, instead of there being two r-bit multiplicative inverses in $GF(2^r)$ for the whole design, there could be $2t$ inverses in $GF(2^s)$, as is presented in Fig. 11.9 for $t = 2$. Since the number of two input gates to implement the inverse is proportional to the square of the number of bits at its input, a modification where $t = 2$ would result in roughly in the 50 % decrease overhead associated with the architecture based on robust codes. As a consequence this also results in a slight decrease in the level of robustness and in the introduction of errors which are detected with different probabilities.

Table 11.2 Comparison of protection architectures of the linear block of AES based on different alternatives

Code[1]	Number of 2-input gates Predictor	EDN	Overhead (%)	$\lvert K_d \rvert^2$	Q_m^3
Linear parity	31	32	30	2^{32}	0
Robust parity	185	32	100	0	0.5
Min. dist. robust [245]	196	64	120	0	0.5
Hamming	253	80	153	2^{32}	0
Gen. Vasil'ev [245]	292	116	188	2^{6}	0.5
$(x, (Px)^3)$ [245]	432	266	322	2^{26}	2^{-5}

[1] All codes have 32 information bits [2] $\lvert K_d \rvert$ is the number of undetectable errors [3] Q_m is the maximum error masking probability of detectable errors

11.6.2.1 Protection of the Linear Block of AES Based on Minimum Distance Robust and Partially Robust Codes

Protection architectures for the linear block of AES based on minimum distance robust and partially robust codes can be found in [245]. It was shown that for slow fault injection mechanisms, where the attacker cannot change the injected faults at every clock cycle, minimum distance robust and partially robust codes can provide better fault detection capabilities than linear codes and traditional robust codes.

The hardware overhead in terms of the number of two-input gates required for the implementation of the predictor and the EDN for different alternatives are compared in Table 11.2. Compared to architectures based on linear codes, architectures based on robust or partially robust codes have better protection against strong fault injection attacks at the cost of higher hardware overhead. The architecture based on $(x, (Px)^3)$ partially robust codes requires more than 300 % hardware overhead for the protection of a linear block. Since the nonlinear block of AES is much larger than its linear block, the overall percentage hardware overhead of architectures based on robust or partially robust codes is much smaller than that in the data presented in Table 11.2. In [335], it was shown that architectures based on $(x, (Px)^3)$ partially robust codes require less than 80 % overhead to protect the whole AES device. For more information about minimum distance robust codes and their applications, please refer to [245, 414, 415].

11.7 Secure FSM Design Based on Nonlinear Codes

Protecting the data path of cryptographic algorithms is the focus of existing countermeasures against active fault attacks. However, even if the data path is protected with the most secure scheme, the unprotected control unit may create a serious vulnerability in the system. For instance, by injecting specifically chosen errors into the part of the IC that implements the control units, the adversary may bypass the encryption

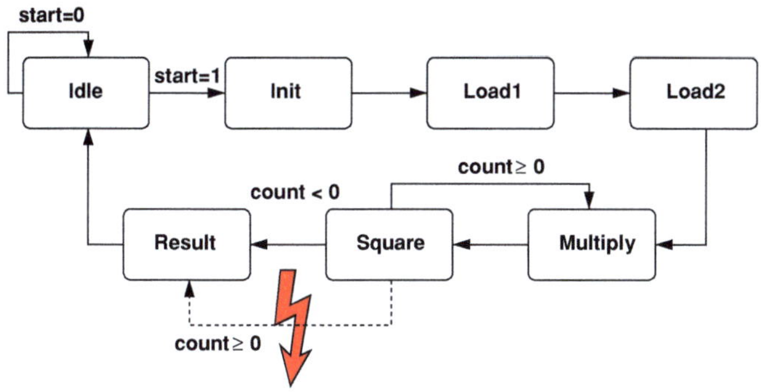

Fig. 11.10 Fault injection example on the control unit of the Montgomery ladder Algorithm with Point of Attack indicated by the dashed transition [157]

states in an FSM to conveniently gain access to secret information. Similarly, in a cryptographic authorization protocol the state which checks the validation of login information can be skipped with minimal effort. Thereby, the adversary can directly impersonate a valid user. Such attacks are indeed practical since states in an FSM are implemented using flip-flops, which can be easily attacked using bit-flips realized through fault injections. For example, Fig. 11.10 shows an example attack scenario on the FSM of the Montgomery ladder algorithm [157]. In their paper, Gaubatz et al. show that the attacker can recover the secret exponent using this attack with mild effort. As a result, secure FSM design becomes an important problem.

11.7.1 The Error Detection Technique

In this section, we describe how to use the nonlinear error detection codes for securing the next-state logics of FSMs. In order to achieve this, we first provide the details of a specific nonlinear coding structure which we propose to use as the main building block of our security scheme.

The initial version of the robust codes defined in [216] does not preserve arithmetic and hence cannot be applied to protect arithmetic structures against fault attacks. As a solution to this problem, Gaubatz et al. proposed a new type of nonlinear arithmetic codes called "robust quadratic codes" in [156]. The following definition from [156] rigorously defines this particular class of binary codes.

Definition 11.6 [156] Let C be an arithmetic single residue (n, k) code with $r = n - k$ redundant bits. Then $C_p = \{(x, w)|x \in \mathbb{Z}_{2^k}, w = (x^2 \bmod p) \in GF(p)\}$ where $2^k - p < \epsilon$ and $r = k$.

In this definition, ϵ is used to represent a relatively small number. As will be explained in Theorem 11.3, ε determines how secure the coding scheme is. As ε gets smaller, the utilized code becomes more secure. In addition, in this definition, $\mathbb{Z}_{2^k}$ is used to represent k-bit integers. While quantifying the error detection capability of C_p, [156] makes the following uniformity assumption:

Assumption 11.1 The values of $x \in \mathbb{Z}_{2^k}$ in C_p are uniformly distributed and each value of x is equally likely.

Under this assumption, reference [156] provides the following bound on the error masking equation which quantifies the error detection performance of the nonlinear code C_p.

Theorem 11.3 *[156] For C_p, $\max_{e \neq 0}(Q(e)) = 2^{-k} \cdot \max(4, 2^k - p + 1)$.*

Our protection scheme is built on top of this coding structure. The main idea is to encode the variables of the FSM using the nonlinear robust code (n, k) as in Definition 11.6, and to use the error detection capability of this coding scheme for fault detection. However, a direct implementation of this coding scheme for an FSM would cause a serious security problem. Note that the specific error detection technique proposed by Gaubatz et al. works under the uniformity assumption, which states that all the codewords are observed with the same probability (Assumption 11.1). This is a valid assumption if the code is applied to an arithmetic structure (such as an adder or a multiplier) where the inputs, and hence the outputs, tend to be uniformly distributed. However, when the FSMs are involved, this assumption becomes invalid because

1. depending on the FSM, some states may be visited more than others while the device is in operation, and
2. the number of inputs and states in an FSM are usually relatively small over a large domain.

Due to this nonuniform behavior, the security level provided by Theorem 11.3 does not apply if this nonlinear coding scheme is directly applied to an FSM. In this case, the error detection probabilities of Theorem 11.3 will be much smaller because the number of valid codewords (fault-free next-state values) determine the value of $M = |C_p|$ in the error masking probability of (11.1).

If the whole state space is being used uniformly by the valid next-state values (information carried by the code), then $|C_p| = 2^k$ as in this theorem. However, in the nonuniform case of FSMs, the value of $|C_p| = t$, where t is the number of states in the FSM. This alone dramatically increases the error masking probability of the detection scheme when $t \ll 2^k$ (Note that the security level provided by the applied nonlinear code is directly proportional to the value of k. The error detection probability of the scheme decreases as k gets smaller. As a result, for reasonable security levels, k should be reasonably large. However, when this is the case, we will have $t \ll 2^k$ because the number of states in an FSM is usually minimal). In addition, if there are some states that are visited more than others, this will decrease the effective value of $|C_p|$ even further, and hence will also cause an increase in the error masking probability.

Our solution to this problem is built around two innovative ideas:

- Arithmetic formulation of the next-state logic using Lagrange interpolation.
- Randomized embedding to solve the nonuniformity problem discussed before. In this method, each state value will have multiple images and this will provide unpredictability and uniformity.

More specifically, we first propose making state transitions work according to an arithmetic formula in a nonredundant field GF(q), i.e. $s' = f(s, i)$ where s' is the next-state value, i is the input, and s is the current-state value. In this setting, s', s, and $i \in GF(q)$.

However, this is not sufficient for solving the previously discussed nonuniformity problem because even though the state transitions are now working according to an arithmetic formula, the same nonuniform behavior is observed at the state registers. To solve this problem, we need unpredictability. In other words, the state machine variables must be practically unpredictable to the attacker. As a result, as the second step of the solution, we propose embedding the nonredundant field $GF(q)$ into a larger redundant ring $\mathbb{R} = \mathbb{Z}_M$ using a transformation $\phi : GF(q) \mapsto \mathbb{R}$ (or $\phi : \mathbb{R} \mapsto \mathbb{R}$) where ϕ is a homomorphism. Therefore, all the arithmetic operations defined for GF(q) can also be carried out in $\mathbb{R}$. In our case, randomized embedding is achieved by defining the ring modulus M and the scaling factor S with M=$S \times q$ where q is the prime defining the field GF(q). In this setting, we define the function ϕ as $\phi(s) = s + R \times q (\text{mod } M)$ where R is a randomly chosen value from $\mathbb{Z}_S$. This scaling effectively partitions the ring $\mathbb{R}$ into co-sets, of which only one contains the nonredundant codewords. As a result of this scaling, each state and input value now has S images. In other words, the same state and input will now be represented by S different values in the ring $\mathbb{R}$ depending on the value of the random R. This will increase the uniformity of the state values observed at the output of the next-state logic, and essentially increase the error detection capability of the nonlinear coding technique we propose in this section. Note that when the state value is reduced using q, we obtain the nonredundant value of the state. Note that the state and input values observed in this FSM will randomly be one of the S images of their nonredundant values. A numerical application of this protection scheme is shown in Example 11.4. Note that in this example $q = 11$. This means that the nonredundant value of each state and input value will be one of $\{0, 1, 2, 3, 4, 5, 6, 7, 8, 9, 10\}$ (i.e. integers modulo 11). This is the only co-set that includes the nonredundant state and input values. However, due to the randomization process, each nonredundant state and input value will have $S = 390451572$ different images ($2 \equiv 13 \equiv 24$, and so on).

We now formally define a randomized robust code by merging this embedding with the robust code definition introduced by Gaubatz et al. [156] in Definition 11.6 as follows.

Definition 11.7 [10] We define the co-set randomized robust code (n, k) with $r = n - k$ redundant bits as $C = \{(x, w) | x = y + R \times q (\text{mod } M), \forall x \text{ and } \forall y \in \mathbb{Z}_M, w = x^2 (\text{mod } p) \in GF(p), R \in \mathbb{Z}_S\}$ where $r = k$, $k \geq \max(\lceil \log_2 M \rceil, \lceil \log_2 p \rceil)$.

After the randomization of the next-state function f is achieved, we can use individually robust arithmetic circuits such as multipliers, adders, and so on to build the proposed robust FSM. To achieve this, we utilize the nonlinear code explained in Definition 11.7 to encode the state variables and inputs. Next, we use the individually robust arithmetic circuits that will work in the ring $\mathbb{R}$ to implement the next-state function f. Robust adder and multiplier implementation examples that work under the utilized code are provided in [156].

In the following theorem, we establish the error detection probability of our scheme. A detailed proof of this theorem can be found in [10].

Theorem 11.4 *[10] For the nonlinear code C that is defined as in Definition 11.8, the error masking probability will be upper-bounded by* $S^{-1} \cdot \max(4, 2^k - p + 1)$, *where* $S > \max(4, 2^k - p + 1)$.

Example 11.4 Consider the FSM of Fig. 11.10. As discussed before, this FSM bounces between the "Square" and "Multiply" states most of the time. If we select $GF(q) = GF(11)$ and $S = 390451572$, then $M = 4294967292$ and we will need a $(64, 32)$-code with $p = 4294967291$. In this case, injected errors are detected with a probability of at least $1 - S^{-1} \cdot \max(4, 2^k - p + 1) = 1 - (2^{32} - 4294967291 + 1)/(2 \times 390451572) \simeq 1 - 2^{-27}$. The denominator in this case becomes $2 \times S$ because the "Square" and "Multiply" states and their images will be uniformly distributed.

11.7.2 Case Study

In this section, we present the application of the proposed technique to the example FSM shown in Fig. 11.10. The first step is to use two-level Lagrange interpolation to generate an algebraic polynomial for the next state logic. This polynomial represents the next-state s' as a function of the input and current state variables i and s, respectively. For the example FSM we investigate here, the following polynomial (see [10] for more details on the computation process) shows the result of the two-level Lagrange interpolation:

$$\begin{aligned} s' = f(s, i) &= (4i)s^6 + (2 + 4i)s^5 \\ &+ (3 + 6i)s^4 + (1 + 2i)s^3 + (5 + 3i)s^2 + (5 + i)s + i. \end{aligned} \quad (11.7)$$

This function can be implemented in an efficient way. The idea is to add time-redundancy and reuse the expensive hardware modules (e.g. the multiplier and the adder) in a serial manner. This can be achieved by rearranging the next-state polynomial of (11.7) using Horner's method. In this case, the next-state equation becomes

$$s' = ((((((4i)s+(2+4i))s+(3+6i))s+(1+2i))s+(5+3i))s+(5+i))s+i. \quad (11.8)$$

Once this polynomial is computed, the next step is to encode the input and current state values using the co-set randomized code of Definition 11.7. In this case, the input will be $(i, |i^2|_p)$ and the current state will be $(s, |s^2|_p)$ where $i, s \in Z_M$. Next,

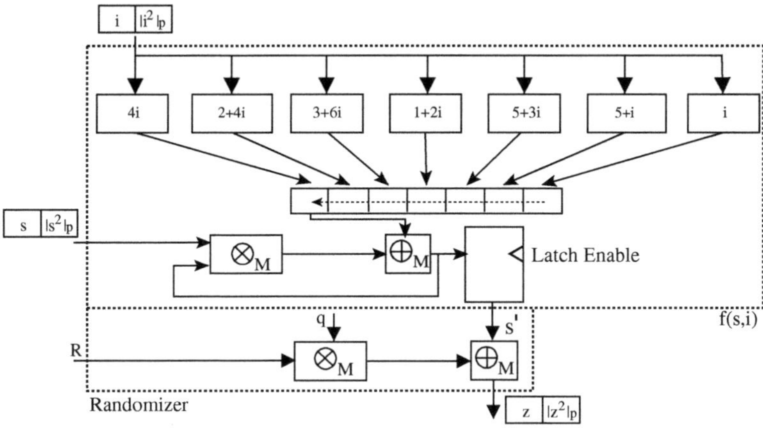

Fig. 11.11 Time redundant (serial) arithmetic hardware implementation of the next-state logic

using the computed function $f(s, i)$, we compute $(s', |(s')^2|_p)$. In this computation, all the main data path operations are conducted over modulo M. Also note that each arithmetic unit (in both function f and the randomizer unit) is a robust one which implements nonlinear error detection individually. For example, all the multipliers in function f and the randomizer unit compute the output and its redundant checksum using the inputs and their redundant checksums. These individual components can detect an injected fault and raise an error signal. Next, using the randomizer unit, we compute $z = s' + R \times q (\text{mod } M)$. These operations are conducted using robust arithmetic units and are over modulo M as well. The resulting randomized next-state value is then fed as the current state value into the next-state logic in the next iteration. Note that we do not need to recover s' for the following next state computation. Since the function $f(s, i)$ and randomizer unit work modulo M, the next-state value will always be a randomized image of the correct next state value. The nonredundant form of the next-state value ($\in$ GF(q)) can always be computed by reducing the randomized images modulo q. The resulting implementation is shown in Fig. 11.11 for this specific example. For details on FSM security using nonlinear codes, the reader is referred to [10].

11.7.3 Implementation Results

In order to measure the performance of our scheme, we implemented the following cases for the FSM, shown in Fig. 11.10:

1. regular FSM with no error detection or any type of redundancy.
2. FSM protected with (10, 5, 5) co-set randomized robust codes (our solution).

Table 11.3 Hardware implementation results

	Gate count	Time delay (ns)
Non-red FSM	80	0.64
(10, 5, 5)	8,305	29.99

Table 11.4 Gate counts and hardware overhead caused by our protection scheme for different Montgomery multipliers

	Non-red FSM	Robust FSM	Robust data Path	FSM overhead (%)
MM(256,2,64)	3,843	269,010	436,625	37.58
MM(256,1,128)	3,812	266,882	705,560	27.05
MM(512,1,128)	3,841	268,870	760,090	25.76
MM(512,1,256)	7,307	511,532	2,500,949	16.74

We implemented these techniques using VHDL and synthesized them using the tcb013lvhptc and DW (Design Ware) libraries of the Synopsys design tool. We report the gate count and time delay results in Table 11.3.

Even though the area overhead and time delay caused by our scheme seems relatively high initially, we argue that this is the price that needs to be paid to secure FSMs against strong adversaries. Remember that FSM security is a difficult problem, especially against an advanced attacker. Also note that even though our protection technique is not appropriate for low-power cryptographic hardware, it can be applied to circuits that include large and parallel arithmetic units in the data path. Low-power cryptographic units have relatively large FSMs due to their serial nature. In this case, large area overhead of our FSM protection scheme might affect the circuit area and performance in an unacceptable way. However, we argue that the overhead of our technique will be reasonable for cryptographic hardware with parallel arithmetic units (i.e. adders, multipliers, etc.) in the data path. These kinds of circuits have relatively small FSMs and hence the overhead of our protection scheme becomes minimal.

In order to clarify this idea, we investigated the effect of our FSM protection scheme on the overall circuit area. First of all, we analyzed four different Montgomery multipliers (MMs) with different parameters. MM(x,y,z) indicates a Montgomery multiplier with different parameters where x is the size of the operands, y is the number of pipeline stages, and z is the word size. The results are shown in Table 11.4. Note that the overhead caused by our scheme gets smaller for circuits with large and parallel data paths. For example, our FSM protection scheme causes an area overhead of approximately 16 % for the MM(512,1,256) case.

Furthermore, we argue that time delay caused by our scheme will not have a big impact on the circuit performance because FSMs are usually not in the critical path of a circuit, critical path usually goes through the data path, which includes various arithmetic operations (i.e. multiplications, divisions, etc.). As a result, the time delay caused by our technique will be masked and the throughput of the circuit will not be affected by this protection technique.

11.8 Secure ECC Based on Nonlinear Codes

Elliptic Curve Cryptosystems (ECCs) have also been a target of active fault attacks. In [44], Biehl et al. showed that using fault injection, ECC point multiplication can be forced to be computed over a less secure elliptic curve. As a result, it becomes relatively easy to solve the discrete log problem on which the ECC is based. They also proposed implementing bit faults during random moments of a multiplication operation and showed that it is possible to reveal the secret key d in a bit-by-bit fashion. In [90], the authors relaxed the assumptions of Biehl et al. in terms of the location and precision of the injected faults. Even with this new attacker model, their attack essentially recovers the (partial) secret in the ECC discrete log problem. Similarly, in [54], Blomer et al. proposed the so-called "Sign Change Fault" attacks on the ECC-based systems. Using this attack, they recover the secret scalar in polynomial time. In this section, we will discuss how nonlinear robust codes can be used to protect ECC operations.

11.8.1 Elliptic Curve Cryptography Overview

This section briefly describes the elliptic curve discrete logarithm problem (ECDLP) and ECC formulations over finite fields of prime characteristics. A point P of order $\#n$, selected over an elliptic curve E defined over a finite field GF(p), can be used to generate a cyclic subgroup $\langle P\rangle = \{\infty, P, 2P, 3P, \ldots, (\#n-1)P\}$ of $E(GF(p))$.

The ECDLP is the underlying number theoretical problem used by ECC, and it is defined as determining the value $k \in [1, \#n-1]$, given a point $P \in E(GF(p))$ of order $\#n$, and a point $Q = kP \in \langle P\rangle$. In a cryptosystem, the private key is obtained by selecting an integer k randomly from the interval $[1, \#n-1]$. Then the corresponding public key is the result of scalar point multiplication $Q = kP$, which is computed by a series of point additions and doublings.

ECC can be built upon two curves: (1) Weierstrass and (2) Edwards. In this chapter, we focus on Edwards curves. However, note that the technique we propose is a generic one that can be applied to all elliptic curve structures (Weierstrass and Edwards) and all coordinate systems (projective and affine).

11.8.1.1 Edwards Formulation for Elliptic Curves

An elliptic curve E defined over a prime field $GF(p)$ (with $p > 3$) can be written in the Edwards normal form as

$$E(GF(p)) : x^2 + y^2 = c^2(1 + dx^2y^2), \tag{11.9}$$

where the parameter c can be chosen as 1 without loss of generality. The addition of two points $P = (x_1, y_1)$ and $Q = (x_2, y_2)$ in $E(GF(p))$, resulting in a third point $P + Q = (x_3, y_3)$ in $E(GF(p))$, can be computed as

$$x_3 = \frac{x_1 y_2 + y_1 x_2}{1 + d x_1 x_2 y_1 y_2} \pmod{p}, \tag{11.10}$$
$$y_3 = \frac{y_1 y_2 - x_1 x_2}{1 - d x_1 x_2 y_1 y_2} \pmod{p}.$$

This equation is valid even if $P = Q$, and it never results in a point at infinity. An Edwards elliptic curve defined as in (11.9) is converted to homogeneous projective coordinates as $E(\mathrm{GF}(p)) : X^2 + Y^2 = Z^4 + dX^2Y^2$ where $X = xZ, Y = yZ$. The following formulas compute the unified point addition and doubling (11.11) and optimized doubling (11.12) operations with projective coordinates [33]:

$$X_3 = Z_1Z_2(X_1Y_2 + Y_1X_2)(Z_1^2Z_2^2 - dX_1X_2Y_1Y_2) \pmod{p}, \tag{11.11}$$
$$Y_3 = Z_1Z_2(Y_1Y_2 - X_1X_2)(Z_1^2Z_2^2 + dX_1X_2Y_1Y_2) \pmod{p},$$
$$Z_3 = (Z_1^2Z_2^2 - dX_1X_2Y_1Y_2)(Z_1^2Z_2^2 + dX_1X_2Y_1Y_2) \pmod{p};$$

$$X_3 = 2X_1Y_1(X_1^2 + Y_1^2 - 2Z_1^2) \pmod{p}, \tag{11.12}$$
$$Y_3 = (X_1^2 - Y_1^2)(X_1^2 + Y_1^2) \pmod{p},$$
$$Z_3 = (X_1^2 + Y_1^2)(X_1^2 + Y_1^2 - 2Z_1^2) \pmod{p}.$$

11.8.2 The Error Detection Technique

We mainly propose applying nonlinear codes to secure operations conducted over elliptic curves, i.e. point addition and doubling operations against active fault injection attacks. In this chapter, we are focusing on ECC structures based on prime fields GF(p), yet a similar idea can be applied to protect elliptic curves that are defined over binary fields as well.

The main idea is to encode the coordinates of elliptic curve points using the systematic nonlinear (n, k)-code of Definition 11.6. This code essentially uses redundancy for error detection. We define the following error check function on a point coordinate $X \in \mathrm{GF}(p)$ to obtain a nonlinear error check-sum:

$$w = h(X) = X^2 (\mathrm{mod}\, p) \in GF(p). \tag{11.13}$$

Consequently, the point coordinate X is encoded as $(X, h(X))$. We now formally define a robust code by embedding the nonlinear code definition introduced by Gaubatz, Sunar, and Karpovsky [156] into elliptic curves as follows.

Definition 11.8 [9] define the prime field robust code (n, k) with $r = n-k$ redundant bits as $C = \{(x, w)|x \in GF(p), w = x^2(\text{mod} p) \in GF(p)\}$ where $r = k$.

In the nonredundant case, a point P on an elliptic curve E is represented as $P = (X, Y, Z)$ (assuming projective coordinates). However, with the new robust code definition we have, each point will be represented as $P = (X, X_w, Y, Y_w, Z, Z_w)$, where the subscript w is used to show the checksum portions.

The following theorem establishes the security level provided by the nonlinear code described in Definition 11.8 for any elliptic curve E. The detailed proof of this theorem can be found in [9].

Theorem 11.5 *[9] For the nonlinear code C of Definition 11.8, the error masking probability is upper-bounded by* $max(4, 2^k - p + 1) \cdot (p + 1 - 2\sqrt{p})^{-1}$.

Example 11.5 Consider the NIST recommended prime field curve P-192. For this curve, $(2^k - p + 1) = 18446744073709551618 \approx 2^{64}$. In this case, according to Hasse's theorem, the number of valid points on this curve will be at least $(p + 1 - 2\sqrt{p}) \approx 2^{192}$. As a result, the minimum error detection capability proposed by our scheme in this case will be $1 - 2^{64}/2^{192} = 2^{-128}$.

11.8.3 Proposed Point Addition/Doubling Construction

In this section, we provide the secure implementation of the unified point addition/doubling operation for Edwards curves. This implementation utilizes the error detection technique described in Sect. 11.8.2. The main idea of the nonlinear error detection is to create two computation paths that are nonlinear to each other. As the first step to achieving this, the coordinates of the input points in an operation (point addition or doubling) are encoded using the nonlinear code described in Definition 11.8. One of the nonlinear paths is the original nonredundant data path. The second path, which is called the "predictor" block, runs in parallel to the nonredundant path, and essentially predicts the checksum of the results of the original computation. At this point, it is important to note that we do not simply replicate the original hardware to implement the predictor. For each data path, the total operation count is expressed in terms of multiplications, divisions and additions and subtractions, where **M** stands for multiplication, **D** stands for division, and **A** stands for addition or subtraction.

For Edwards curves that are using projective coordinates, the unified point addition operation computes the point $P3 = (X_3, Y_3, Z_3)$ using the input points $P1 = (X_1, Y_1, Z_1)$ and $P2 = (X_2, Y_2, Z_2)$. The explicit formula that implements the point addition is shown in (11.11). In the following, we show how the predictor block works. It mainly computes the expected X_{3w}, Y_{3w}, Z_{3w} using the inputs and their checksums. More specifically, the expected checksums should be

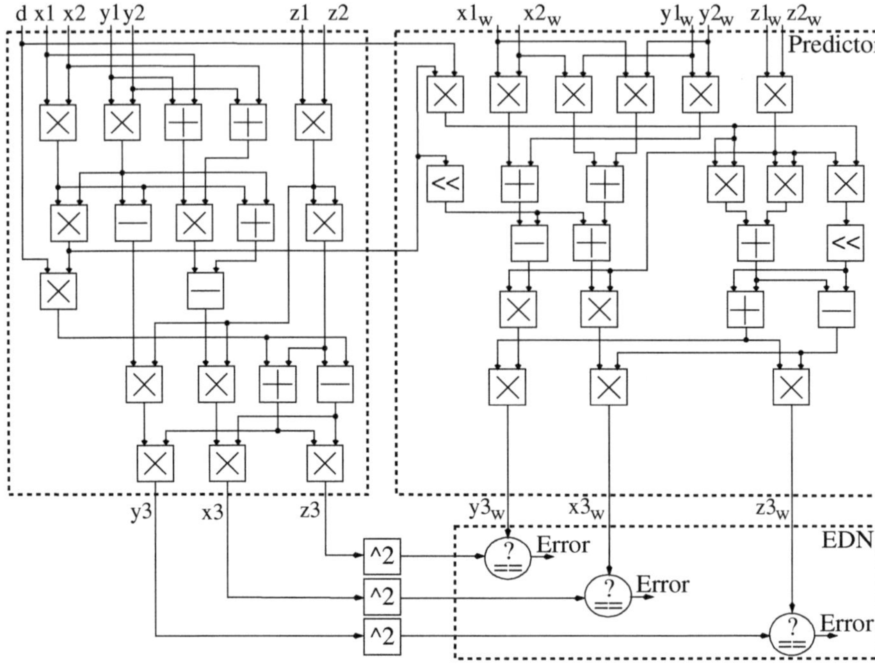

Fig. 11.12 Secure Edwards projective unified point addition

$$
\begin{aligned}
X_{3w} &= (X_3)^2 = [Z_1Z_2(X_1Y_2 + Y_1X_2)(Z_1^2Z_2^2 - dX_1X_2Y_1Y_2)]^2,\\
Y_{3w} &= (Y_3)^2 = [Z_1Z_2(Y_1Y_2 - X_1X_2)(Z_1^2Z_2^2 + dX_1X_2Y_1Y_2)]^2,\\
Z_{3w} &= (Z_3)^2 = [(Z_1^2Z_2^2 - dX_1X_2Y_1Y_2)(Z_1^2Z_2^2 + dX_1X_2Y_1Y_2)]^2.
\end{aligned}
$$

Next, we express the terms on the right-hand side as a function of the inputs and their checksums. As an example, we show how to achieve this for X_3. The same method can also be applied to the Y and Z coordinates as well.

$$
\begin{aligned}
X_{3w} &= [Z_1Z_2(X_1Y_2 + Y_1X_2)(Z_1^2Z_2^2 - dX_1X_2Y_1Y_2)]^2\\
&= Z_1^2Z_2^2(X_1^2Y_2^2 + Y_1^2X_2^2 + 2X_1Y_2Y_1X_2)(Z_1^4Z_2^4 - 2Z_1^2Z_2^2dX_1X_2Y_1Y_2\\
&\quad + d^2X_1^2X_2^2Y_1^2Y_2^2)\\
&= Z_{1w}Z_{2w}(X_{1w}Y_{2w} + Y_{1w}X_{2w} + 2\alpha)(Z_{1w}^2Z_{2w}^2 - 2Z_{1w}Z_{2w}d\alpha\\
&\quad + d^2X_{1w}X_{2w}Y_{1w}Y_{2w}),
\end{aligned}
$$

where $\alpha = X_1X_2Y_1Y_2$. After some algebra, we get the following equation array for each coordinate of the resulting point $P3$. These equations mainly represent the function implemented by the predictor unit in our design.

$$\alpha = X_1X_2Y_1Y_2;$$
$$A = Z_{1w}Z_{2w}; \quad B = X_{1w}Y_{2w}; \quad C = Y_{1w}X_{2w}; \quad D = X_{1w}X_{2w};$$
$$E = Y_{1w}Y_{2w}; \quad F = d\alpha; \quad G = B + C + 2\alpha; \quad H = D + E - 2\alpha;$$
$$K = A^2 + F^2; \quad L = AF; \quad M = K - 2L; \quad N = K + 2L;$$

$$X_{3w} = AGM; \quad Y_{3w} = AHN; \quad Z_{3w} = MN;$$

The total operation count for this predictor unit will be 14**M** + 7**A**. Note that all the operations in this setup are modulo p, where p is the prime that generates the finite field the elliptic curve is defined over.

The hardware implementation of this technique is shown in Fig. 11.12. In this figure, the block on the left is the original, nonredundant data path that computes the unified point addition. The predictor block mainly implements the X_{3w}, Y_{3w}, and Z_{3w} computations defined above. Next, the output coordinates are squared to compute their checksums. Finally, the EDN compares the results of these two paths. If all the results match, this means that the conducted operation is fault-free. However, if there is a mismatch in any one of the coordinate comparisons, this points to an injected fault. Hence, an error signal is asserted. Once the error signal is asserted, either the secret can be flushed or the device can be reset. For details on ECC security using nonlinear codes, the reader is referred to [9].

11.8.4 Area Estimation for the Proposed ECC Protection Technique

The area overhead caused by the application of our scheme is dependent on the areas of arithmetic unit implementations in a particular system. Without knowing the relative area ratios of division, multiplication and addition/subtraction units, it is not possible to provide an exact overhead measure. However, given the higher complexity of divisions and multiplications with respect to additions/subtractions, it is reasonable to ignore additions and subtractions to obtain an estimation of the overhead. Also, we assume that the area of a multiplication unit is on the order of a division unit. This is a reasonable assumption because it does not make sense to have an affine system where the area of a divider is much larger than the area of a multiplier.

Having made these assumptions, the estimated percentage overheads of the nonlinear error detection schemes we propose for point operations are presented in Table 11.5. Note that this table provides results for both Weierstrass and Edwards curves for different coordinate systems. We observe that the Weierstrass-based elliptic curve systems can be protected with reasonable area overhead. Note that the application of our scheme causes 175, 169, and 130 % overheads for the affine point doubling, Jacobian point addition, and Jacobian point doubling operations, respectively. The worst case for securing Weierstrass operations is the affine point addition, which causes an area overhead of 300 %. In addition, we also observe that the predic-

Table 11.5 Overhead analysis for the error detection technique proposed in this chapter

ECC system	Nonredundant data path	Our predictor data path	Our complete data path	Our estimated % overhead
Weierstrass affine point addition	1 **D**+ 2**M**+ 6**A**	1**D**+8**M** + 13**A**	2**D**+ 10**M**+19**A**	300
Weierstrass affine point doubling	1**D**+3**M**+ 5 **A**	1**D**+6**M**+ 9**A**	2**D**+9**M** + 14**A**	175
Weierstrass Jacobian point addition	16**M**+7**A**	27**M**+13**A**	43**M**+ 20**A**	169
Weierstrass Jacobian point doubling	10 **M**+ 5**A**	13**M**+8**A**	23**M**+ 13**A**	130
Edwards affine unified point addition	2 **D**+5**M**+ 7**A**	2**D**+6**M**+8 **A**	4**D**+11**M**+15**A**	114
Edwards projective unified point addition	12**M**+7**A**	14**M**+ 7**A**	26**M**+14**A**	117

tor unit for point additions incurs more overhead than point doublings in Weierstrass systems. This is also an expected result since point addition operations are more complex than point doubling operations.

On the other hand, the balanced normal form of the Edwards formulation provides simpler predictor designs with less overhead compared to the Weierstrass systems. In other words, the Edwards formulation is more appropriate for the nonlinear error detection technique that is proposed in this paper. The secure unified point addition in affine coordinates require an overhead of 114 %, while in projective coordinates, the overhead ratio is 117 %.

11.9 Conclusion

Application of nonlinear error detection codes is one of the most effective solutions to active fault injection attacks. In this chapter, we provided a comprehensive analysis of nonlinear codes by investigating various constructions and their applications. More specifically, we discussed protection of the AES data path, FSMs, and ECC using nonlinear robust codes.

Acknowledgments The authors would like to express their thanks to Dr. Deniz Karakoyunlu for his support and valuable discussions during the development of Sect. 11.8.

Chapter 12
Lattice-Based Fault Attacks on Signatures

Phong Q. Nguyen and Mehdi Tibouchi

Abstract Since the introduction of the LLL algorithm in 1982, lattice reduction has proved to be one of the most powerful and versatile tools of public key cryptanalysis. In particular, it has sometimes been combined with fault injection to break physical implementations of public key cryptosystems. We present several examples of lattice-based fault attacks against DSA and RSA signatures, together with the necessary mathematical background.

12.1 Introduction

A lattice is a regular arrangement of points in space which can be described as the set of integral linear combinations of certain collections of linearly independent vectors $\mathbf{b}_1, \ldots, \mathbf{b}_d$ called bases. Lattices have infinitely many bases, but some are more interesting than others; finding interesting lattice bases, consisting of short and nearly orthogonal vectors, is a mathematical problem with a long history, dating back to the work of Lagrange and Gauss on quadratic forms in the late eighteenth and early nineteenth centuries.

However, it was not until around 1980, with the advent of algorithmic number theory, that general procedures for obtaining such interesting bases and solving related lattice problems were proposed. The starting point of modern lattice reduction is the seminal 1982 paper by Lenstra et al. [251] introducing the algorithm that became known as LLL.

This algorithm and its later refinements were quickly recognized as powerful cryptanalytic tools: they were used by Adleman as early as 1983 [1] in a generalization of Shamir's attack on the Merkle-Hellman knapsack-based cryptosystem [370]. They

---

P. Q. Nguyen (✉) · M. Tibouchi
Département d'informatique, École Normale Supérieure,
Paris, France
e-mail: pnguyen@ens.fr

M. Joye and M. Tunstall (eds.), *Fault Analysis in Cryptography*,
Information Security and Cryptography, DOI: 10.1007/978-3-642-29656-7_12,

have been the foundation of many more cryptanalytic results ever since, including attacks on linear congruential generators [147, 386], RSA signatures with constant padding [71, 290], implementations of El Gamal signatures [302] and the numerous applications of Coppersmith's techniques for finding small roots of algebraic equations [101].

In this chapter, we present several recent examples of attacks where these powerful lattice methods were used in conjunction with fault injection to break physical implementations of cryptographic signature schemes:

- against DSA (Sect. 12.3), an attack by Naccache et al. [300], involving the search for the lattice point closest to a target point in space;
- against the RSA signature scheme ISO/IEC 9796-2 (Sect. 12.4), a polynomial attack by Coron et al. [104] based on results similar to Coppersmith's, and another attack by Coron et al. [108] based on a rather different technique, namely the so-called orthogonal lattice technique introduced by Nguyen and Stern in 1997 [307].

12.2 Preliminaries on Lattices

We start with some preliminary definitions and results about lattices. A reader unfamiliar with these notions may want to skim through this section in a first reading and refer back to it later as needed. A similar presentation with a larger scope and more details can be found in [303, Sect. 6].

12.2.1 Notation and Background

We will consider $\mathbb{R}^n$ endowed with its usual structure as an Euclidean vector space. Bold letters will denote vectors, usually in row notation. If $\mathbf{x} = (x_1, \dots, x_n)$ and $\mathbf{y} = (y_1, \dots, y_n)$ are two vectors, their Euclidean inner product is denoted by

$$\langle \mathbf{x}, \mathbf{y} \rangle = \sum_{i=1}^{n} x_i y_i .$$

The corresponding Euclidean norm is denoted by

$$\|\mathbf{x}\| = \sqrt{\langle \mathbf{x}, \mathbf{x} \rangle} = \sqrt{x_1^2 + \cdots + x_n^2}.$$

For any subset S of $\mathbb{R}^n$, we define the linear span of S, denoted by $\operatorname{span}(S)$, as the minimal vector subspace of $\mathbb{R}^n$ containing S.

Let $\mathbf{b}_1, \dots, \mathbf{b}_m$ be in $\mathbb{R}^n$. The vectors $\mathbf{b}_i$ are said to be *linearly dependent* if there exist scalars $x_1, \dots, x_m \in \mathbb{R}$ which are not all zero and such that:

$$\sum_{i=1}^{m} x_i \mathbf{b}_i = 0.$$

Otherwise, the $\mathbf{b}_i$'s are said to be *linearly independent*.

The Gram determinant of $\mathbf{b}_1, \ldots, \mathbf{b}_m \in \mathbb{R}^n$, denoted by $\Delta(\mathbf{b}_1, \ldots, \mathbf{b}_m)$, is by definition the determinant of the Gram matrix $\left(\langle \mathbf{b}_i, \mathbf{b}_j \rangle\right)_{1 \leq i,j \leq m}$. This real number $\Delta(\mathbf{b}_1, \ldots, \mathbf{b}_m)$ is always ≥ 0, and it turns out to be 0 if and only if the $\mathbf{b}_i$'s are linearly dependent. The Gram determinant is invariant by any permutation of the m vectors, and by any integral linear transformation of determinant ± 1 such as adding to one of the vectors a linear combination of the others. The Gram determinant has a very useful geometric interpretation: when the $\mathbf{b}_i$'s are linearly independent, $\sqrt{\Delta(\mathbf{b}_1, \ldots, \mathbf{b}_m)}$ is the m-dimensional volume of the parallelepiped spanned by the $\mathbf{b}_i$'s.

12.2.2 Lattices and Lattice Bases

For our purposes, a lattice will be any subgroup L of $(\mathbb{Z}^n, +)$ for some n (such a subgroup is sometimes called an integral lattice, to distinguish it from more general definitions of a lattice). Examples include $\mathbb{Z}^n$ itself, the zero lattice $\{\mathbf{0}\}$, and the set $a\mathbb{Z} + b\mathbb{Z}$ of linear combinations of any two integers $a, b \in \mathbb{Z}$ (which is simply $\gcd(a, b)\mathbb{Z}$). For another example, consider n integers $a_1, \ldots, a_n$, together with a modulus M. Then the set of all $(x_1, \ldots, x_n) \in \mathbb{Z}^n$ such that $\sum_{i=1}^{n} a_i x_i \equiv 0 \pmod{M}$ is a lattice, as it is clearly a subgroup of $\mathbb{Z}^n$.

Let $\mathbf{b}_1, \ldots, \mathbf{b}_m$ be arbitrary vectors in $\mathbb{Z}^n$. Denote by $L(\mathbf{b}_1, \ldots, \mathbf{b}_m)$ the set of all integral linear combinations of the $\mathbf{b}_i$'s:

$$L(\mathbf{b}_1, \ldots, \mathbf{b}_m) = \left\{ \sum_{i=1}^{m} n_i \mathbf{b}_i : n_1, \ldots, n_m \in \mathbb{Z} \right\}.$$

This set is a subgroup of $\mathbb{Z}^n$, and hence a lattice L, called the lattice *generated* or *spanned* by the $\mathbf{b}_i$'s. When the $\mathbf{b}_i$'s are linearly independent, we say that $(\mathbf{b}_1, \ldots, \mathbf{b}_m)$ is a *basis* of the lattice L, in which case each lattice vector decomposes itself uniquely as an integral linear combination of the $\mathbf{b}_i$'s.

Bases and sets of generators are useful for representing lattices, and for performing computations. One will typically represent a lattice by some lattice basis, which can itself be represented by a matrix with integer coefficients. If $(\mathbf{b}_1, \ldots, \mathbf{b}_m)$ is a basis of L, with $\mathbf{b}_i = (b_{i,1}, \ldots, b_{i,n})$, we will represent the lattice $L = L(\mathbf{b}_1, \ldots, \mathbf{b}_m)$ by the following matrix:

$$\begin{pmatrix} b_{1,1} & \cdots & b_{1,n} \\ \vdots & & \vdots \\ b_{m,1} & \cdots & b_{m,n} \end{pmatrix}$$

whose rows are the coordinates of the $\mathbf{b}_i$'s.

We define the *dimension* or *rank* of a lattice L, denoted by $\dim(L)$, as the dimension d of the vector space $\text{span}(L)$. The dimension is the maximal number of linearly independent lattice vectors. Any lattice basis of L must have exactly d elements. It is clear that there exist d linearly independent lattice vectors, but such vectors do not necessarily form a basis, as in to the case of vectors spaces. However, it is in fact always possible to find a basis of L: in other words, lattices of dimension d in $\mathbb{Z}^n$ are exactly all subsets of $\mathbb{Z}^n$ of the form

$$L = L(\mathbf{b}_1, \ldots, \mathbf{b}_d)$$

for some linearly independent vectors $\mathbf{b}_1, \ldots, \mathbf{b}_d \in \mathbb{Z}^n$.

12.2.3 Lattice Volume

Let $(\mathbf{b}_1, \ldots, \mathbf{b}_d)$ and $(\mathbf{c}_1, \ldots, \mathbf{c}_d)$ be two bases of a lattice L in $\mathbb{R}^n$. An elementary result states that there exists a $d \times d$ integral matrix $U = (u_{i,j})_{1 \le i,j \le d} \in \mathcal{M}_d(\mathbb{Z})$ of determinant ± 1 such that $\mathbf{c}_i = \sum_{j=1}^{d} u_{i,j}\mathbf{b}_j$ for all $1 \le i \le d$. It follows that the Gram determinant of those two bases are equal:

$$\Delta(\mathbf{b}_1, \ldots, \mathbf{b}_d) = \Delta(\mathbf{c}_1, \ldots, \mathbf{c}_d) > 0.$$

The *volume* of the lattice L is then defined as

$$\text{vol}(L) = \Delta(\mathbf{b}_1, \ldots, \mathbf{b}_d)^{1/2},$$

which is independent of the choice of the lattice basis $(\mathbf{b}_1, \ldots, \mathbf{b}_d)$. It corresponds to the d-dimensional volume of the parallelepiped spanned by any basis. In the important case of full-dimensional lattices where $\dim(L) = n = \dim(\mathbb{R}^n)$, the volume is also equal to the absolute value of the determinant of any lattice basis.

If an explicit basis is known, computing the lattice volume amounts to computing a determinant. However, in some cases, we may not know any explicit lattice basis; in such as cases, the following elementary result may be useful for computing the volume nonetheless. If L_1 and L_2 are two lattices in $\mathbb{Z}^n$ of the same dimension such that $L_1 \subset L_2$, then L_2/L_1 is a finite group of order denoted by $[L_2 : L_1]$ which satisfies

$$\text{vol}(L_1) = [L_2 : L_1] \cdot \text{vol}(L_2).$$

For example, consider n integers $a_1, \ldots, a_n$, together with a modulus M. We have seen in Sect. 12.2.2 that the set L of all $(x_1, \ldots, x_n) \in \mathbb{Z}^n$ such that $\sum_{i=1}^{n} a_i x_i \equiv 0 \pmod{M}$ is a lattice in $\mathbb{Z}^n$, but it is not obvious how to explicitly write down a basis of L. However, note that $L \subset \mathbb{Z}^n$ and that the dimension of L is n because L contains $M\mathbb{Z}^n$. It follows that $\text{vol}(L) = [\mathbb{Z}^n : L]$. Furthermore, the definition of L says that it is the kernel of the group homomorphism $\mathbb{Z}^n \to \mathbb{Z}/M\mathbb{Z}$ sending $(x_1, \ldots, x_n)$ to $\sum_{i=1}^n a_i x_i$. The image of this morphism is the subgroup generated by $\gcd(M, a_1, a_2, \ldots, a_n)$, and hence

$$\mathrm{vol}(L) = [\mathbb{Z}^n : L] = \frac{M}{\gcd(M, a_1, a_2, \ldots, a_n)}.$$

12.2.4 Lattice Reduction

As previously noted, any lattice has a basis. In fact, any lattice of dimension 2 or more has infinitely many bases. However, some of these bases are "better" than others, in the sense that they give more compact descriptions of the lattice and make it easier to answer various questions about the lattice.

In the case of Euclidean vector spaces, it is usually convenient to work with orthogonal bases. Lattices, on the other hand, do not always admit orthogonal bases. Nevertheless, it can be shown that one can always find so-called *reduced bases*, consisting of "relatively short" vectors that are "not too far" from being orthogonal.

Giving precise definitions for such notions as "short vectors" and "reduced bases", showing that such vectors or such bases exist, and describing methods to find them is the purpose of the theory of lattice reduction.

Minkowski's Successive Minima

In order to explain what a reduced basis is, we need to define what is meant by short lattice vectors. Let L be a lattice of dimension ≥ 1 in $\mathbb{R}^n$. Since any closed hyperball centered at 0 contains finitely many vectors of L, there exists a nonzero vector in L of minimal length. The Euclidean norm of that shortest nonzero vector is called the *first minimum* of L, and is denoted by $\lambda_1(L) > 0$. Any vector $\mathbf{v}$ in L of norm $\lambda_1(L)$ is called a shortest vector of L. It is never unique, since $-\mathbf{v}$ is also a shortest vector.

For that reason, one must be careful when defining the *second-to-shortest* vector of a lattice. Minkowski [289] defined the other minima as follows. For all $1 \leq i \leq \dim(L)$, the ith *minimum* $\lambda_i(L)$ is defined as the minimum of $\max_{1\leq j\leq i} \|\mathbf{v}_j\|$ over all families of i linearly independent lattice vectors $\mathbf{v}_1, \ldots, \mathbf{v}_i \in L$. Clearly, the minima are increasing: $\lambda_1(L) \leq \lambda_2(L) \leq \cdots \leq \lambda_d(L)$. And it is not difficult to see that there always exist linearly independent lattice vectors $\mathbf{v}_1, \ldots, \mathbf{v}_d$ reaching simultaneously the minima, that is $\|\mathbf{v}_i\| = \lambda_i(L)$ for all i.

As a side note, perhaps surprisingly, it is not the case in general that such vectors form a basis of L. In fact, one can find lattices (of dimension ≥ 5) in which there is no basis at all formed by vectors of length equal to the successive minima.

Random Lattices

It is possible to give a relatively precise description of how Minkowski's minima behave "generically", namely for so-called random lattices. The precise definition of a random lattice is somewhat technical but the idea is that there is a natural way to pick a lattice "uniformly at random" (see, e.g., [6] for details).

It is proved in [7] that a random lattice of rank n satisfies asymptotically, with overwhelming probability,

$$\forall\, 1 \leq i \leq n, \quad \lambda_i(L) \approx \sqrt{\frac{n}{2\pi e}}\, \mathrm{vol}(L)^{1/n} \tag{12.1}$$

This means that all minima are close to each other and to the radius of an n-dimensional ball of volume $\mathrm{vol}(L)$.

Furthermore, [7] also shows that asymptotically, in a random n-rank lattice L, there exists with overwhelming probability a lattice basis $(\mathbf{b}_1, \ldots, \mathbf{b}_n)$ such that

$$\forall\, 1 \leq i \leq n, \quad \|\mathbf{b}_i\| \approx \sqrt{\frac{n}{2\pi e}}\, \mathrm{vol}(L)^{1/n}. \tag{12.2}$$

Such a basis consists of very short vectors, since their norms are close to the successive minima. Thus, random lattices typically have nice bases.

In applications, lattices are often not picked at random in the previous sense but constructed in a particular way. Such nonrandom lattices can sometimes behave quite differently from random lattices: for example, the first minimum can be arbitrarily small compared to $\mathrm{vol}(L)^{1/n}$ for lattices constructed in an ad hoc way. However, it is often a reasonable heuristic to infer the "typical" behavior of lattices in a given problem from the case of random lattices, and assume that the previous properties hold most often nonetheless.

Reduction Notions and LLL

We just mentioned that random lattices always have nice bases: what about general lattices? The goal of lattice reduction is to prove the existence of nice lattice bases in every lattice, and to describe procedures to find such bases. These nice bases are referred to as reduced.

There are multiple definitions of a reduced basis: usually, one first defines a notion of reduction, then shows that there exist bases which are reduced in this sense, and finally proves that bases which are reduced in this sense have interesting properties, mathematical (how short are the vectors of a reduced basis?) and computational (how easy is it to compute the basis?).

One classical notion of reduction, which will be enough for our purposes, is the Lenstra–Lenstra–Lovász reduction [251] (LLL for short). It is not very strong (in the sense that vectors of an LLL-reduced basis are not necessarily very short) but it is computationally inexpensive: we can compute LLL-reduced basis of lattices of relatively large dimension quickly.

We will not need a precise definition of LLL reduction, but the following few properties will come in handy. All lattices admit LLL-reduced bases, and any LLL-reduced basis $(\mathbf{b}_1, \ldots, \mathbf{b}_d)$ of a lattice L satisfies

1. $\|\mathbf{b}_1\| \leq 2^{(d-1)/4}(\mathrm{vol}\, L)^{1/d}$.
2. For all $1 \leq i \leq d$, $\|\mathbf{b}_i\| \leq 2^{(d-1)/2}\lambda_i(L)$.
3. $\|\mathbf{b}_1\| \times \cdots \times \|\mathbf{b}_d\| \leq 2^{d(d-1)/4}\, \mathrm{vol}\, L$.

The vectors of an LLL-reduced basis are thus at most exponentially far from the minima.

12.2.5 Lattice Problems in Practice

There are many computational problems related to lattices. Some of them are easy: for example, given bases of two lattices L_1, L_2 in $\mathbb{Z}^n$, it is easy to decide if $L_1 \subset L_2$, or to find a basis of $L_1 \cap L_2$. However, many lattice problems are believed to be computationally hard for lattices of high dimension.

The most famous lattice problem is the Shortest Vector Problem (SVP for short): given a basis of a lattice L, find $\mathbf{v} \in L$ such that $\|\mathbf{v}\| = \lambda_1(L)$. A related problem is the Closest Vector Problem (CVP for short): given a basis of a lattice L and a target vector $\mathbf{t} \in \mathbb{Z}^n$, find $\mathbf{v} \in L$ minimizing $\|\mathbf{t} - \mathbf{v}\|$, that is, such that $\|\mathbf{t} - \mathbf{v}\| \leq \|\mathbf{t} - \mathbf{w}\|$ for all $\mathbf{w} \in L$.

There are NP-hardness results for both SVP and CVP [285], so efficient algorithms for solving these problems in large lattice dimensions are unlikely to exist. The best known algorithms are exponential or worse [8, 210].

The best that can be done with more efficient algorithms is finding approximate solutions: for example, γ-SVP is the problem of finding a vector $\mathbf{v} \in L$ such that $\|\mathbf{v}\| \leq \gamma\lambda_1(L)$, and becomes easier as γ grows with respect to lattice dimension d. In particular, LLL and its variants provide polynomial algorithms for a^d-SVP (and the similarly defined a^d-CVP) for certain constants $a > 1$; thus, SVP and CVP can be approximated within an exponential factor in polynomial time.

In practice, in low dimension, say ≤ 100, as is the case in most instances of the attacks presented hereafter, the most important lattice problems become easy: for instance, exact SVP and CVP can be quickly solved using existing tools. The main reason is that lattice reduction algorithms behave better than their worst-case bounds: see, for instance, [306] for the case of LLL, and [152] for the case of BKZ.

However, when the lattice dimension becomes very high, it is difficult to predict experimental results in advance. Several factors seem to influence the result: the lattice dimension, the input basis, the structure of the lattice, and, in the case of CVP, the distance of the target vector to the lattice. What is always true is that one can quickly approximate SVP and CVP up to exponential factors with an exponentiation base very close to 1 (see [152] for concrete values of the exponentiation bases), but in high dimension, such exponential factors may not be enough for cryptanalytic purposes, depending on the application. If better approximation factors are required, one should perform experiments to see if existing algorithms are sufficient. If the lattice and the input basis are not exceptional, there is no reason to believe that exact SVP can be solved in very high dimension (say $\geq$400), although one can always give it a try. Furthermore, if the target vector is not unusually close to the lattice, there is also no reason to believe that exact CVP could be solved in very high dimension (say $\geq$400).

12.3 Fault Attacks on DSA Signatures

At PKC 2005, Naccache et al. [300] presented a fault attack against smart card implementations of the DSA signature standard. The attack combines a glitch attack on signature generation that produces DSA signatures where a few bits of the nonce are known, and a previously introduced lattice-based technique [304] that will recover the DSA secret key from sufficiently many such signatures with partially known nonces.

In practice, for DSA signatures using a 160-bit subgroup, it was found that 27 faulty signatures with one zeroed byte were sufficient to recover the secret key. Moreover, the technique generalized to other El Gamal-type signature schemes, such as ECDSA [305] and Schnorr signatures.

The attack is based on the well-known fact that the security of DSA signatures crucially relies on the nonce used in signature generation being chosen uniformly at random. Statistical biases can often be exploited for key recovery. This fact has been illustrated by many attacks, the latest example being Sony's PlayStation 3 gaming console: hackers have apparently been able to retrieve ECDSA secret keys [134] because some nonces have allegedly been reused.

12.3.1 The DSA Signature Scheme

DSA is an El Gamal-like signature algorithm specified by NIST in their Digital Signature Standard [141]. It can be described as follows.

Parameters and Keys

The system parameters are three integers p, q, and g where p and q are prime, q divides $p - 1$, and $g \in \mathbb{Z}_p^*$ has order q, as well as a cryptographic hash function H of output length equal to the byte size of q. The private key is an element $\alpha \in \mathbb{Z}_q^*$, and the public key is $\beta = g^\alpha \pmod{p}$.

Signature

To sign a message m, the signer picks k uniformly at random in $\{0, 1, \ldots, q-1\}$ and sets

$$r := (g^k \bmod p) \pmod q \quad \text{and} \quad s := \frac{H(m) + \alpha r}{k} \pmod q.$$

The signature is the pair (r, s).

Verification

The verifier considers (r, s) as a valid signature of m if

$$r \stackrel{?}{=} (g^{wh} \beta^{wh} \bmod p) \pmod q$$

where $h = H(m)$ and $w = 1/s (\bmod q)$.

Recommended Parameters

The original Digital Signature Standard required p to be between 512 and 1,024 bits long, q to be 160 bits long, and the hash function H to be SHA-1. The current version also allows parameter sizes of (2,048, 224), (2,048, 256) and (3,072, 256) with SHA-2 as a hash function.

12.3.2 Attack Model

Faulty signatures are valid DSA signatures (r_i, s_i) where the ℓ least significant bits of the corresponding nonces k_i are all 0.

Such faulty signatures are obtained in practice by causing a glitch in the signing device during the generation of the nonce: since k is generated by loading a series of random bytes into memory, fault injection makes it possible to skip part of the loop involved in that generation, resulting in clear least significant bits. A timing analysis of the power trace then makes it possible to check whether the generation has actually been faulty, by examining if it is shorter than a normal execution or not.

12.3.3 Description of the Attack

Given sufficiently many faulty signatures, the secret key α can be recovered using a technique based on lattices [185, 304]. The idea is to use the congruence

$$\alpha r \equiv sk - H(m) \pmod{q}.$$

This relation reveals nothing about α when k is chosen truly at random in $\mathbb{Z}_q$, but when some bits of k are known, some information about α is leaked since all other parameters are publicly known. Intuitively, knowing ℓ bits of k should reveal ℓ bits of information about α, so if q is N bits long, α should be recoverable from about N/ℓ faulty signatures. Moreover, the relation is affine in both α and k: that is why we can hope to attack the problem with lattices.

Let us describe in more details how one can take advantage of the leaked information. In our case, we know that the ℓ least significant bits of k are 0, so we can write $k = 2^\ell b$ for some integer $b \geq 0$. Writing $h = H(m)$, we get

$$\alpha r \cdot 2^{-\ell} s^{-1} \equiv b - h \cdot 2^{-\ell} s^{-1} \pmod{q}.$$

Now let

$$t = \lfloor r \cdot 2^{-\ell} s^{-1} \rceil_q \quad \text{and} \quad u = \lfloor -h \cdot 2^{-\ell} s^{-1} \rceil_q$$

where for any integer z, $\lfloor z \rceil_q$ denotes the only integer $0 \leq a < q$ such that $a \equiv z \pmod{q}$. Both t and u can be computed from public information, and the additional information we have, namely that $b < q/2^\ell$, can be expressed in terms of t and u as

$$0 \leq \lfloor \alpha t - u \rceil_q < q/2^\ell.$$

Therefore, if we denote by $|\cdot|_q$ the distance to $q\mathbb{Z}$, i.e. $|z|_q = \min_{a \in \mathbb{Z}} |z - aq|$, we have

$$|\alpha t - u - q/2^{\ell+1}|_q \leq q/2^{\ell+1}.$$

In other words, we know an approximation of αt modulo q:

$$|\alpha t - v/2^{\ell+1}|_q \leq q/2^{\ell+1}$$

where v is the integer $2^{\ell+1} u + q$.

Given a number of faulty signatures (r_i, s_i) of various messages m_i, say d of them, the same method yields pairs of integers (t_i, v_i) such that

$$|\alpha t_i - v_i/2^{\ell+1}|_q \leq q/2^{\ell+1}.$$

The goal is to recover α from this data. This problem is very similar to the hidden number problem considered by Boneh and Venkatesan in [58], and is approached by transforming it into a lattice closest vector problem.

More precisely, consider the $(d+1)$-dimensional lattice L spanned by the rows of the following matrix:

$$\begin{pmatrix} 2^{\ell+1}q & 0 & \cdots & 0 & 0 \\ 0 & 2^{\ell+1}q & \ddots & \vdots & \vdots \\ \vdots & \ddots & \ddots & 0 & \vdots \\ 0 & \ldots & 0 & 2^{\ell+1}q & 0 \\ 2^{\ell+1}t_1 & \ldots & \ldots & 2^{\ell+1}t_d & 1 \end{pmatrix}.$$

The inequality $|v_i/2^{\ell+1} - \alpha t_i|_q \leq q/2^{\ell+1}$ implies the existence of an integer c_i such that

$$|v_i - \alpha \cdot 2^{\ell+1}t_i - 2^{\ell+1}qc_i| \leq q. \qquad (12.3)$$

Note then that the row vector

$$\mathbf{c} = \left(\alpha \cdot 2^{\ell+1}t_1 + c_1 \cdot 2^{\ell+1}q, \ldots, \alpha \cdot 2^{\ell+1}t_d + c_d \cdot 2^{\ell+1}q, \alpha\right)$$

belongs to L, since it can be obtained by multiplying the last row vector by α and then subtracting appropriate multiples of the first d row vectors. Since the last coordinate of this vector discloses the hidden number α, we call $\mathbf{c}$ the *hidden vector*. The hidden vector is very close to the (publicly known) row vector $\mathbf{v} = (v_1, \ldots, v_d, 0)$. Indeed, by (12.3), the distance from $\mathbf{c}$ to $\mathbf{v}$ (in the Euclidean norm, say) is bounded by

$$\|\mathbf{v} - \mathbf{c}\| \leq q\sqrt{d+1}.$$

On the other hand, one has $\mathrm{vol}(L) = 2^{(\ell+1)d} \cdot q^d$. Therefore, assuming heuristically that L behaves like a random lattice, Eq. (12.1) suggests that the length of the shortest vector in L should be

$$\lambda_1(L) \approx \sqrt{\frac{d+1}{2\pi e}}\,\mathrm{vol}(L)^{1/d+1} = \sqrt{\frac{d+1}{2\pi e}} \cdot 2^{(\ell+1)d/(d+1)}q^{d/(d+1)}.$$

If $\|\mathbf{v} - \mathbf{c}\|$ is much shorter than this length, we expect $\mathbf{c}$ to be the closest vector to $\mathbf{v}$ in L. This is realized when

$$q\sqrt{d+1} \ll \sqrt{\frac{d+1}{2\pi e}} \cdot 2^{(\ell+1)d/(d+1)}q^{d/(d+1)}$$

or equivalently

$$q \ll \left(\frac{2^\ell}{\sqrt{\pi e/2}}\right)^d.$$

If q is N bits long, this is satisfied for

$$d \gtrsim \frac{N}{\ell - \log_2 \sqrt{\pi e/2}}.$$

Provided that we have that many faulty signatures, if we can find the closest vector to $\mathbf{v}$ in the lattice L, it is heuristically likely to be the hidden vector $\mathbf{c}$, which makes it possible to recover the private key α.

For $N = 160$ and $\ell = 8$ (one byte zeroed out), we get $d \gtrsim 23$. Practical experiments in [300] indicate that, in this setting, α is retrieved with probability about 10 % for $d = 23$, 63 % for $d = 25$ and 100 % for $d = 27$. In all those cases, the lattice dimension is small enough that solving the closest vector problem is easy in practice.

The above estimate is heuristic, but it is possible to give parameters for which attacks of this kind can be proved rigorously [304]. Such attacks apply to DSA but also to any El Gamal-type signature scheme (see, for instance, [305] for the case of ECDSA).

12.3.4 Proposed Countermeasures

Contrary to what happens in many other fault attacks, these faulty DSA signatures are actually valid signatures. In particular, checking their validity after signature generation will not help thwart the attack.

Several countermeasures are proposed in [300] to prevent the attacker from successfully generating signatures with partially zeroed nonces, including execution randomization (e.g. randomize the order in which the buffer containing the nonce is filled, using a linear congruential generator) and repeated refreshment (e.g. generate several random numbers with random delay in between, and use the XOR of these values as the nonce).

12.4 Fault Attacks on Randomized RSA Signatures

One of the best-known examples of a fault attack is the so-called Bellcore attack [56, 199] on RSA-CRT signatures. Let us first recall it succinctly.

To sign a message m with RSA, a signer computes $\sigma = \mu(m)^d \bmod N$, where N is the public modulus, d is the private exponent and μ is a certain encoding function. Since this computation is time-consuming, it is very common to use the Chinese Remainder Theorem (CRT) to obtain a four fold speedup by first evaluating

$$\sigma_p = \mu(m)^{d \bmod p-1} \pmod p \quad \text{and} \quad \sigma_q = \mu(m)^{d \bmod q-1} \pmod q$$

and then deducing σ from σ_p and σ_q with the CRT.

This method is vulnerable to fault attacks: if an attacker can inject a fault during the computation of σ_q, the whole computation will produce a faulty signature σ' satisfying

$$\sigma' \equiv \mu(m)^d \pmod p \quad \text{and} \quad \sigma' \not\equiv \mu(m)^d \pmod q,$$

which allows the attacker to factor N by computing

$$\gcd\left((\sigma')^e - \mu(m) \bmod N, N\right) = p.$$

Clearly, Boneh et al.'s fault attack applies to any deterministic RSA encoding μ, such as the Full Domain Hash [29], where μ is a full-length hash function. The attack is also applicable to probabilistic signature schemes where the random nonce used to generate the signature is sent along with the signature.

However, if the nonce is only recovered when verifying the signature, or if some part of the message is unknown, the attack does not apply directly. For example, with a probabilistic encoding of the form $\mu(m) = m\|r$, where a random nonce r is simply concatenated with the message m, the nonce r is only recovered when verifying a correct signature. Given a faulty signature σ', the attacker cannot retrieve r nor infer $(m\|r)$ which would be necessary to compute $\gcd\left((\sigma')^e - \mu(m) \bmod N, N\right) = p$. It is not advisable to use this particular toy encoding (for example, one can clearly forge signatures if r is allowed to be as large as $N^{1-1/e}$), but more serious probabilistic encodings in which the randomness is recovered as part of signature verification (such as PSS [31]) are similarly immune to the Bellcore attack.

Coron et al. [104] have shown how, for certain padding schemes, the case of unknown nonces or partially unknown messages can be tackled nonetheless if the unknown part is not too large. The attack uses a technique by Herrmann and May [179] for finding small roots of linear equations modulo an unknown factor p of a public modulus N. This technique is in turned based on Coppersmith's lattice-based method for finding small roots of polynomials [101].

The main application considered by Coron et al. was RSA signatures using the ISO/IEC 9797-2 standard padding scheme [191], where the message is partially unknown, a common situation in smart card applications, especially EMV smart cards [132]. The ISO/IEC 9797-2 encoding function is of the form

$$\mu(m) = \texttt{6A}_{16} \,\|\, m[1] \,\|\, H(m) \,\|\, \texttt{BC}_{16}$$

where $m = m[1]\|m[2]$ is split in two parts, and H is a hash function, typically SHA-1. Only $m[2]$ is transmitted together with the signature $\sigma = \mu(m)^d \bmod N$. To recover the whole message and verify the signature, the recipient first computes $\mu(m) = \sigma^e \bmod N$, from which he deduces $m[1]$, and then checks that the hash value $H(m)$ is correct.

In the cases considered in [104], the message prefix $m[1]$ is partially known to a fault attacker (because messages are formatted in a particular way, as is common in applications) but some variable part is unknown, and the hash value $H(m)$ is unknown, as well as is a result. Yet, if the unknown part of $m[1]$ is not too large (up to about 160 bits for a 2,048-bit modulus), then, in Boneh et al.'s attack, the private key can be recovered with a single fault.

In addition, the same paper presents an extension of this attack to the case when multiple faults are available, which makes it theoretically possible to deal with larger unknown message parts. However, this variant involves lattice reduction in dimensions exponential in the number of faults (using heuristic, multivariate versions of Coppersmith's method), and becomes quickly impractical.

Later, Coron et al. presented another multiple fault attack in the same setting [108] eschewing Coppersmith-like techniques entirely. It is based on orthogonal lattices, as used in earlier cryptanalytic results such as [307, 308]. The lattice sizes involved are moderate and make the attack quite practical for relatively large unknown message parts. For a particular EMV use case highlighted in [108], and well beyond the reach of the attack in [104], ten faulty signatures are sufficient to recover the private key almost instantaneously.

12.4.1 The ISO/IEC 9797-2 Signature Scheme

ISO/IEC 9797-2 is an encoding standard allowing partial message recovery [190, 191], and is used in many embedded applications, including EMV smart cards [132]. The encoding can be used with hash functions H of various digest sizes k_h. When k_h, the message size and the size of N (denoted k) are all multiples of 8, the ISO/IEC 9797-2 encoding of a message $m = m[1] \parallel m[2]$ is

$$\mu(m) = \texttt{6A}_{16} \parallel m[1] \parallel H(m) \parallel \texttt{BC}_{16}$$

where $m[1]$ consists of the $k - k_h - 16$ leftmost bits of m and $m[2]$ represents the remaining bits of m. In [191] it is required that $k_h \geq 160$.

The ISO/IEC 9797-2 encoding itself is deterministic, but it is often used with messages containing parts which are either secret or picked at random upon signature generation. This is particularly the case in EMV smart cards. The message to be signed can be put in the following general form: $m = m[1] \parallel m[2]$ with

$$m[1] = \alpha \parallel r \parallel \alpha' \quad \text{and} \quad m[2] = \text{DATA}$$

where r is an unknown bit string (e.g. a random nonce), α and α' are bit strings known to the attacker, and DATA is a possibly unknown bit string. Thus, the ISO/IEC 9797-2-encoded message is

$$\mu(m) = \texttt{6A}_{16} \parallel \alpha \parallel r \parallel \alpha' \parallel H(\alpha \parallel r \parallel \alpha' \parallel \text{DATA}) \parallel \texttt{BC}_{16}$$

where both r and the hash value are unknown. In typical use cases, the hash value is a 160-bit long SHA-1 digest, and N is an RSA modulus of around 1,024 bits. The unknown string r can be of various sizes depending on the nature of the signed message; the shorter r is, the easier the attack becomes.

More generally, we will consider encoded messages of the form

$$\mu(m) = A + B \cdot x_0 + C \cdot y_0$$

where B and C are public constants (in the ISO/IEC 9797-2 case, suitable powers of 2), A is a number representing the known part of the encoding, and x_0 and y_0 are the unknown parts (equal to r and $H(m)$ respectively). The signature is then computed as $\sigma = \mu(m)^d \bmod N$ as usual, using the CRT.

12.4.2 Attack Model

A fault is injected into the exponentiation modulo q part of the RSA-CRT signature generation, resulting in a faulty signature σ satisfying

$$\sigma^e \equiv A + B \cdot x_0 + C \cdot y_0 \pmod{p} \quad \text{and} \quad \sigma^e \not\equiv A + B \cdot x_0 + C \cdot y_0 \pmod{q}.$$

Dividing by B and subtracting the left-hand side, the faulty signature yields a relation of the form

$$a + x_0 + c \cdot y_0 \equiv 0 \pmod{p} \quad \text{and} \quad a + x_0 + c \cdot y_0 \not\equiv 0 \pmod{q}$$

where $a = B^{-1}(A - \sigma^e) \bmod N$ is a value known to the attacker, and $c = B^{-1} \cdot C \bmod N$ is a public constant.

In other words, the fault attack provides the attacker with an integer a such that for a certain unknown pair (x_0, y_0) of bounded size, the following holds:

$$a + x_0 + c \cdot y_0 \equiv 0 \pmod{p} \quad \text{and} \quad a + x_0 + c \cdot y_0 \not\equiv 0 \pmod{q}.$$

We will write the bounds on x_0 and y_0 as $0 \le x_0 < N^\gamma$ and $0 \le y_0 < N^\delta$. The total fraction of unknown bits in the encoded message is thus $\gamma + \delta$.

Intuitively, in both of the attacks described below, the attacker takes advantage of the affine relation in x_0 and y_0 mod p to recover the unknown part with lattice techniques, and use it to factor N.

In practice, the fault can be injected using, for example, voltage spikes during the computation modulo q (see [104, Sect. 4]).

The resulting value modulo q (of the faulty signature, or equivalently of $a+x+cy$) is usually modeled as a random element in $\mathbb{Z}_q$; this is the random fault model.

12.4.3 Single-Fault Attack

After a single fault, the attacker obtains a bivariate linear polynomial $f(x, y) = a + x + c \cdot y$ such that $f(x_0, y_0) \equiv 0 \pmod{p}$ and $f(x_0, y_0) \not\equiv 0 \pmod{q}$ for some unknown pair (x_0, y_0) satisfying known bounds. They can then apply the following Coppersmith-like result by Herrmann and May.

Theorem 12.1 (Herrmann–May [179]) *Let N be a sufficiently large composite integer with a divisor $p \geq N^{\beta}$. Let $f \in \mathbb{Z}[x, y]$ be a bivariate linear polynomial. Assume that $f(x_0, y_0) = 0 \bmod p$ for some (x_0, y_0) such that $|x_0| \leq N^{\gamma}$ and $|y_0| \leq N^{\delta}$. Then for any $\varepsilon > 0$, under the condition*

$$\gamma + \delta \leq 3\beta - 2 + 2(1 - \beta)^{3/2} - \varepsilon \tag{12.4}$$

one can find linearly independent polynomials $h_1, h_2 \in \mathbb{Z}[x, y]$ such that $h_1(x_0, y_0) = h_2(x_0, y_0) = 0$ over $\mathbb{Z}$, in time polynomial in $\log N$ *and* ε^{-1}.

In our case, N is a (presumably balanced) RSA modulus, so $\beta = 1/2$ and condition (12.4) becomes

$$\gamma + \delta < \frac{\sqrt{2} - 1}{2} \approx 0.207. \tag{12.5}$$

Under that condition, we obtain two linearly independent polynomials $h_1, h_2 \in \mathbb{Z}[x, y]$ such that (x_0, y_0) is a root of both.

Suppose further that h_1, h_2 are in fact *algebraically* independent. It is then easy to compute the finite list of their common roots, e.g. by taking the resultant with respect to y and solving for x, and thus to find (x_0, y_0). Once (x_0, y_0) is computed, one can proceed as in the Bellcore attack to factor N:

$$\gcd\big(f(x_0, y_0) \bmod N, N\big) = p.$$

The algebraic independence assumption makes the attack heuristic, but the experimental validation carried out in [104] suggests that it works well for $\gamma + \delta$ not too large. For example, with a 2,048-bit modulus and 160-bit hash function, a 158-bit long nonce r is recovered in less than a minute with a single fault. The total fraction of unknown message bits is then $\gamma + \delta = (158 + 160)/2{,}048 \approx 0.155$.

However, condition (12.5) is quite restrictive, especially when N is small. Indeed, for a 1,024-bit modulus with a 160-bit hash function, the theoretical maximum nonce size that can be recovered is 58 bits, and the algorithm is only manageable in practice for much smaller sizes. As a result, the attack performs worse than exhaustive search for r for this modulus size.

To tackle larger unknown message parts, it is necessary to take advantage of several faulty signatures. This Coppersmith-based attack admits a natural extension to multiple faults [104, Sect. 2.4], but the lattice dimensions involved grow exponentially with the number of faulty signatures, so that the extension is only practical for a very small number of faults, and does not extend the scope of the single-fault attack by a large margin.

12.4.4 Multiple-Fault Attack

We now describe the multiple-fault attack of [108], which uses lattices in a very different way. We are now given ℓ faulty signatures, and in particular a vector $\mathbf{a} = (a_1, \ldots, a_\ell)$ of integers such that, for two short unknown vectors $\mathbf{x} = (x_1, \ldots, x_\ell)$ and $\mathbf{y} = (y_1, \ldots, y_\ell)$, we have:

$$\mathbf{a} + \mathbf{x} + c \cdot \mathbf{y} \equiv \mathbf{0} \pmod{p}. \tag{12.6}$$

The attack then proceeds in three steps.

Linearization

The first step is to eliminate $\mathbf{a}$ in (12.6) to obtain a linear relation. To do so, we try to find vectors $\mathbf{u}_j$ orthogonal to $\mathbf{a}$ modulo N, and take the inner products of (12.6) with the $\mathbf{u}_j$'s.

The set $L_{\mathbf{a}} = \{\mathbf{u} \in \mathbb{Z}^\ell : \langle \mathbf{a}, \mathbf{u} \rangle \equiv 0 \pmod{N}\}$ of vectors $\mathbf{u}$ orthogonal to $\mathbf{a}$ modulo N is a lattice in $\mathbb{Z}^\ell$ of rank ℓ and volume N, as seen in Sect. 12.2.3. It is possible to find a reduced basis of it by standard orthogonal lattice techniques [307] as follows. Denote by $L_{\mathbf{a}}^{(0)}$ the lattice in $\mathbb{Z}^{\ell+1}$ generated by the rows of the matrix

$$\begin{pmatrix} \kappa N & 0 & \cdots & \cdots & 0 \\ \kappa a_1 & 1 & 0 & \cdots & 0 \\ \kappa a_2 & 0 & 1 & \ddots & \vdots \\ \vdots & \vdots & \ddots & \ddots & 0 \\ \kappa a_\ell & 0 & \cdots & 0 & 1 \end{pmatrix} \tag{12.7}$$

where κ is a large scaling constant. Now for any vector $\mathbf{u}^{(0)} = (u_0, u_1, \ldots, u_\ell)$ in this lattice, one has $u_0 = \kappa(a_1 u_1 + \cdots + a_\ell u_\ell + mN)$ for some integer m. In particular, if $|u_0| < \kappa$, then $u_0 = 0$, and $\mathbf{u} = (u_1, \ldots, u_\ell)$ must satisfy $\langle \mathbf{a}, \mathbf{u} \rangle \equiv 0 \pmod{N}$. Thus, any vector $\mathbf{u}^{(0)}$ in $L_{\mathbf{a}}^{(0)}$ whose norm is less than κ gives rise to a vector $\mathbf{u} \in L_{\mathbf{a}}$. Conversely, if $\mathbf{u}$ is a point in $L_{\mathbf{a}}$, then $\mathbf{u}^{(0)} = (0, u_1, \ldots, u_\ell)$ is in $L_{\mathbf{a}}^{(0)}$.

Since it admits $N\mathbb{Z}^\ell$ as a sublattice, the lattice $L_{\mathbf{a}}$ in $\mathbb{Z}^\ell$ contains a linearly independent family consisting of ℓ vectors of length at most N. Taking the corresponding vectors in $L_{\mathbf{a}}^{(0)}$, this implies that $\lambda_j(L_{\mathbf{a}}^{(0)}) \leq N$ for $1 \leq j \leq \ell$.

Consider then an LLL-reduced basis $(\mathbf{u}_1^{(0)}, \ldots, \mathbf{u}_{\ell+1}^{(0)})$ of $L_{\mathbf{a}}^{(0)}$. Recall from Sect. 12.2.4 that a property of LLL reduction is that for $1 \leq j \leq \ell$,

$$\|\mathbf{u}_j^{(0)}\| \leq 2^{\ell/2} \lambda_j(L_{\mathbf{a}}^{(0)}) \leq 2^{\ell/2} N.$$

Hence, if one chooses κ large enough, all vectors $\mathbf{u}_j^{(0)}$, $1 \le j \le \ell$, are of norm less than κ and give rise to independent vectors $\mathbf{u}_j$ that are orthogonal to $\mathbf{a}$ modulo N (it is in fact an LLL-reduced basis of the lattice of such orthogonal vectors). In particular, taking the inner product of (12.6) with $\mathbf{u}_j$, we get, for $1 \le j \le \ell$,

$$\langle \mathbf{x}, \mathbf{u}_j \rangle + c\langle \mathbf{y}, \mathbf{u}_j \rangle \equiv 0 \pmod{p}. \tag{12.8}$$

Moreover, we can give a heuristic estimate the size of the vectors $\mathbf{u}_j$. Assuming that $L_{\mathbf{a}}$ behaves like a random lattice, the length of these vectors should be around $\sqrt{\ell/2\pi e} \cdot N^{1/\ell}$ since $\mathrm{vol}(L_{\mathbf{a}}) = N$. Neglecting the small multiplicative factor, we can heuristically expect that $\|\mathbf{u}_j\| \lesssim N^{1/\ell}$.

Orthogonalization

For all j, let $\alpha_j = \langle \mathbf{x}, \mathbf{u}_j \rangle$ and $\beta_j = \langle \mathbf{y}, \mathbf{u}_j \rangle$. We expect to have, for all j except the last few, $|\alpha_j| \lesssim N^{\gamma+1/\ell}$ and $|\beta_j| \lesssim N^{\delta+1/\ell}$. Now recall from (12.8) that

$$\alpha_j + c \cdot \beta_j \equiv 0 \pmod{p}.$$

If $(\alpha_j, \beta_j) \ne (0, 0)$, this amounts to writing $-c$ as the modular ratio $\alpha_j/\beta_j \bmod p$. But this is only possible in general if the sum of the sizes of α_j and β_j is at least the size of p. In other words, if $N^{\gamma+1/\ell} \cdot N^{\delta+1/\ell} \ll p$, that is

$$\gamma + \delta + \frac{2}{\ell} \lesssim \frac{1}{2}, \tag{12.9}$$

we should have $\alpha_j = \beta_j = 0$ for all j except perhaps the last few. This means that $\mathbf{u}_j$ is orthogonal to $\mathbf{x}$ and $\mathbf{y}$ in $\mathbb{Z}^\ell$.

Assume that this does in fact hold for $1 \le j \le \ell - 2$, and consider the lattice of vectors in $\mathbb{Z}^\ell$ orthogonal to $\mathbf{u}_j$ for $1 \le j \le \ell - 2$. This is a bidimensional lattice containing $\mathbf{x}$ and $\mathbf{y}$, and we can find a basis $(\mathbf{x}', \mathbf{y}')$ of it with LLL. This is done by computing the LLL-reduction of the following lattice in $\mathbb{Z}^{\ell-2+\ell}$, generated by the rows of

$$\begin{pmatrix} \kappa' u_{1,1} & \cdots & \kappa' u_{\ell-2,1} & 1 & & 0 \\ \vdots & & \vdots & & \ddots & \\ \kappa' u_{1,\ell} & \cdots & \kappa' u_{\ell-2,\ell} & 0 & & 1 \end{pmatrix}$$

where κ' is a suitably large constant [307].

Factoring

Finally, using LLL again as in the linearization step, we can construct a short vector $\mathbf{v}$ orthogonal to both $\mathbf{x}'$ and $\mathbf{y}'$ modulo N. Then $\mathbf{v}$ is also orthogonal mod N to all $\mathbb{Z}$-linear combinations of $\mathbf{x}'$ and $\mathbf{y}'$, in particular $\mathbf{x}$ and $\mathbf{y}$. Therefore, taking the inner product of (12.6) with $\mathbf{v}$, we get

$$\langle \mathbf{a}, \mathbf{v} \rangle \equiv 0 \pmod{p}$$

and we can thus factor N as required:

$$\gcd(\langle \mathbf{a}, \mathbf{v} \rangle, N) = p.$$

Summary and Experiments

If condition (12.9) is satisfied, the following algorithm should heuristically find a nontrivial factor of N:

1. Find an LLL-reduced basis $(\mathbf{u}_1, \ldots, \mathbf{u}_\ell)$ of the lattice of vectors in $\mathbb{Z}^\ell$ orthogonal to $\mathbf{a}$ modulo N.
2. Find a basis $(\mathbf{x}', \mathbf{y}')$ of the lattice of vectors in $\mathbb{Z}^\ell$ orthogonal to $\mathbf{u}_j$, $1 \leq j \leq \ell - 2$.
3. Find a short vector $\mathbf{v}$ orthogonal to $\mathbf{x}'$ and $\mathbf{y}'$ modulo N (for example, the first vector of an LLL-reduced basis of the lattice formed by such orthogonal vectors).
4. Return $\gcd(\langle \mathbf{a}, \mathbf{v} \rangle, N)$.

The attack is heuristic because it might be the case that $\mathbf{u}_j$ is not actually orthogonal to $\mathbf{x}$ and $\mathbf{y}$ in $\mathbb{Z}^\ell$ for all $j \leq \ell - 2$.

Experimentally, however, success rates are very high when condition (12.9) is verified. For example, for $\gamma + \delta = 0.33$, condition (12.9) says $\ell \gtrsim 12$. Simulations from [108] with a 1,024-bit balanced RSA modulus N find a 13 % success rate for $\ell = 12$ and a 100 % success rate for $\ell = 13$. Note by the way that a total unknown message part of $\gamma + \delta = 0.33$ is out of reach of the single-fault attack from [104], and its Coppersmith-based multi-fault variant is totally inapplicable for such parameter sizes, whereas this attack is computationally very easy.

12.4.5 Proposed Countermeasures

Since faulty signatures are invalid, a possible countermeasure is to check the signature after generation. Since signature verification is significantly cheaper than signature generation in cases of interest (because of a low public exponent), this is a reasonable option, and preferable to incurring the four fold performance penalty of not using the Chinese Remainder Theorem at all. More generally, other usual countermeasures against RSA-CRT fault attacks, as proposed, for example, by Shamir [201, 372] or

Giraud [162], apply to this setting. However, see [226] for an indication that such strategies may prove ineffective in more elaborate attack models.

Furthermore, it should be noted that ad hoc signature paddings have no proof of security even against standard attacks. The ISO/IEC 9797-2 padding scheme, in particular, has a number of known vulnerabilities [107, 109]. It may be advisable to use a probabilistic RSA signature scheme like PSS [31] instead, which is actually secure in the random fault model considered here, as proved by Coron and Mandal [106].

Chapter 13
Fault Attacks on Pairing-Based Cryptography

Nadia El Mrabet, Dan Page and Frederik Vercauteren

Abstract Over the last ten years, the use of bilinear maps or "pairings" as building block primitives within cryptographic schemes has become commonplace. This trend has been supported by insight into their security properties and methods for efficient evaluation; the latter aspect has provided results that now allow even embedded devices to execute pairing-based schemes. However, this raises questions relating to physical security in the same way as for RSA- and ECC-based schemes. Specifically, the secure deployment of a pairing-based scheme necessitates the study of related fault attacks. This chapter attempts to survey the state of the art in this respect; it aims to describe the main results in this area, and give an overview of potential countermeasures.

13.1 Introduction

Fault Attacks

The rich suite of physical attack techniques now available represents a clear and present danger to devices that execute cryptography; various real-world examples of their threat include recent attacks on the KeeLoq range of RFID devices used for

N. El Mrabet
Department of Computer Science, Université de Caen,
Caen, France

D. Page (✉)
Department of Computer Science, University of Bristol,
Bristol, UK
e-mail: page@cs.bris.ac.uk

F. Vercauteren
Department of Electrical Engineering, Katholieke Universiteit Leuven,
Leuven, Belgium

M. Joye and M. Tunstall (eds.), *Fault Analysis in Cryptography*,
Information Security and Cryptography, DOI: 10.1007/978-3-642-29656-7_13,

automotive and building access control [130]. Such attacks are often categorized as follows:

1. a side-channel attack passively monitors execution, and recovers information through analysis of collected profiles, and
2. a fault attack actively influences execution and hopes to recover information as a result of unintended behavior.

Typically the goal is to recover security-critical information embedded in the device; this might mean recovery of key material that subsequently allows the attacker to construct a clone device, for example.

Although the field of side-channel analysis now includes remote methods for execution monitoring (e.g., network-based attacks on SSL [74]), fault attacks require physical access to the device under attack. On the one hand, with such access (potentially after the device is depackaged) fault induction methods are numerous [21]. For example, one might try to "glitch" the clock signal to alter execution control flow, or alter memory content by exposing it to intense light or a laser beam; in either case, the fault that results may be permanent or transient. On the other hand, the same requirement limits the class of devices that fault attacks can target. Typically, this class is limited to embedded and mobile devices, the quintessential example being smart cards that are carried into, and used within, an adversarial environment.

Pairing-Based Cryptography

First mooted by Shamir [371] in 1984, identity-based cryptographic schemes represent a conceptually ideal partner for smart cards since the latter are often used as identity-aware tokens. The seminal Identity Based Encryption (IBE) scheme of Boneh and Franklin [57] functions by making constructive use of bilinear maps, or "pairings"; this approach is now popular within a wider context.

We expand on the details later, but at a high level a pairing

$$e : \mathbb{G}_1 \times \mathbb{G}_2 \longrightarrow \mathbb{G}_T$$

simply maps inputs from the additive groups $\mathbb{G}_1$ and $\mathbb{G}_2$ onto a result in the multiplicative group $\mathbb{G}_T$. A crucial feature of said mapping is the bilinear property that ensures

$$e(a \cdot A, b \cdot B) = e(A, B)^{a \cdot b}.$$

Efficient realizations parametrize $\mathbb{G}_1$ and $\mathbb{G}_2$ using an elliptic curve, and $\mathbb{G}_T$ using a finite field; note that parametrization should ensure intractable Discrete Logarithm Problems (DLPs) in all groups, in part meaning that inversion of the pairing [150] is also intractable.

Given two inputs, the evaluation of a pairing is typically performed via an algorithm derived from the so-called "Miller loop" [286]; for an overview of related optimizations, see the description of Scott [364]. Two features are pertinent:

1. the security-critical information in a pairing-based cryptographic scheme is often one of the inputs to the pairing, and
2. a suitably implemented pairing uses no data-dependent branches (or at least none based on security-critical information).

Pairing evaluation is now computationally efficient enough to be deployed on smart cards [366], thus posing the question of resilience to fault attacks that take the above features into account.

Chapter Overview

A wide range of fault attacks against Elliptic Curve Cryptography (ECC) exist [44, 54, 90, 143]; Otto [315, Chap. 4] gives a concise overview. However, although ECC underpins the concrete use of pairing-based cryptography (by virtue of using elliptic curve groups), direct application of ECC-oriented attacks is usually ineffective. In part, this is because said attacks focus on recovering a scalar multiplier; in the context of pairings, the security-critical information is instead an input point. As such, one is typically required to focus on the security of pairing evaluation itself rather than on scalar multiplication.

This approach presents both a problem and an opportunity. The opportunity for attack is made wider by the complex parametrizations and algorithms used in pairing-based cryptography. One can, for example, easily identify the potential for faults in

1. precomputed values or parameters, for example, group parameters such as the order or generator,
2. inputs to the pairing, for example, the input points, and may not reside in the correct group,
3. intermediate values, for example, the so-called "Miller variable", which acts as an accumulator during the pairing computation.

Our focus in this chapter is fault attacks on pairing-based cryptography: we aim to describe the state of the art in relation to the challenges outlined above, and describe open problems and issues. Note that we strictly consider fault attacks only, i.e., numerous results relating to side-channel attack are, although sometimes related, out of our scope.

13.2 Background and Notation

Let E be an elliptic curve over a finite field $\mathbb{F}_q$, with $\mathcal{O}$ denoting the identity element of the associated group of rational points $E(\mathbb{F}_q)$. For a positive integer $r | \#E(\mathbb{F}_q)$ coprime to q, let $\mathbb{F}_{q^k}$ be the smallest extension field of $\mathbb{F}_q$ which contains the rth roots of unity in $\overline{\mathbb{F}}_q$; the extension degree k is called the security multiplier or embedding degree. Let $E(\mathbb{F}_q)[r]$ denote the subgroup of $E(\mathbb{F}_q)$ of all points of order dividing r, and similarly for the degree k extension of $\mathbb{F}_q$.

To support a description of the attacks we outline parametrizations of, and algorithms for several, concrete pairings. However, we let

$$e : \mathbb{G}_1 \times \mathbb{G}_2 \longrightarrow \mathbb{G}_T$$

generically denote such a pairing, whose specific type is determined by the context; $\overline{e}$ denotes the same pairing wherein some fault occurs or is injected. The two groups $\mathbb{G}_1$ and $\mathbb{G}_2$ will be subgroups of elliptic curve groups, whereas $\mathbb{G}_T$ is a subgroup of the multiplicative group of a finite field.

All currently known pairings evaluate a combination of so-called "Miller functions", written $f_{n,P}(\cdot)$ for a given scalar n and a point P, in either $\mathbb{G}_1$ or $\mathbb{G}_2$. Such a function is defined as a ratio of two bivariate polynomials, and can be evaluated at an elliptic curve point Q to yield a resulting finite field element. As is the case for ratios of univariate polynomials, such a function is uniquely determined (up to multiplication by a constant) by its zeroes and poles and their respective multiplicities. This information is precisely contained in the divisor of a function, namely the formal sum of the zeroes and poles with multiplicities. A Miller function $f_{n,P}$ satisfies

$$\text{div}(f_{n,P}) = n(P) - ([n]P) - (n-1)\mathcal{O},$$

i.e., the function has a zero at P of order n, a pole of order 1 at the point $[n]P$ and a pole of order $n-1$ at $\mathcal{O}$.

Miller [286] described an efficient procedure to compute any function $f_{n,P}$ and evaluate it at an elliptic curve point. The crucial ingredient is a simple "double-and-add" strategy based on the following observation: let i and j be positive integers; then

$$f_{i+j,P} = f_{i,P} \cdot f_{j,P} \cdot \frac{l_{[i]P,[j]P}}{v_{[i+j]P}}, \tag{13.1}$$

where $l_{[i]P,[j]P}$ is the equation of the line through $[i]P$ and $[j]P$ (or the tangent line when $[i]P = [j]P$), and $v_{[i+j]P}$ is the equation of the vertical line through $[i+j]P$. The resulting algorithm is described in Algorithm 13.1. Scott [364] gives an overview various approaches to optimization of the algorithm, an important example being the denominator elimination approach due to Barreto et al. [25]; this allows removal (assuming a suitable parametrization) of the computationally expensive inversions otherwise required during updates to the Miller variable f.

13.2.1 The Weil Pairing

The Weil pairing was introduced by Weil [287, 418] in 1940 in his proof of the Riemann hypothesis for curves. The simplified definition is as follows: let E be an elliptic curve over $\mathbb{F}_q$ and let $P, Q \in E(\mathbb{F}_{q^k})[r]$ with $P \neq Q$; then

Algorithm 13.1: Miller's algorithm to compute $f_{n,P}(Q)$

Input: $P = (x_P, y_P) \in \mathbb{G}_1$, $Q = (x_Q, y_Q) \in \mathbb{G}_2$ and $n = \sum_{i=0}^{s} n_i 2^i$
Output: $f_{n,P}(Q) \in \mathbb{F}_{q^k}$

$f \leftarrow 1$
$T \leftarrow P$
for $i = s - 1$ **downto** 0 **do**
  $f \leftarrow f^2 \cdot l_{T,T}(Q)/v_{2 \cdot T}(Q)$
  $T \leftarrow [2]T$
  if $n_i = 1$ **then**
    $f \leftarrow f^2 \cdot l_{T,P}(Q)/v_{T+P}(Q)$
    $T \leftarrow T + P$
  end
end
return f

$$e_W(P, Q) = (-1)^r \frac{f_{r,P}(Q)}{f_{r,Q}(P)}.$$

The most important properties of the Weil pairing are that the pairing is bilinear, nondegenerate, i.e., for every non-zero $P \in E(\mathbb{F}_{q^k})[r]$, there exists $Q \in E(\mathbb{F}_{q^k})[r]$ such that $e_W(P, Q) \neq 1$, and that it is reduced, i.e., $e_W(P, Q)$ is an rth root of unity in $\mathbb{F}_{q^k}$. Note that the Weil pairing requires two evaluations of Miller functions.

13.2.2 The Tate Pairing

The Tate pairing [396], or more correctly the Tate-Lichtenbaum [145, 146, 255] pairing, is defined as follows: let $P \in E(\mathbb{F}_{q^k})[r]$ and $Q \in E(\mathbb{F}_{q^k})/rE(\mathbb{F}_{q^k})$ with $P \neq Q$; then

$$e_T(P, Q) = f_{r,P}(Q) \cdot (\mathbb{F}^*_{q^k})^r.$$

The result of the Tate pairing is only defined up to an rth power. In order to obtain a unique value, the output is raised to the power $(q^k - 1)/r$, which results in the reduced Tate pairing, namely

$$\hat{e}_T(P, Q) = f_{r,P}(Q)^{\frac{p^k-1}{r}}.$$

The computation of $f_{r,P}(Q)$ is typically called the Miller loop, whereas the subsequent computation of $f_{r,P}(Q)^{\frac{p^k-1}{r}}$ is called the final powering (or final exponentiation).

The main properties are similar to those of the Weil pairing. All other pairings discussed in this chapter are simply efficient methods to compute a fixed power of the

Algorithm 13.2: The Duursma–Lee algorithm

Input: $P = (x_P, y_P) \in \mathbb{G}_1$ and $Q = (x_Q, y_Q) \in \mathbb{G}_2$
Output: $e(P, Q) \in G_T$

$f \leftarrow 1$
for $i = 1$ **upto** m **do**
  $x_P \leftarrow x_P^3,\ y_P \leftarrow y_P^3$
  $\mu \leftarrow x_P + x_Q + b$
  $\lambda \leftarrow -y_P y_Q \sigma - \mu^2$
  $g \leftarrow \lambda - \mu\rho - \rho^2$
  $f \leftarrow f \cdot g$
  $x_Q \leftarrow x_Q^{1/3},\ y_Q \leftarrow y_Q^{1/3}$
end
return f^{q^3-1}

reduced Tate pairing when the arguments are restricted to a cleverly chosen domain, i.e., the eigenspaces of Frobenius.

Duursma and Lee [129] introduced an approach to evaluation using a family of hyperelliptic curves that includes supersingular curves over finite fields of characteristic 3. For $\mathbb{F}_q$ with $q = 3^m$ and $k = 6$, suitable curves are defined by an equation of the form

$$E : y^2 = x^3 - x + b$$

with $b = \pm 1 \in \mathbb{F}_3$. If $\mathbb{F}_{q^3} = \mathbb{F}_q[\rho]/(\rho^3 - \rho - b)$, and $\mathbb{F}_{q^6} = \mathbb{F}_{q^3}[\sigma]/(\sigma^2 + 1)$, then the distortion map $\phi : E(\mathbb{F}_q) \rightarrow E(\mathbb{F}_{q^6})$ is defined by $\phi(x, y) = (\rho - x, \sigma y)$. Then, setting $\mathbb{G}_1 = \mathbb{G}_2 = E(\mathbb{F}_{3^m})$ and $\mathbb{G}_T = \mathbb{F}_{q^6}$, Algorithm 13.2 computes an admissible, symmetric pairing.

13.2.3 The η and η_G Pairings

Barreto et al. [26] introduced the η pairing by generalizing the Duursma–Lee approach to allow use of supersingular curves over finite fields of any small characteristic; Kwon [246] independently used the same approach, and in both cases characteristic 2 is of specific interest. The η pairing already has a simple final powering, but work by Galbraith et al. [151] (see [312, Sect. 5.4]) demonstrates that it can be eliminated entirely; the crucial difference is the lack of normal denominator elimination, which is enabled by evaluation of additional line functions. Interestingly, analysis of the approach demonstrates no negative security implication in terms of pairing inversion and so on. We follow Whelan and Scott [420] by terming this approach the η_G pairing.

For $\mathbb{F}_q$ with $q = 2^m$ and $k = 4$, suitable curves are defined by an equation of the form

Algorithm 13.3: The η algorithm

Input: $P = (x_P, y_P) \in \mathbb{G}_1$ and $Q = (x_Q, y_Q) \in \mathbb{G}_2$
Output: $e(P, Q) \in G_T$

$f \leftarrow 1$
for $i = 1$ **upto** m **do**
  $x_P \leftarrow x_P^2,\ y_P \leftarrow y_P^2$
  $\mu \leftarrow x_P + x_Q$
  $\lambda \leftarrow \mu + x_P x_Q + y_P + y_Q + b$
  $g \leftarrow \lambda + \mu t + (\mu + 1)t^2$
  $f \leftarrow f \cdot g$
  $x_Q \leftarrow x_Q^{1/2},\ y_Q \leftarrow y_Q^{1/2}$
end
return f^{q^2-1}

Algorithm 13.4: The η_G algorithm

Input: $P = (x_P, y_P) \in \mathbb{G}_1$ and $Q = (x_Q, y_Q) \in \mathbb{G}_2$
Output: $e(P, Q) \in G_T$

$f \leftarrow 1$
$T \leftarrow P$
for $i = 1$ **upto** m **do**
  $\lambda \leftarrow x_T^2 + 1$
  $u \leftarrow (y_Q + y_T + \lambda(x_Q + x_T + 1)) + (\lambda + x_Q + q)t + (\lambda + x_Q + 1)t^2$
  $v \leftarrow (x_Q + x_T + 1) + t + t^2$
  $f \leftarrow f^2 \cdot \frac{u}{v}$
  $T \leftarrow [2]T$
end
return f

$$E : y^2 + y = x^3 + x + b$$

with $b \in \mathbb{F}_2$. If $\mathbb{F}_{q^2} = \mathbb{F}_q[s]/(s^2 + s + 1)$ and $\mathbb{F}_{q^4} = \mathbb{F}_{q^2}[t]/(t^2 + t + s)$ then the distortion map $\phi : E(\mathbb{F}_q) \to E(\mathbb{F}_{q^4})$ is defined by $\phi(x, y) = (x + s^2, y + sx + t)$. Note that $s = t^5$ and that t satisfies $t^4 = t + 1$, so we can also represent $\mathbb{F}_{q^4}$ as $\mathbb{F}_q[t]/(t^4 + t + 1)$. Then, by setting $\mathbb{G}_1 = \mathbb{G}_2 = E(\mathbb{F}_q)$ and $\mathbb{G}_T = \mathbb{F}_{q^4}$, Algorithms 13.3 and 13.4 compute admissible symmetric pairings.

13.2.4 The Ate Pairings

The ate pairing was introduced by Hess et al. [181] and generalises the η-pairing to ordinary elliptic curves. The main difference between the Tate and η-pairings is that the arguments are swapped, i.e., the ate pairing is defined on $\mathbb{G}_2 \times \mathbb{G}_1$. Furthermore, $\mathbb{G}_2$ is now instantiated with a very specific subgroup of $E(\mathbb{F}_{q^k})[r]$, namely the q-eigenspace of Frobenius, i.e.,

$$\mathbb{G}_2 = \{(x, y) \in E(\mathbb{F}_{q^k})[r] \mid (x^q, y^q) = [q](x, y)\}.$$

The definition then is very similar to the reduced Tate pairing: let $T = q - \#E(\mathbb{F}_q)$; then the reduced ate pairing is

$$\hat{e}_a(P, Q) = f_{T,Q}(P)^{\frac{q^k-1}{r}}.$$

After its invention, several other variants of the ate pairing were introduced in [170, 250, 274, 434], finally culminating in the notion of optimal pairing [180, 407] but all have the same structure: all compute a (product of) Miller functions followed by a final powering.

13.3 Attacks

In the following, we outline (in chronological order) attacks that represent the state of the art in this field; the attacks can be roughly categorized as injecting faults into either

1. the parametrization, specifically the Miller loop bound, or
2. intermediate computation, specifically the Miller variable.

13.3.1 Attack 1

Page and Vercauteren [318] attack the Duursma–Lee algorithm using a fault that changes the Miller loop bound: the fault model assumes that a random fault is induced in m, and this creates an observable difference in the number of iterations within the Miller loop. For two fixed inputs, the approach computes two results (one valid and one faulty); the security-critical input can be extracted from their quotient since this effectively cancels out terms not influenced by the fault.

13.3.1.1 Recovering a Point

Consider the Duursma–Lee algorithm invoked with inputs $P = (x_P, y_P)$ and $Q = (x_Q, y_Q)$ where P is a security-critical point and Q is selected by the attacker; assume for the moment that the final powering (i.e., line 9) does not exist. Let $\overline{e}_\Delta$ denote the invocation of Algorithm 13.2 where an attacker induces some transient fault that replaces the loop bound m (in Line 2) with Δ. Given this ability, instead of producing a product of polynomials of the form

$$\prod_{i=1}^{m}\left[(-y_P^{3^i}\cdot y_Q^{3^{m-i+1}}\sigma-(x_P^{3^i}+x_Q^{3^{m-i+1}}+b)^2)-(x_P^{3^i}+x_Q^{3^{m-i+1}}+b)\rho-\rho^2\right]$$

the algorithm instead produces

$$\prod_{i=1}^{\Delta}\left[(-y_P^{3^i}\cdot y_Q^{3^{m-i+1}}\sigma-(x_P^{3^i}+x_Q^{3^{m-i+1}}+b)^2)-(x_P^{3^i}+x_Q^{3^{m-i+1}}+b)\rho-\rho^2\right]$$

for some value Δ (remembering that for now, we ignore the final powering). If the attacker is able to induce a fault so that $\Delta = m + 1$, then recovering P is trivial: compute the two results

$$R_1 = \overline{e}_m(P, Q) \quad \text{and}$$
$$R_2 = \overline{e}_{m+1}(P, Q);$$

i.e., R_1 is correct and R_2 is faulty. Using $g_{(i)}$ to denote the ith factor of a product produced by the algorithm, dividing the two results produces a single factor

$$R=\frac{R_2}{R_1}=g_{(m+1)}=(-y_P^{3^{m+1}}\cdot y_Q\sigma-(x_P^{3^{m+1}}+x_Q+b)^2)-(x_P^{3^{m+1}}+x_Q+b)\rho-\rho^2.$$

Given that $\forall z \in \mathbb{F}_q, z^{3^m} = z$, the attacker can easily extract x_P or y_P based on knowledge of x_Q and y_Q.

However, the ability to selectively induce a fault so that $\Delta = m + 1$ is a little tenuous; it is far more realistic to assume that each fault induces $\Delta = m \pm d$ for a random, unknown disturbance d. As such, it is reasonable to assume the attacker must compute the two results

$$R_1 = \overline{e}_{m\pm d}(P, Q) \quad \text{and}$$
$$R_2 = \overline{e}_{m\pm d+1}(P, Q),$$

again producing a single term, $g_{(m\pm d+1)}$; this allows the same approach as above, but demands that the attacker know d so as to correct the differing powers of x_P, y_P, x_Q and y_Q. Recovery of d is, however, reasonable: since the algorithm is straight-line and takes a fixed number of operations, the execution time leaks the number of loop iterations performed fairly directly. This approach requires only a modest number of invocations due to an argument similar to that of the birthday paradox; we simply keep provoking random faults until we recover an R_1 and R_2 that satisfy our requirements.

13.3.1.2 Reversing the Final Powering

The remaining problem is the reversal of the final powering: given the result $R = e(P, Q)$ we want to recover S, the value that was computed by the algorithm before the final powering (i.e., before line 9) so that $R = S^{q^3-1}$. It is clear that given R, the value of S is only determined up to a non-zero factor in $\mathbb{F}_{q^3}$. Indeed, for non-zero $c \in \mathbb{F}_{q^3}$ we have $c^{q^3-1} = 1$. Furthermore, given one solution $S \in \mathbb{F}_{q^6}$ to $X^{q^3-1} - R = 0$, all others are of the form $c \cdot S$ for non-zero $c \in \mathbb{F}_{q^3}$.

Given the description above it should be clear that the attacker does not need to reverse the powering of a full product of factors, but rather a single factor with a fairly special form. That is, given

$$R = \frac{R_2}{R_1} = \frac{\overline{e}_{m\pm d+1}(P, Q)}{\overline{e}_{m\pm d}(P, Q)} = g_{(m\pm d+1)}^{q^3-1}$$

the attacker aims to recover $g_{(m\pm d+1)}$, which will, in turn, allow recovery of the correct x_P and y_P. This requires

1. a method to compute one valid root of $R = g^{q^3-1}$ for some factor g, and
2. a method to derive the correct value of g from among all possible solutions.

The first problem can be solved very efficiently as follows: multiply the equation $X^{q^3-1} - R = 0$ by X to obtain

$$X^{q^3} - R \cdot X = 0,$$

and note that the operator $X^{q^3} - R \cdot X$ is a linear operator on the two-dimensional vector space $\mathbb{F}_{q^6}/\mathbb{F}_{q^3}$. Since $\mathbb{F}_{q^6} \cong \mathbb{F}_{q^3}[\sigma]/(\sigma^2 - 1)$, we can write $X = x_0 + \sigma x_1$ and $R = r_0 + \sigma r_1$, with $x_0, x_1, r_0, r_1 \in \mathbb{F}_{q^3}$. Using this representation, we see that the above equation is equivalent to

$$M \cdot X = \begin{pmatrix} 1 - r_0 & r_1 \\ r_1 & 1 + r_0 \end{pmatrix} \begin{pmatrix} x_0 \\ x_1 \end{pmatrix} = 0.$$

The kernel of the matrix M is a one-dimensional vector space over $\mathbb{F}_{q^3}$ and thus provides all the solutions to $X^{q^3-1} - R = 0$.

To choose the correct root from among all $q^3 - 1$ possibilities, one can use the specific form of the factors in the computed product. Each factor g is of the form

$$g = g_0 + g_1\rho - \rho^2 + g_2\sigma,$$

with $g_0, g_1, g_2 \in \mathbb{F}_q$. To recover g from $R = g^{q^3-1}$, one first obtains $g' = c \cdot g$ for some $c \in \mathbb{F}_{q^3}$ using the root finding approach above, and then computes c^{-1} and thus g itself. Again this boils down to a simple linear system of equations: by multiplying g' with an appropriate factor in $\mathbb{F}_{q^3}$, we can assume g' is of the form

$g' = 1 + (g'_0 + g'_1\rho + g'_2\rho^2)\sigma$. Let $d = c^{-1} = g/g' \in \mathbb{F}_{q^3}$, then d clearly is of the form $d = d_0 + d_1\rho - \rho^2$. To determine $d_0, d_1 \in \mathbb{F}_q$, we use the fact that the terms $\rho\sigma$ and $\rho^2\sigma$ do not appear in g. This finally gives the following linear system of equations:

$$\begin{pmatrix} g'_1 & g'_0 + g'_2 \\ g'_2 & g'_1 \end{pmatrix} \begin{pmatrix} d_0 \\ d_1 \end{pmatrix} = \begin{pmatrix} g'_1 + g'_2 \\ g'_0 + g'_2 \end{pmatrix}.$$

13.3.1.3 Extensions of the Attack Technique

El Mrabet [131] extends the idea of using a fault attack on the Miller loop bound to a wider (and arguably more useful) class of pairings. The disadvantage of the paper is the fault model: the paper focuses on recovery of points assuming that the final powering will be removed using a secondary fault attack or leakage via the scan chain [350] (i.e., the pre-powering result is given to the attacker, rather than the attacker being forced to recover it cryptanalytically from the post-powering result).

However, assuming this is possible, the main contribution is, development of various explicit attacks against variants of Miller's algorithm, i.e., Algorithm 13.1. For example, the paper shows it is possible to attack a pairing that employs Jacobian projective coordinates more or less as easily as one using affine coordinates once the final powering is eliminated.

13.3.2 Attack 2

Conceptually, Whelan and Scott [420] again adopt an approach similar to that of to Page and Vercauteren [318]: for two fixed inputs, they compute two results (one valid and one faulty). However, their attack differs significantly: the fault model assumes the Miller variable f, or the line function used to update it, has a random fault induced. Various pairings with different final powerings are analysed; the conclusion is that pairings with no final powering (e.g., η_G pairing) are easier to attack than those with a complex final powering (e.g., Tate pairing).

We adopt the same notation as in the original paper, writing elements of $\mathbb{F}_{q^k}$ as "cells" throughout. That is,

$$z \in \mathbb{F}_{q^k} = \sum_{i=0}^{k-1} z_i X^i \mapsto [z_0][z_1]\cdots[z_{k-1}]$$

where the right-hand side should be interpreted as a vector of elements in the base field, so each $z_i \in \mathbb{F}_q$. The main assumption is that a fault can target any of the cells in memory.

13.3.2.1 Excluding the Final Powering

One of the specific examples discussed is the η_G pairing described by Algorithm 13.4. The authors assume an attacker can inject a transient fault in the first cell $[v_0]$ of the function v computed in line 6 of Algorithm 13.4 during the last round of the Miller loop (i.e., when $i = m$). Denote by $\bar{\eta}_G(P, Q)$ the output of this faulty pairing; then dividing $\eta_G(P, Q)$ by this value gives

$$\frac{\eta_G(P, Q)}{\bar{\eta}_G(P, Q)} = \frac{u/v}{u/\bar{v}} = \frac{\bar{v}}{v} = \frac{[\bar{v}_0][v_1][v_2][v_3]}{[v_0][v_1][v_2][v_3]} = [N_0][N_1][N_2][N_3],$$

where the $N_i \in \mathbb{F}_{2^m}$ are the cells of the division of the two pairings. From Algorithm 13.4 it follows that $v_1 = 1$, $v_2 = 1$ and $v_3 = 0$. By multiplying the last equality with $[v_0][v_1][v_2][v_3]$, and using the fact that $v_1 = 1$, one finally obtains a simple expression for the correct v_0, namely

$$v_0 = \frac{N_0 + N_2 + 1}{N_1}.$$

Since v_0 is explicitly given by $x_Q + x_{[2^{m-1}]P} + 1$, the coordinate $x_{[2^{m-1}]P}$ can be easily extracted (since the adversary knows x_Q) and from this it is easy to recover $\pm P$ (simply multiply by the inverse of 2^{m-1} modulo the group order), and thus P by recomputing the pairing.

13.3.2.2 Including the Final Powering

The authors also investigate a more general situation by introducing a targeted fault during computation of the reduced Tate pairing. Since r is an odd prime, we know that $r_0 = 1$; this implies that in the last iteration of Algorithm 13.1 (i.e., when $i = 0$), line 7 will always be executed. Assuming we can inject a fault into the computation of $l_{T+P}(Q)$, we obtain

$$\frac{\bar{\hat{e}}_T(P, Q)}{\hat{e}_T(P, Q)} = \left(\frac{\bar{l}_{[r-1]P,P}(Q)}{l_{[r-1]P,P}(Q)}\right)^{(q^k-1)/r} = \left(\frac{\bar{v}_P(Q)}{v_P(Q)}\right)^{(q^k-1)/r},$$

where the last equality follows from the fact that P has order r. The problem, however, is that one would need to reverse the final powering, and find the correct root from amongst all r possibilities. This is a special instance of the so-called Hidden Root Problem [406], which currently seems impossible for exponents of the form $(q^k - 1)/r$ with r a proper divisor of $\Phi_k(q)$ (and Φ_k the kth cyclotomic polynomial).

13.4 Countermeasures

A number of potential countermeasures hope to foil the attacks described in Sect. 13.3; some have a dual use within the context of side-channel attacks. Said countermeasures can be roughly categorized as either

1. acting to detect an injected fault, and thereby aborting computation before security-critical results are made accessible to the attacker, or
2. randomizing the input or intermediate computation so that even if an effective attack is mounted, it cannot recover the security-critical information (without knowledge of the randomness used).

13.4.1 Duplicate Computation

A standard algorithm-neutral approach to fault detection is to duplicate the computation, and check that the same result is produced in all cases. In the context of pairings, this means one can compute

$$\begin{aligned} R_1 &= e(P, Q) \quad \text{and} \\ R_2 &= e(P, Q) \end{aligned}$$

and check that $R_1 = R_2$. Extending this approach, bilinearity allows one to instead compute

$$\begin{aligned} R_1 &= e(P, Q) \quad \text{and} \\ R_2 &= e(\alpha \cdot P, \beta \cdot Q) \end{aligned}$$

for random α and β, and then check that $R_1^{\alpha \cdot \beta} = R_2$. This has the advantage that even if the attacker can inject the same fault in both cases, the check will still detect the fault with high probability.

The disadvantage in both cases is the implied overhead. If the pairing is implemented in hardware, the countermeasure implies a two fold increase in space, whereas a software approach implies at least a two fold increase in execution time.

13.4.2 Checking Intermediate Results

Duplicate computation can be viewed as a course-grained approach mirrored by a fine-grained alternative that checks each intermediate result; the former can abort only after computation has finished, which is disadvantageous if the fault was injected soon after the computation started, whereas the latter can abort as soon as the fault is detected. The frequency of checks now determines the overhead, although this can be "tuned" to suit the level of security required.

Algorithm 13.5: A randomised Duursma–Lee algorithm 159

Input : $P = (x_P, y_P) \in \mathbb{G}_1$ and $Q = (x_Q, y_Q) \in \mathbb{G}_2$.
Output: $e(P, Q) \in G_T$.

$r_0 \in_R \mathbb{F}_{q^6}, r_1 \in_R \mathbb{Z}$
$f_0 \leftarrow r_0,\ f_1 \leftarrow 1$
$m' \leftarrow m + r_1$
for $i = 1$ **upto** m' **do**
  $x_P \leftarrow x_P^3,\ y_P \leftarrow y_P^3$
  $\mu \leftarrow x_P + x_Q + b$
  $\lambda \leftarrow -y_P y_Q \sigma - \mu^2$
  $g \leftarrow \lambda - \mu\rho - \rho^2$
  $f_1 \leftarrow f_1 \cdot g$
  if $i = m$ **then** $f_0 \leftarrow f_1$ **else** $f_0 \leftarrow f_0$
  $x_Q \leftarrow x_Q^{1/3},\ y_Q \leftarrow y_Q^{1/3}$
end
return $f_0^{q^3-1}$

For example, one can check that intermediate elliptic curve points satisfy the curve equation; this guards against faults in their coordinates. The Tate pairing offers an attractive extension of this approach; the final value of the accumulator point (i.e., the value of T after execution of the Miller loop) should be $T = \mathcal{O}$. However, similar checks on finite field elements (e.g., the Miller variable) are more problematic. With a large prime characteristic base field, one might make use of

1. the encoding strategy of Gaubatz, Sunar and Karpovsky [156], which lends itself to modular multi-precision arithmetic such as the approaches of Montgomery [293] or Barrett [27],
2. or the checksum-oriented "wooping" approach of Bos [59, Chap. 6].

Ozturk [316, Chap. 5] discusses fault-resilient arithmetic with direct consideration of the Tate pairing; the work expands on Ozturk et al. [317], and uses $\mathbb{F}_{3^m}$ (and extensions thereof) as an example. Although generally applicable, the focus in on the extension field in particular since this helps to minimize latency.

13.4.3 Randomised or Fault-Resilient Miller Loop Counter

Ozturk [316, Sect. 5.3] discusses the use of fault-resilient counters to foil attacks focused on changing the Miller loop bound; suggested instantiations include those of Gaubatz et al. [156].

Ghosh et al. [159, Sect. 4] take a different approach, opting to randomize the loop bound as detailed in Algorithm 13.5. The idea is to perform $m' > m$ iterations in total, storing the correct result in the iteration where $i = m$. Although the implications for performance are a potential disadvantage, this approach makes recovery of d in the attack of Page and Vercauteren [318], for example, more difficult.

13.4.4 Input Randomization and Blinding

Some of the attacks described in Sect. 13.3 rely on the fact that the attacker has knowledge of, and potentially control over, one of the input points; assume this point is Q. Several potential countermeasures focus on this fact: the idea is that by randomizing Q or the Miller variable, the attacker is unable to recover information from a faulty result.

1. Page and Vercauteren [318] note that since
$$e(a \cdot A, b \cdot B) = e(A, B)^{a \cdot b},$$
one can randomize P and Q by selecting random α and β such that $\alpha \cdot \beta = 1 (\bmod r)$, and then computing
$$e(\alpha \cdot P, \beta \cdot Q) = e(P, Q).$$
2. Page and Vercauteren [318] note that since
$$e(A, B + C) = e(A, B) \cdot e(A, C),$$
one can randomize Q by precomputing a random point R and $S = e(P, R)^{-1}$, and then computing
$$e(P, Q + R) \cdot S = e(P, Q).$$
3. Scott [365] discusses the fact that one can randomize the Miller variable (during each iteration) via multiplication by some random $\omega \in \mathbb{F}_q$; this has no effect on the result since the randomization is eliminated by the final powering.
4. Kim et al. [230] discuss the possibility of using randomized projective coordinates; the idea is to operate on the affine point $Q = (x_Q, y_Q)$ using the projective representative
$$(\lambda \cdot x_Q, \lambda \cdot y_Q, \lambda)$$
for some random $\lambda \in \mathbb{F}_q$ instead. This dictates use of a projective (re)formulation of the pairing algorithm.
5. Shirase et al. [376] describe a scheme for randomizing input points via addition of a random field element, and then removing the effect via reformulation of the pairing algorithm.

An important note is that Ghosh et al. [159, Sect. 3] incorrectly rule out the first two approaches; their argument is based on the mistaken assumption that the bilinearity of a normal pairing is retained by a faulty pairing. For the attack of Page and Vercauteren [318] in particular, they assume

$$\overline{e}_{\Delta}(a \cdot A, b \cdot B) = \overline{e}_{\Delta}(A, B)^{a \cdot b},$$

which is not true unless $\Delta = m$, i.e., there is no fault.

13.5 Conclusion

The field of pairing-based cryptography is moving fast; in under ten years it has grown from an interesting aside into an industrially supported and soon to be standardized endeavour. On the one hand, study of physical security, fault attacks specifically, has not yet caught up with the vast range of parametrizations, algorithms and use cases. On the other hand, selected attacks using reasonable fault models have already been demonstrated; it seems likely that as implementations mature and proliferate, practical study of what faults can reasonably be induced will yield further attacks.

Even so, the range of approaches is far smaller than in the context of block ciphers. In part this is due to the mathematical complexity of pairings. This fact suggests future attack opportunities might just as likely result from permanent faults in software (e.g., through errors in implementation) as from more hardware-oriented fault induction. In the case of ECC, this is already an issue[1]; it is interesting that pairing-based cryptography seems to exacerbate the underlying problem.

[1] See http://www.mail-archive.com/openssl-dev@openssl.org/msg23208.html.

Part IV
Miscellaneous

Chapter 14
Fault Attacks on Stream Ciphers

Alessandro Barenghi and Elena Trichina

Abstract In this chapter, we provide an outlook on fault attack techniques aimed at breaking stream ciphers. The chapter will start with an overview of the possible targets for fault attacks among this class of encryption algorithms, and subsequently provide two in-depth case studies on enhancing impossible cryptanalysis against RC4 and differential cryptanalysis against Trivium. After the analysis the chapter will end with a broader overview of the attacks on other notable stream ciphers and provide directions for future research in the field.

14.1 Introduction

Prior to the appearance of fast block ciphers, such as DES, stream ciphers ruled the world of encryption. The classical examples of such ciphers are rotor machines, among which Enigma occupies the most prominent place. Although the methods of design and analysis of block and stream ciphers are quite different, the distinction between them nowadays is somewhat vague. One of the main reasons why modern state-of-the-art stream ciphers seem to be not as well defined as block ciphers is because of the great variety of their constructions and because, with the advent of sophisticated computers, communications do not happen bit-by-bit anymore but in multi-bit packets. This trend has pushed the use of fast software-oriented block ciphers in a stream-like direction, thanks to special modes of operations, such as Counter, Output FeedBack, Cipher Block Chaining and Galois Counter modes. A confirmation of this trend is the adoption of the KASUMI stream cipher as a 3GPP

A. Barenghi (✉)
Dipartimento di Elettronica e Informazione,
Politecnico di Milano, Milan, Italy
e-mail: barenghi@.elet.polimi.it

E. Trichina
STMicroelectronics, Rousset, France

M. Joye and M. Tunstall (eds.), *Fault Analysis in Cryptography*, Information Security and Cryptography, DOI: 10.1007/978-3-642-29656-7_14,

standard for encryption. In fact, KASUMI is a version of a block cipher named MISTY1 running in a Counter mode and is employed to encrypt streams of data in cellphone communications.

The two main parts of a stream cipher are the state-transition function, which, given an old state, computes a new state, and a filter, which, given the actual state, produces the output. The output of a stream cipher is a random-looking stream of bits (or digits) which are typically XORed with a plaintext resulting in a ciphertext. Thus stream ciphers can be viewed as a computational analogy of a one-time pad cipher, replacing a long, perfectly random, secret key with a short secret "seed" from which a stream of bits is generated pseudorandomly. For practical purposes it is fundamental that the pseudorandom stream be computationally indistinguishable from a stream of truly random bits.

Many stream ciphers are still based on Feedback Shift Registers (FSRs), where these registers are combined in a variety of ways: LFSR with nonlinear filters, irregular stepping functions, nonlinear feedback functions, irregular decimation/shrinking and any kind of nonlinear transform. (L)FSR-based ciphers are up to this day the main workhorses when there is a need to encrypt large quantities of fast streaming data. A number of secret military hardware-oriented ciphers belong to this group, as do"civil" ciphers E0 (Bluetooth), designed for a short-range LAN, and one of the most widely used (by the virtue of being included in the GSM cipher suite for ensuring over-the-air privacy protection) ciphers A5/1.

A new, popular modern trend uses parts of block cipher-like rounds mixed with LFSR-like structures; an example of such ciphers is MUGI [417], which has been recommended by the Japanese e-government initiative CRYPTREC. In [50] this approach has been generalized by turning any standard block cipher into a stream cipher by a combination of an appropriate mode and a "leak extraction" phase, i.e. outputting some parts of the internal state of a cipher at certain rounds, thus increasing the rate at which the keystream is produced.

A champion in simplicity and robustness of modern stream ciphers, RC4 is based on a 256-byte permutation table, somewhat similar to an S-Box, where a state transition function makes a pseudorandom swap of two elements of the table. The value of one of the indices of the swapped elements is sequentially incremented and thus is known, while the value of the other is computed as a sum of the previous value and a result of a lookup in the table with the first index. RC4 with a 128-bit key is a defacto Internet standard and is also recommended by the CRYPTREC initiative. One of the finalists of the EU-funded network of excellence ECRYPT's competition eSTREAM, HC-128 has a similar internal structure, but it uses a much larger state made of two 512×32-bit word tables with rather complex state transition functions and a nonlinear output filter.

We refer readers to [133] for a journey to a fascinating world of modern stream ciphers, their making and breaking. In this chapter we will concentrate on so-called fault attacks, i.e. attacks based on the simple discovery that by injecting faults in hardware during execution of cryptographic algorithms, one can get privileged information from within the encryption process, information which otherwise would have been hidden from an attacker who is trying to analyze a cipher based only on its mathematical specification. It is clear that in theory fault attacks are much more powerful than regular cryptographic attacks: indeed, consider the (hypothetical) situation when an injected fault results in the bypassing of the initialization phase of a stream cipher, causing a direct leak of a key material, or when it changes the address of the output register, pointing instead to the register containing a current state!

Fortunately, not all theoretical fault models can be easily achieved in practice. Nonetheless, the state of the art in physical fault attacks is constantly improving, and a large number of spectacular successes in breaking CRT-RSA and AES have been reported in the literature. Interestingly enough, until now there seem to be no practical implementations of fault attacks on stream ciphers, although, starting with Hoch and Shamir [182], where they developed a method of exploiting perturbations of LFSR-based stream ciphers to recover the key, the number of publications describing "realistic" theoretical attacks on stream ciphers has been steadily growing.

In this chapter we survey techniques of differential fault analysis of a number of stream ciphers, with special emphasis on some leaders of the eSTREAM competition. However, to give a flavor of fault attacks, we will start with an "impossible" fault analysis of RC4 that is easy to describe and to understand, followed by a more traditional differential fault analysis of the same cipher. We proceed with a fault analysis of the LFSR-based cipher Trivium in Sect. 14.3 and demonstrate how far differential fault analysis can be pushed by describing an attack on a markedly different and much more complex cipher, HC-128, in Sect. 14.4. A brief survey of publications on fault analysis on other stream ciphers, such as Grain, Rabbit, and SNOW 2.0 is given in Sect. 14.5. In the conclusion we address some aspects of the practical implementation of fault attacks and countermeasures for software and hardware implementations of stream ciphers and how they differ from those of block ciphers and public key algorithms.

14.2 Impossible Cryptanalysis Enhanced: Faults on RC4

14.2.1 Cipher Description and Properties

The Alleged RC4 cipher is a stream cipher designed by Ron Rivest in 1987 and made public in 1995 by an anonymous researcher who analyzed the implementation provided by RSA Security and posted the source code on a newsgroup. After the public disclosure of its inner workings, the cipher has been widely used in a large range of applications, ranging from document encryption (the PDF file format employs RC4 as a standard protection) to network traffic protection (IEEE 802.11b wireless equivalent privacy). One of the key reasons for this widespread adoption is the simplicity of the cipher structure, which results in ease of implementation. The cipher state is characterized by two indices i and j which range from 0 to 255, and may thus be contained in a single byte variable, and an array of 256 bytes which is properly initialized by the key scheduling procedure. The values contained in the array are always the same (all the integers ranging from 0 to 255) and are permuted by the algorithm based on the values of the two counters i and j. In order to initialize the inner state of the cipher, RC4 performs a first permutation of the S array (described in Algorithm 4.1) according to the key supplied, which may be from 5 to 256 bytes long. The initialization operates 256 swaps among the variables, employing an eight-bit slice of the key together with the value of j to determine the index of the array cells to swap.

Algorithm 14.1: The RC4 key schedule

Input: $K = (k_0, k_1, \ldots, k_l)$: cipher key l: key length in bytes
Output: i, j, $S = (s_0, \ldots, s_{255})$: initialized inner state

begin
  $S \leftarrow (0, 1, 2, \ldots, 255)$
  $j \leftarrow 0$
  $i \leftarrow 0$
  for $i < 255$ **do**
    $j \leftarrow (j + s_j + k_{i \bmod l}) \bmod 256$
    $temp \leftarrow s_i$
    $s_i \leftarrow s_j$
    $s_j \leftarrow temp$
    $i \leftarrow i + 1$
  end
  $i \leftarrow 0$
  $j \leftarrow 0$
  return i, j, S
end

Algorithm 14.2: The RC4 step function

Input: i, j, $S = (s_0, \ldots, s_{255})$ the inner state of the cipher
Output: o the output byte

$i \leftarrow i + 1 \bmod 256$
$j \leftarrow (j + s_i) \bmod 256$
$temp \leftarrow s_i$
$s_i \leftarrow s_j$
$s_j \leftarrow temp$
return $o \leftarrow s_{s_i + s_j}$

After the initialization step is completed, the cipher is able to output an eight-bit value per step, obtaining it as described by Algorithm 4.2. At first, the value of i, which acts as a counter, is incremented by 1, and employed to update the value of j based on the value of the ith cell of the state. After the updates, the values of i and j are used to perform a swap between two values of the inner state and output a single byte based on their contents. Due to the nature of the updating algorithm, it is possible that particular conditions of the two indices lead to extremely short and regular permutation cycles of the inner state. In particular, Finney [140] observed in 1994 that, in the case where the two conditions $j = i + 1$ and $S[j] = 1$ hold, the RC4 cipher enters a short cycle of size $256 \cdot 255$. During this cycle, every 256 steps the cipher shifts its internal state one byte to the left and outputs a fixed byte of the state. This in turn implies that the whole ordered content of the state is output as a keystream during the short cycle in which the cipher is operating. A further observation is the fact that, if the cipher is set in one of the states belonging to a short cycle, it will never exit it, thus enabling an attacker to recognize the peculiar condition through observing the periodicity of the output. During regular working conditions, RC4 will never be in one of these states, since the initial key setup procedure sets both i and j to zero, thus placing the cipher outside the short cycles.

14.2.2 Impossible States and Faults

Whilst RC4 cannot enter the short cycles on its own, it is possible to force it into one of the states belonging to them through a proper fault injection. In 2005, Biham, Granboulan and Nguyen [45] proposed a fault attack against RC4 which exploits this possibility. The fault assumption made is that the attacker is able to modify the value of either i or j at random during the execution of the RC4 algorithm. No particular assumptions are made on which fault pattern is injected into the values except that only one of them is modified. The attack technique is thus split into two parts: understanding how probable it is that the cipher enters a short cycle and recognizing the short cycling output with the smallest number of faulty bytes in order to reconstruct the inner state with the smallest possible amount of keystream information.

The probability that a random fault in either i or j will force the cipher into an otherwise impossible state is 2^{-16}. This is because of the fact that the new value set will match the first condition ($j = i + 1$) with probability 2^{-8}, given two indices over eight bits and the probability that $S[j] = 1$ for a randomly selected value of j is exactly 2^{-8} because the state holds 256 different values. This observation implies that the attacker is able to obtain an exploitable fault by injecting on average 2^{16} faults during the cipher operation, taking care to restart the enciphering device after each unsuccessful trial.

In order to detect if the cipher has entered one of Finney's states, the most naïve approach implies recording twice the bytes of a short cycle and acknowledging the effective periodicity of the output, thus requiring the attacker to obtain $2 \cdot 255 \cdot 256 = 2^{17}$ bytes of the keystream. In order to reduce the quantity of keystream to be gathered each time before a reliable check can be done, it is useful to observe the frequency of the repetitions of the output byte values.

In regular working conditions, the cipher selects the output bytes pseudorandomly, thus resulting in a repetition in the output roughly every $256^{\frac{1}{2}} = 16$ bytes because of the Birthday paradox. On the other hand, when cycling in Finney's states, the selection of the output bytes $o = s_{s_i+s_j}$ is driven only by the value of s_i, since the value of s_j is fixed to 1 because of the peculiarity of the state. This in turn implies that the selection of the output byte will be driven by different indices except for the case where a value of the array is swapped twice. The authors of [45] determined that this results experimentally in a duplicate output byte occurring once approximately every 80 bytes, which enables an attacker to successfully discard a faulty output if duplications in the output occur more frequently than that (the authors suggest that every keystream with a duplicated byte occurring more frequently than once every 30 bytes can be safely discarded).

After obtaining the exploitable faulty keystream an attacker needs to reconstruct the inner state of the cipher. The simplest way to do this is to select a random byte and read the state by skipping 255 bytes at a time in the obtained keystream, exploiting the fact that, while the location taken from the state array is the same every 256 outputs, the content of the state is rotated a single byte to the left every 256 cycles. A quicker way to obtain the state is to exploit the information leaked by the order of the output bytes, which enables the attacker to infer a couple of consecutive values in the inner state, thus reducing the possible state space, together with the fact that, when the output byte is equal to 1, $s_i + s_j = i$.

The authors of the attack report that the average number of keystream bytes to be obtained before a Finney state is correctly entered and the inner state is recovered is around 2^{21} on average, considering also the faulty keystream which gets discarded by the aforementioned duplicated output frequency test.

In addition to the presented impossible cryptanalysis on RC4, the authors of the same paper also propose a common differential cryptanalysis on the same cipher. The key point of the second method is to inject faults precisely one per inner state byte and then deduce from the position and value of the faulty outputs which cell has been damaged. Whilst this attack requires significantly less faults than the other one, the required fault model is rather stringent, since it implies both perfect timing and quite selective locality in injecting the faults.

14.3 Differential Fault Analysis of Trivium

14.3.1 Cipher Description

Trivium is a stream cipher selected by the eSTREAM project as one of the most promising ones for hardware implementation. Proposed by De Cannière and Preneel in [116], this cipher was to be as simple as possible without sacrificing security. As a consequence, the architecture of the keystream generation algorithm does not involve many components: it only employs three registers and a couple of two-input one-output logical XOR and AND gates.

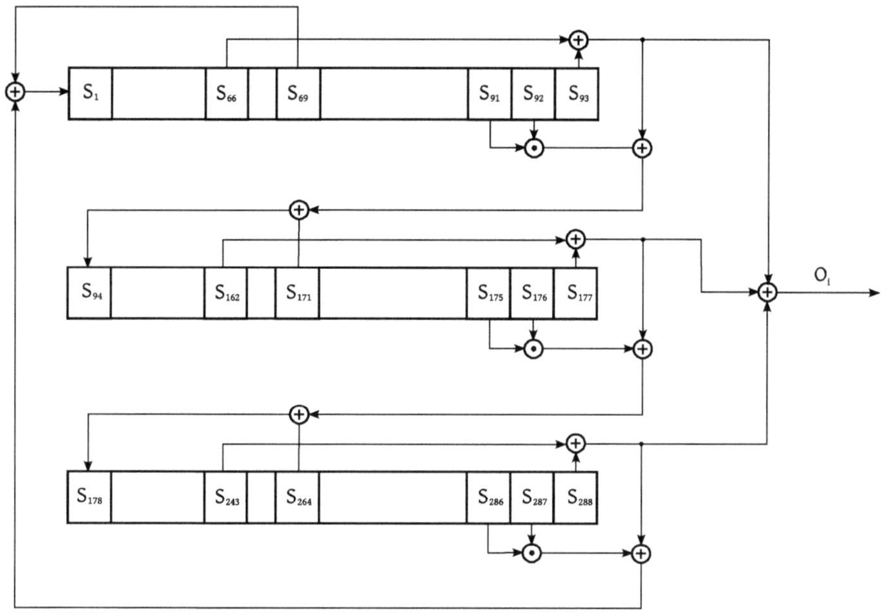

Fig. 14.1 Trivium block diagram

Figure 14.1 provides a block diagram of the cipher. As depicted in Fig. 14.1, the output of the cipher is the result of the linear combination of three different values, each obtained from a different internal register of the state. From now on, we will be denoting the state bits by $s_1 \dots s_{288}$.

A key feature of the feedback function, which will also be used to lead the fault attacks, is the fact that the degree of the three feedbacks into the three registers is 2 (since there is a single AND gate in the feedback loop). Moreover, the two variables involved in the second degree term of the feedback are contained in two consecutive bits of the same register, while the third variable, employed in the feedback circuit of the ith is taken from the $(i + 1) \bmod 3$rd one in order to provide inter-register diffusion.

In order to initialize the keystream generator, the algorithm requires an initialization vector and a key, both 80 bits long. These values are used to initialize the first bytes from left to right of the first two registers: in particular, the key fills the state bits from s_1 to s_{80} and the initialization vector fills the state bits from s_{94} to s_{173}. After this assignment phase, all the other bits of the internal state are set to 0, with the exception of the last three bits of the third register (s_{286}, s_{287} and s_{288}), which are set to 1. The initialization phase is completed through clocking the whole system a total of 1,152 times (i.e. four times the width of the inner state), without outputting anything. After this initialization, the bits contained in the inner state of the cipher have strongly nonlinear dependence on the key and IV bits, which is the key security feature of Trivium.

After the initialization phase, the cipher outputs one bit per clock cycle, which will be denoted by o_{cycle} in the remainder of this chapter. For the sake of clarity, the keystream obtained during an execution in which a fault has been injected will be denoted by $\bar{o}_{cycle}$.

The goal of an attacker is the reconstruction of the complete inner state of the cipher, i.e. the contents of the three state registers: once these data are recovered, it is possible to reverse-clock the cipher and roll back the inner state to the start of the initialization. It is possible to determine when the cipher has been fully rolled back thanks to the peculiar initialization padding, and thus extract the recovered IV and key bits compromising the security of the scheme.

14.3.2 Attack Technique

The attack which will be presented in this chapter relies on the assumption that the attacker is able to inject a single bit-flip fault into the internal state of the Trivium cipher during the a precise clock cycle t_0. After the fault injection, the attacker collects some of bits of the faulty keystream (some hundreds) and then restarts the encrypting device before injecting a new fault.

The cryptanalysis of Trivium starts with the observation that the relation between the inner state and the output is purely linear. This implies that, every time an attacker obtains a single bit of the keystream, he implicitly gains information on the inner state of the cipher in the form of a linear equation of the state bits.

In particular, the output bit at the ith clock cycle can be written as

$$s_{66} \oplus s_{93} \oplus s_{162} \oplus s_{177} \oplus s_{243} \oplus s_{288} = o_i,$$

thus yielding 66 linear equations binding the first 66 output bits with the internal state bits. The following 82 keystream bits have quadratic dependencies with the inner state and the degree of the relation increases quickly afterwards, preventing an attacker from directly solving the system by brute force.

In order to successfully attack the cipher, an attacker will need more linear equations involving the state bits; this is achievable by analyzing the differences between the correct and a faulty keystream. The effect of a single bit-flip fault assumed to be the fault model employed by the attacker is to substitute an inner state term in the equations s_i with $\bar{s}_i = s + 1$. This in turn implies that the difference between the correct and faulty output bit is equal to the difference between two linear equations: one representing the correct output, and the other the same equation with the correct term substituted with the one taking into account the fault.

The only missing point for applying this substitute and subtract technique is that the attacker must be able to determine which position of the state has been hit by the bit flipping fault in order to do the proper substitution.

The key to spotting which bit has been altered is observing the different distances between the two bits involved in the linear output combination for each register. In fact, the two tapped bits are respectively 27, 15 and 45 positions apart for the first, second and third registers. The simplest case in determining the fault position occurs when the faulty bit $\bar{s}_e$ is placed between the beginning of a register and the first tap (i.e. $e \in [1; 66] \cup [94; 162] \cup [178; 243]$). In this case, the second nonzero bit in the keystream difference appears after 27, 15 or 45 clock cycles from the first, revealing that the faulty bit is being employed by the second output tap of a specific register. Combining the second appearance with the position of the first nonzero bit in the keystream difference it is possible to correctly reconstruct the register and the specific bit in which the flip occurred.

In case a fault occurs outside the aforementioned interval, it is still possible to infer the position of the fault, albeit the second degree effects of the AND gates on the feedback must also be taken into account, since they may generate other differences in the output keystream.

Summing up, the first basic attack to Trivium may be possible by obtaining enough linear equations in the state bits, adding to the 66 which are straightforward from the cipher structure those which bind bit differences between the correct and faulty ciphertexts. Once a whole system of 288 equations in the state bits is obtained, it is possible to solve it through Gaussian elimination and obtain all the values of the inner state at a specific time. By clocking back the Trivium cipher, if desired, the attacker may also recover the original key and the IV employed.

A way to further improve the efficiency of the attack is to precompute a pool of linear equations for all the possible positions where the fault may hit and store them in a lookup table. Since the equations depend only on the cipher structure, this precomputation effort need be done only once. The practical attack to the cipher thus is reduced to the fault injection and keystream collection parts alone.

The authors of [45] provide an estimate of the average number of faults needed to recover the whole cipher state, which is around 380. This figure can be significantly reduced if the attacker is willing to brute-force a small number of the state bits: for instance, by guessing only eight bits, the number of faults can be reduced to 270, and by guessing 28 bits the number of faults can be cut in half with respect to the whole state retrieval.

In the aforementioned article the authors also propose a way to improve even more upon the number of linear equations obtained from a single fault. The key idea behind the improvement is the observation of the fact that most of the quadratic equations of Trivium involve only terms of the form $s_i s_{i+1}$. These equations may be efficiently represented by adding a new variable for each quadratic term and thus needing only twice the amount of memory (288 variables for the state, 287 variables for all the possible pairs of adjacent bits).

Once all the output equations involving only such quadratic and linear terms have been precomputed, the attack starts by injecting faults in the same way as in the basic case. The key difference is that, as soon as it is possible to obtain the value of a variable from the system, the solver replaces all the occurrences of it with the actual constant value, thus possibly reducing the degree of some of the quadratic equations back to 1.

This technique leads to a significant reduction in the number of faults which need to be injected in order to recover the whole state: 43 faults are sufficient in order to obtain the whole set of $\{s_1 \ldots s_{288}\}$ variables. On the other hand, with this attack improvement, trying to guess a small number of bits through brute-force does not involve a significant reduction of the number of faults which need to be injected. In fact, guessing more than half of the state bits (168 out of 288) leads to a reduction in the number of faults to 42, thus saving the attacker a single fault injection.

The number of injected faults may be further reducible if all the possible quadratic equations, involving also nonadjacent bits of the state, are considered. According to the authors of [184], this yields an extra 20 % of equations from the same faults at the cost of using memory efficient techniques to store all the equations, since all the possible generic quadratic terms are $\binom{288}{2}$ instead of just 287.

14.4 An Advanced Case: Differential Fault Analysis of HC-128

14.4.1 Cipher Description

HC-128 is a stream cipher proposed by Wu [423] and is part of the software portfolio of the eSTREAM project. The cipher is based on an internal state composed of two separate tables P and Q, with 512 entries, each 32 bits wide, which are employed to generate the keystream, and two indices which drive the generation process.

The keystream generation process is driven by an index, i, which acts as both a counter up to 1 , 024 and a switch between the table of the cipher that is refreshed with the feedback function and the table that is not. In particular, when i assumes a value between 0 and 511 P undergoes diffusion, while when its value is $512 < i < 1023$, Q is modified. The i index also selects how the output depends on the inner state, employing a LUT to add complexity to the relationship binding state and outputs.

As HC-128 is (almost) symmetric, Fig. 14.2 provides a block diagram of the table under refreshment for a fixed value of i. The other "half" of the cipher can easily be obtained by swapping the role of P and Q and changing the rotation constants and directions of the rotations.

As depicted in Fig. 14.2, the nonlinear feedback function, which is driven by $j = i \bmod 512$, employs four values from the selected state table (P in the figure), rotates three of the four 32-bit values and recombines them through a mixture of XOR ($\oplus$ in figure) and addition modulo 2^{32} ($\boxplus$ in figure) operations. The recombined value is then combined with the old value of the cell to obtain the one which will be replacing it. The positions of the first three elements are determined from the index j, respectively adding to them modulo 512 (the actual length of the table) three constant values: -10, -3 and $+1$.

After the table update operation is performed through the feedback function, the cipher computes the output 32-bit word through the nonlinear output function which

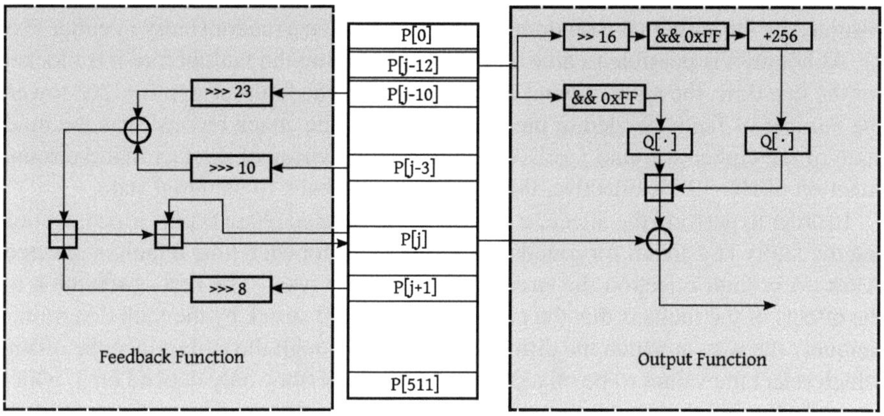

Fig. 14.2 A diagram of one of the two symmetric halves of HC-128

employs two values from the P state table. The first one is employed to perform a lookup on the other state table Q, by using two of its four bytes to look up two values, one in the lower, the other in the upper half of Q. These two values are then combined through addition and the result is employed to mask the value of the P table which had just been refreshed before being output as the output word for the cycle.

After outputting the word, the cipher increments the value of the counting index i by 1 and resumes the keystream generation.

The key scheduling strategy of the cipher is based on initializing the P and Q tables with the key material and expanding it until both tables are filled. After the whole state is filled with key material, the cipher is clocked 1,024 times to avoid the production of low-degree keystream in output. After these 1,024 rounds, the cipher is ready to generate a usable keystream.

As can be noticed, the bit-wise relations among the cipher, the counters and the internal state are highly nonlinear and the cipher also has a wide temporal memory implemented through the use of the values from the table at rest as a mask for the output. These features significantly reduce the possibility of breaking this stream cipher using classic cryptanalysis, since it is very difficult to obtain exploitable linear or near-linear relations.

14.4.2 Attack Description

In order to attack the HC-128 cipher through the use of faults, the authors of [184] propose a two-phase technique akin to the one employed to break RC4. In the first phase the position where the fault hit the internal state of the cipher is recovered, while in the second phase the recovery of the inner state takes place. The fault model assumed by the authors is a random 32-bit wide fault on a random entry of either P or Q. Although it is possible to attack the cipher injecting the fault before it is clocked for the first time, the authors point out that injecting the faults at step $i = 268$ lowers the number of faults needed to break the scheme. The attack reconstructs the inner state of the cipher at round $i = 1{,}024$ and then, by virtue of the fact that the round function of HC-128 is bijective, the cipher is rolled back to its initial state.

In order to perform the attack, at first a fault collection phase is performed, recording the faulty keystream for rounds $i \in [268, 1535]$ for each time a fault is injected. After every fault injection the stream cipher is fully reset. The first observation on the effects of the faults is that the position of the word struck by the fault determines uniquely the way in which the difference spreads through the table, since the offsets which select the values to be mixed together are fixed (they only depend on j, which is regularly cycling through the entries as i grows). In the same way, it is possible to track down the effect of the changes up to the produced keystream. The authors suggest transforming the sequence of differences between correct and faulty keystreams into a sequence of bits, where a 0 value means "the two words are the same" and a 1 means "there is a difference between faulty and fault free ciphertexts". It is then possible to uniquely map a sequence obtained in this way to a range of positions in either the P or the Q table where the fault has happened.

After a full recovery of the position of the induced faults, the next step is to understand the actual form of the fault induced by the attacker (i.e. which bits have changed). This is manageable since, from the previous analysis, the attacker is able to distinguish which of the three values of the output function has been affected by a fault, and thus selects the one for which the nonlinear term obtained through the lookup on the other state table is untouched. This in turn implies that the attacker is in possession of a faulty and a fault-free word of the inner state, both masked with the same XOR mask, and is thus able to derive the actual fault injected into the state. By repeating this technique and reusing the fault values obtained in the analysis, it is possible to disclose all the fault patterns injected for the whole cipher. For an in-depth case by case analysis we refer you to the full paper, which provides accurate tables of the differences.

After the fault patterns have been recovered for each faulty keystream, the attacker is now in possession of a vast amount of differential information on the encryptions (namely, a fault for each word is required for the attack to succeed). Due to the additive nature of the output function, it is possible to recover the input values of the output function for all the cycles between 512 and 1023 (which employ one term from the Q table and one from P) and reuse the information to deduce the ones for $1,024 < i < 1,535$ (which, conversely, are built from a single value from P and two lookups from Q). After this phase, the attacker has gathered information on the values before the last XOR performed by the output function and knows the keystream; the last step is thus to build a large system of very sparse linear equations in the state bits (roughly 1.8×10^4) and, by solving it, obtain all the values of the inner state.

From experimental results, obtained by the authors through synthetic simulation, the final equation system has an effective rank of 1,022, thus leaving two bits of the inner state to be guessed. The approach proposed by the authors suggests solving $2^2 = 4$ linear equation systems, and checking which actually produced keystream matches the correct one obtained, to successfully recover the key. The computational bound of this attack technique is determined by the repeated solution of the 1.8×10^4 simultaneous equations four times, which can be achieved in minutes to hours on a typical desktop computer.

14.5 An Overview of Grain, Rabbit and SNOW 3G

The landscape of the attacks to stream cipher does not end with the three attacks mentioned: we will now provide a quick overview of the attack techniques used to break Grain, Rabbit and SNOW: three stream ciphers selected by either the eSTREAM portfolio (Grain, Rabbit) or the 3GPP standard (SNOW 2.0).

Grain is a two state register stream cipher taken into consideration by the eSTREAM project for the hardware implementation portfolio. The design of the cipher involves a linear and a nonlinear feedback function for the two registers, together with a nonlinear filter function to combine their contents and output the keystream. The best classical cryptanalytic techniques have not yet obtained any results against Grain; in [36] the authors report a successful fault attack based on a single bit-flip in the state of the cipher, without knowledge of the flipped bit, though requiring exact (clock-accurate) timing.

Rabbit is a stream cipher which entered both the software and the hardware portfolio of the eSTREAM project thanks to both its reduced size and the vast amount of white papers presented regarding its soundness with respect to a number of cryptanalytic techniques (in particular algebraic, correlation-based and differential). In order to break the ciphers the authors of [40] assume a quite original fault model: the possibility of changing an ADD instruction into a XOR. This assumption allows them to perform a kind of cryptanalysis much akin to the linear approximations of multi-bit additions with simple XORs on the cipher and successfully recover the key within a 2^{34} time complexity.

The SNOW 2.0 cipher has been chosen as one of the standard ciphers to be employed to encrypt 3G cellphone communication by the 3GPP committee. The committee deemed some tweaks to be necessary and renamed the standardized cipher SNOW 3G in order to distinguish it from the previous version.

The stream cipher has its inner state split into two parts: a shifting register with 16 32-bit wide cells and a Finite State Machine composed of three 32-bit registers. At each clock cycle the inner state is updated by performing XORs (represented by the $\oplus$ symbol) and additions modulo 2^{32} ($\boxplus$ in Fig. 14.3) on 32-bit values. In particular, the feedback of the shifting register is obtained by adding together using XOR the value of the third word of the register, s_2 and the values of the other two multiplied by two fixed values, α and α^{-1}, over $\mathbb{Z}_{2^{32}}$. In order to add a nonlinear mask to the output, a Finite State Machine is employed to derive a 32-bit mask to be added to the first value of the shift register, before outputting it as a part of the keystream. The nonlinearity

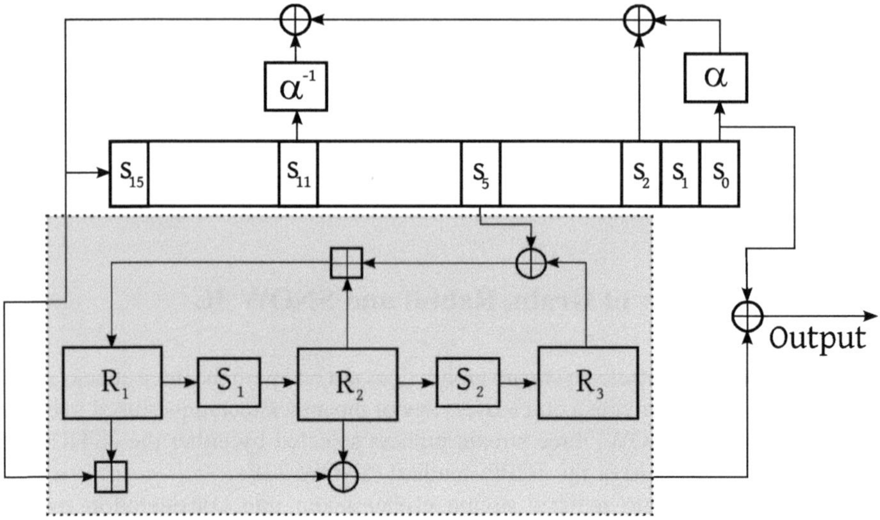

Fig. 14.3 Block diagram of SNOW 3G

in the FSM is obtained by updating the contents of the last two of the three registers through a transformation of the 32-bit value derived from two key primitives of the AES standard: the SubBytes and MixColumns operations. These two primitives are able to map a 32-bit value to another one providing proper, statistical attack immune, diffusion and they can be efficiently implemented in hardware.

In 2009, Debraize et al. [117] proposed a fault attack technique able to successfully break the SNOW 3G cipher. The technique is affine to the one employed to break HC-128, although a couple of extra issues must be addressed. At first, the attacker injects a fault into a word of the inner state s_i, where the position i is known. No particular assumption on the timing of the attack is made, so the fault hypothesis may be reformulated as injecting a fault into an arbitrary location of the state, provided clock-accurate timing is available. After the fault injection, a correct keystream must be collected in order to analyze its differences with the faulty one. Indicating, as in the attack on HC-128, with 1 a difference between two words of the ciphertexts and with 0 two identical words at the same clock cycle, it is possible to build fault fingerprinting sequences. Employing the fault fingerprint, the attacker is able to discover the part of the state in which the fault was injected, and is thus able to write an equation having on its right-hand side the difference between the two keystream values and on its left-hand side, the same difference between the values of the inner state involved in the generation of the output. After collecting a number of different equations, the attacker aims at eliminating the nonlinear terms of the equations (namely, the ones depending on the values of the three registers of the FSM), by subtracting state to state the different equations. Since the construction of the cipher does not allow us to obtain a fully linear system as in the case of HC-128, the authors of [117] employed Gröbner basis decomposition to solve the low degree nonlinear system obtained from these simplifications. The main issue in solving this kind of equation systems is that the problem is known to be NP-hard, so the system of equations must be properly reduced before applying the solving algorithm. The authors report that it is possible to break SNOW 3G after the injection of 22 faults, with a computation time of 20 to 40 seconds on average, which is well within the feasibility of any attacker.

14.6 Conclusion

What conclusion can we draw from our survey of the state of the art of fault attacks on stream ciphers? First of all, that despite a late start, these attacks have proven to be very efficient in theory. A strong feature of these attacks is that they typically use a rather relaxed fault model and rely on traditional stream cipher cryptanalytic techniques. If we look at fault attacks on block and stream ciphers, we notice that the differential fault analysis of block ciphers relies on obtaining differences between two internal states which are not too far in terms of rounds, so, in some sense they act in the same way as reduced round cryptanalysis does: they allow us to get some information which has not yet been fully mixed by the cipher. Moreover, since the fault is localized, the block cipher is also "reduced" in terms of effective block width (for instance, all the attacks on AES aim at brute-forcing a single word).

The part in common for attacks on stream ciphers is the fact that injecting faults still "reduces" the complexity of the inner relations of the cipher, but the problem of exploiting this is tackled in a different way, which is the classical one for stream cipher cryptanalysis, namely, solving equations over $\mathbb{F}_2$, possibly obtaining a group of simultaneous linear equations.

Basically, inducing faults "cuts away" a part of the cipher complexity, but the crypto-analytical techniques for the attack, after the information obtained by injecting faults has been formalized, are the same that as those would be commonly used for cryptanalysis of that cipher type. The advantage of injecting faults in this case is that it is possible to obtain output values generated by very similar equations, thus enabling the attacker to cancel only a part of the involved terms, and hopefully get a purely linear relation which would not leak otherwise.

One notices some similarities between all the papers dedicated to fault attacks on stream ciphers. Among them, one is particularly striking, if not downright disturbing: the almost complete lack of countermeasures. Indeed, only one paper [36] mentions generic countermeasures such as either duplicating the LFSR component of the GRAIN-128 cipher checking internal states for consistency with a comparator, or applying some traditional redundancy-based error detection techniques to the LFSR's internal state. Duplicating the state (and thus computations) would be too expensive for most of the stream ciphers, where the state tends to be the largest component. Duplicating it would basically mean doubling the area requirement for hardware implementations, while for software implementations it would also mean having a 100 % computation time overhead.

Taking into account that typically the key requirements for a stream cipher are high speed and reduced area footprint, such a countermeasure does not look very appealing. In the case of adding redundancy-based techniques, the update operation of the modified LFSR is not a feedback anymore but must be replaced with some more general linear transformation. While this idea may lead to interesting research and generate a number of papers in CHES and FDTC workshops, its practicality is inherently limited by the detection capability of the underlying linear code. Considering that the attack models on most of the stream ciphers tolerate multi-bit faults, implementation of error detection methods based on redundancy is likely to be almost as costly as simple duplication of computations.

Does the absence of countermeasures mean that fault attacks on stream ciphers are more dangerous in practice as they are more difficult to prevent than attacks on block ciphers? We think so. Stream ciphers tend to be vulnerable to fault attacks because the nonlinear mixing effect of the stream cipher is usually obtained in a sort of "incremental fashion", while block ciphers can rely on more effective mixing of the plaintext and key material. It is thus more difficult to disturb the computations of a block cipher "politely", so that the output differences or any kind of relationship between the correct and the faulty outputs make sense for further analysis. This usually involves targeting with a rather high time precision a particular round or a particular bit. One approach to designing a stream cipher that can be more robust against fault attacks is to use a nonlinear function with a high degree as a feedback and never employ linear or almost-linear feedbacks. These technique tend to involve a larger part of the state and create a highly nonlinear bound between the state and the output, which in turn transforms usable differential information into unexploitable relations. This is a direct consequence of the fact that multivariate equation solving over $\mathbb{F}_2$ is NP-complete as soon as the degree of the equation is greater than 2.

A construction akin to "tweakable" ciphers [256] may be helpful since this would quickly raise the degree of the relations binding inputs and outputs. This cipher design proposal suggests using a third input to the cipher, called tweak, which is used to alter the structure of the cipher. The tweak may also be public, but must be changed often, leading to the impossibility of collecting faults with the same cipher structure.

Another idea could be inserting a nonlinear layer which depends on the key (like the S-Boxes in Blowfish [362]), which would also hinder cryptanalysis, since the attacker one would need to consider the differential relations among a family of output functions. The only drawback of this technique is that the function family employed to generate the nonlinear layer should be carefully chosen to avoid having weak members, which in turn would constitute a backdoor. What about the stream ciphers constructed by "leak extraction" from block ciphers, such as LEX [50]? Does protecting them against fault attacks amount to protecting an underlying block cipher? The answer is not that simple since one has to take into account the rounds at which the extraction takes place. The protection will have to be extended to these rounds, instead of being applied only to the last ones, which eventually most likely will result in a full computation duplication as well.

The only attack which can be counted more or less efficiently is the "impossible fault analysis" of RC4, as it is relatively inexpensive to check that the two indices do not fall into the Finney state. However, even this would add two comparisons to each state transition. As an optimization one may want to calculate how often these checks need to be carried out so that the fault is caught before the damage is done. This can be done taking into account that it is possible to detect if the cipher has entered a Finney state only after roughly 20 outputs: this implies that a check every 20 cycles is enough to thwart even the distinguishing capabilities of the attack. If the defender is not concerned with the attacker detecting the Finney state, but only with the state extraction, the check may be delayed further, since the attacker is not in possession of the information concerning the first output of the Finney state, and the only ordering provided by the output values is the one among the interleaved values (which appear once every 255 outputs).

Yet, if stream ciphers are inherently vulnerable to fault attacks, the question is, why there are no publications demonstrating these attacks in practice? The answer to this question illustrates another similarity in reviewed papers: a relatively large number of faulty ciphertexts are required for successful cryptanalysis. Indeed, unlike in fault simulations, injecting faults in real life, no matter what the device or the fault injection means, is never deterministic: not every laser shot leads to an exploitable error and not every glitch produces an expected outcome. Thus to obtain thousands or millions of faulty ciphertexts may take in practice much more time than processing them on a PC. It would be interesting to see experimental results on how long it takes to break RC4 with an impossible fault attack!

Chapter 15
Interaction Between Fault Attack Countermeasures and the Resistance Against Power Analysis Attacks

Francesco Regazzoni, Luca Breveglieri, Paolo Ienne and Israel Koren

Abstract Most of the countermeasures against fault attacks on cryptographic systems that have been developed so far are based on the addition of information redundancy. While these countermeasures have been evaluated with respect to their cost (implementation overhead) and efficiency (fault coverage), little attention has been devoted to the question of the impact their use has on the effectiveness of other types of side-channel attacks, in particular, power analysis attacks. This chapter presents an experimental study whose goal is to determine whether the added information redundancy can increase the vulnerability of a cryptographic circuit to power analysis attacks.

15.1 Introduction

In this chapter we discuss in a comprehensive way the interaction between countermeasures against fault injection attacks and the vulnerability to power analysis attacks, using AES as an example. We focus in particular on the non-linear transformation (S-box) within AES since it is the preferred attack point. Specifically, we concentrate on hardware implementations of AES to which error detection circuits

F. Regazzoni (✉)
Crypto Group, Université Catholique de Louvain, Louvain-la-Neuve, Belgium

F. Regazzoni
ALaRI, University of Lugano, Lugano, Switzerland

L. Breveglieri
Politecnico di Milano, Dipartimento di Elettronica de Informazione (DEI), Milan, Italy

P. Ienne
School of Computer and Communication Sciences, École Polytechnique Fédérale de Lausanne (EPFL), Lausanne, Switzerland

I. Koren
University of Massachusetts, Amherst, MA, USA

M. Joye and M. Tunstall (eds.), *Fault Analysis in Cryptography*, Information Security and Cryptography, DOI: 10.1007/978-3-642-29656-7_15,

have been added. Considered are the basic parity check, double parity, residue checks modulo 3 and 7, complementary parity and a Hamming error correcting code. For all the considered error detection or correction circuits, we analyze the effects that the redundant check bits may have on power analysis attacks using different metrics. These include an information theory-based metric, the success rate of power analysis attacks based on correlation, and the effectiveness of the most common attacks based on difference of means and Hamming weights.

The effects that a specific countermeasure against fault attacks can have on the resistance to power analysis attacks was studied in very few previous publications. Maingot and Leveugle [261, 262] analyzed the impact of four different error detection and correction schemes on power analysis resistance. Their study focused on a register storing the state of the AES encryption, which was enlarged to support the information redundancy necessary for each considered scheme. Using gate-level simulations, they showed how the correlation between the value guessed by the adversary and the value of the register varies depending on the particular error detection code employed. They compared four different error detection codes in search for the best code of secure chips, and based on the correlation, concluded that a complementary parity scheme can improve the circuit's robustness against power-based side-channel attacks as well.

Transistor level-simulations were performed by Regazzoni et al. [339, 338] to compare different error detection codes, including parity codes and residue codes (e.g., mod 3 and 7) using a 180 nm technology. As was done in [261], the authors focused on the output register of the S-box transformation in AES, and they analyzed the impact that the considered codes could have on the resistance against power-based attacks and the role played by measurement noise. Furthermore, they discussed the questions of whether the knowledge of the particular error detection code used in the circuit affects the resistance against power-based side-channel attacks and whether the redundancy helps the attacker even if he is unaware of its presence.

Dual studies concerning the interaction between countermeasures against power analysis attacks and vulnerability to fault attacks were also carried out. Boscher and Handschuh [60] discussed the resistance of masking (randomizing the computation) against fault injection attacks, and concluded that masking does not reduce the effectiveness of differential fault attacks. Selemane et al. [368] and Guilley et al. [171] showed, on the other hand, that WDDL (a dual rail circuit implementations that attempts to make the power consumption uniform and independent of the secret key bits) is intrinsically immune to setup violation fault attacks.

15.2 Considered Error Detection and Correction Circuits

Although we focus in this chapter on the Rijndael [112] block cipher (selected to be the Advanced Encryption Standard [142]), our conclusions are general and applicable to other block ciphers. We concentrate on the S-box step because the output of this non-linear transformation is where the difference between the correct key hypothesis

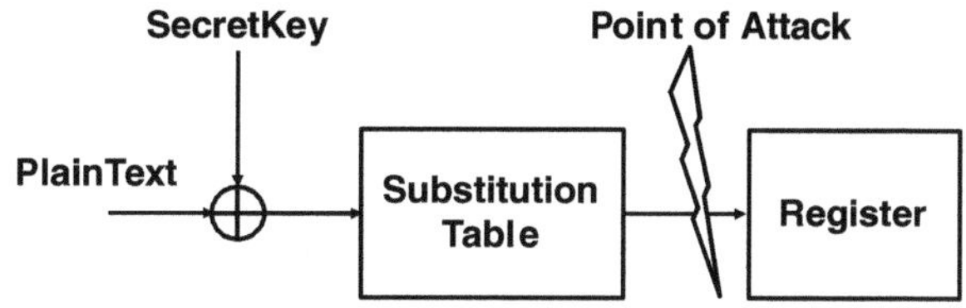

Fig. 15.1 Overview of the considered part of the AES algorithm

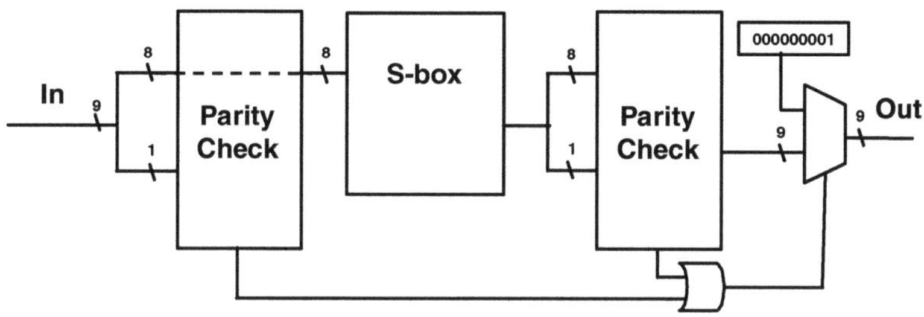

Fig. 15.2 Block diagram of the parity error detection scheme applied to an eight-bit S-box

and the wrong ones is highest, and thus it is the preferred attack point for an adversary [268]. Figure 15.1 depicts the basic configuration used in our experimental power analysis attacks. This configuration is a commonly used simplified implementation of one round of the AES cipher.

The plaintext is added (modulo 2) to the secret key and the result of this XOR operation is used as input to the S-box. The output of the substitution step is stored in a register. In order to always have the same initial condition, a reset signal is applied to the register at the end of each write operation. Although a real implementation of the full algorithm would be somewhat different from this simplified sketch, our purpose is only to estimate the impact of error detection circuits (concatenated to the S-box) on the resistance to power analysis attacks. This approximation (shown in Fig. 15.1) is accepted as sufficiently accurate for analyzing attacks on the most vulnerable portion of the cipher and is therefore adequate for our needs.

Figure 15.2 shows, as an example, an S-box with a parity bit. In this figure, the added check bits are used to detect the presence of errors in two different instances: once at the input and then at the output of the S-box. When new data enters the S-box, the check bits are separated from the data bits and an error detection is performed. If no error is detected, the data bits enter the S-box circuit. The S-box then produces the result of the non-linear transformation plus the corresponding check bits. At this point the second check is performed, again as described before. If no error is detected in both checks, the output of the S-box is forwarded to the next round transformation; otherwise, a faulty output composed of all zeros except for the right most bit is generated to signal the error.

We have implemented several versions of the nonlinear function in the AES S-box, each with a different error detection or correction code. The following circuits are considered:

- **Reference version**. This circuit implements the nonlinear transformation as described in the standard and is used as the reference version.
- **Single parity-based error code**. The single byte parity circuit implements the error detection scheme described by Bertoni et al. [34].
- **Double parity-based error codes**. This code computes two parities: one for the bits with even indices and one for the bits with odd indices.
- **Residue-based codes**. These are the residue codes that use the moduli 3 and 7.
- **Error codes based on complementary parity**. In this scheme, both the even and the odd byte parity bits are computed.
- **Hamming error correcting code**. We consider a (12,4) Hamming code described by the following parity matrix:

$$H = \begin{bmatrix} 1\,1\,1\,0\,0\,0\,1\,1\,1\,0\,0\,0 \\ 1\,0\,0\,1\,1\,0\,1\,1\,0\,1\,0\,0 \\ 0\,1\,0\,1\,0\,1\,1\,0\,0\,0\,1\,0 \\ 0\,0\,1\,0\,1\,1\,0\,1\,0\,0\,0\,1 \end{bmatrix}.$$

15.3 Experimental Setup

Figure 15.3 depicts the complete setup we have used for our evaluation procedure, which is similar to the one presented by Regazzoni et al. [337]. It is composed of a standard Electronic Design Automation (EDA) flow and includes a simulation environment for generating the power consumption traces which are used to provide a measure of the resistance against power analysis attacks. The input to the process is the Register Transfer Level (RTL) description of the S-box and one of the considered error detection/correction circuits. The output is a text file which stores the noise-free instantaneous current consumption of the circuit simulated at a very high resolution of both time and current.

In all our circuit implementations, the S-box module has been described using VHDL at the behavioral level. Because of this, it has been synthesized by the tool as a combinatorial circuit rather than as a memory-based lookup table. It is therefore not necessary to protect the address decoder against injected faults since this component is not present in the synthesized implementation of the substitution function. This approach does not constitute a limitation since it reflects a typical situation when designing a cryptographic unit, where the entire unit is specified using a hardware description language and then synthesized by an EDA tool with no specific implementation constraints imposed. In such cases, the S-box module is often realized as a combinatorial logic.

The VHDL description is synthesized using the STMicroelectronics 90 nm CMOS standard cell library [388] and the Synopsys Design Compiler [391]. If the synthesis tool is set to minimize the circuit's area, it is possible that during the optimization phase of the synthesis process the redundant parts of the circuits (e.g., the circuit generating the complementary parity bit) will be removed. In order to prevent this

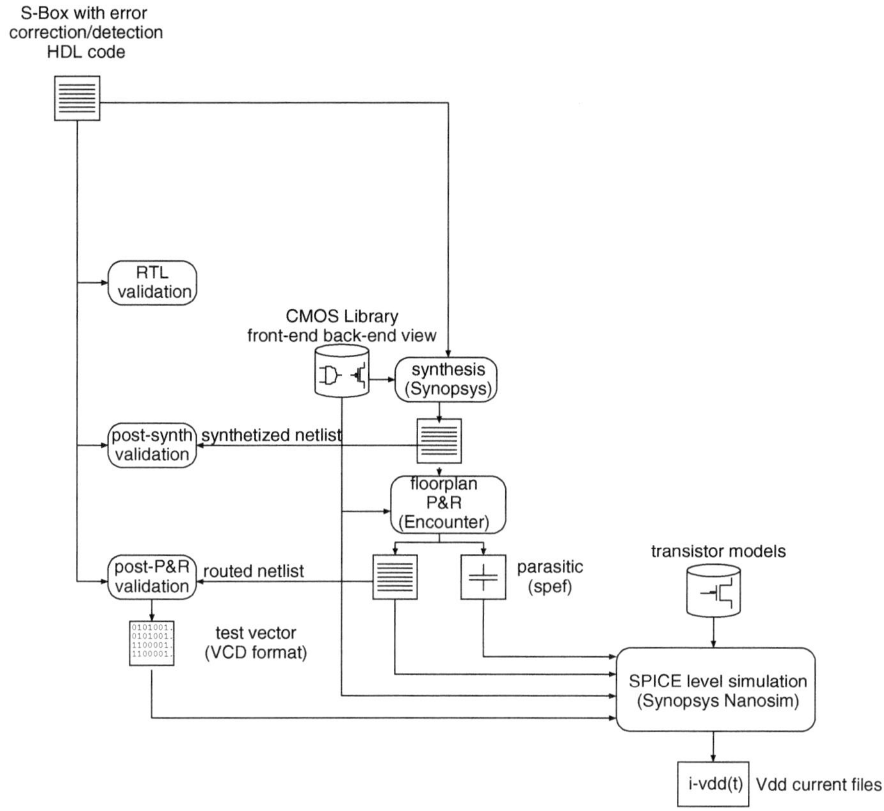

Fig. 15.3 The experimental setup

Table 15.1 Post Synthesis area of each error detection/correction circuit in gate equivalent (GE) area units

S-box type	Circuit area (GE)
Reference	568
Parity	698
Complemented Parity	794
Double Parity	838
Residue Modulo 3	872
Residue Modulo 7	1013
Hamming Code	847

from happening, we first synthesize each component of the circuit separately and then connect the individual components together, forcing the tool to not further optimize the internal design of the individual components. The number of equivalent two-input NAND gates of each circuit is reported in Table 15.1.

The resulting circuit is then placed and routed using Cadence Design Systems SoC Encounter [77]. A parasitics file (in standard parasitic exchange format (SPEF) format) is produced, along with the Verilog netlist of the circuit and an a standard delay format (SDF) file for back annotation of the delays. The flow produces the SPEF and SDF files and the Verilog netlists of the entire design.

Post-place and route simulation is then performed using ModelSim, with the previously generated SDF files to verify the functionality of the circuit and to generate test vectors for transistor-level simulation that will be used to produce the simulated power traces. Synposys' Nanosim is then executed to perform transistor-level simulation, using the SPEF file, the relative Verilog netlist, the SPICE models of the technology cells and the transistor models. This simulation generates vector-based time-varying power profiles which are stored in a text format. This data corresponds to the simulated traces which are later used for mounting the power analysis attacks and the security evaluation.

15.4 Evaluation of the Effects on Power Analysis Resistance

In this section we analyze the possible impact on power analysis resistance of a fault detection circuitry added to a device to protect it against fault injection attacks using state-of-the-art tools and methodologies.

We present below the results of several experiments focusing on the added structural redundancy, attacking the previously described configuration (depicted in Fig. 15.1), where the result of the key addition is passed through the S-box.

In the first and second sets of experiments we explore how the different error detection and correction codes affect the two most common power analysis schemes, namely, the Differential Power Analysis (DPA) based on Kocher's difference of means and the DPA based on Pearson's correlation coefficients which uses the Hamming weight as model. These experiments were carried on using the noise-free traces generated by the SPICE-level simulator.

In the first set of experiments we performed DPA attacks (based on Kocher's difference of means) targeting all the output bits of the S-Boxes. These experiments included attacks on the reference circuit (a straightforward implementation of the AES standard) and on all the error detection codes described in Sect. 15.2. In the second set of experiments we performed attacks which use as model the Hamming weight and as distinguisher Pearson's correlation coefficients, and we considered situations where the adversary is both aware and unaware of the presence of the particular error detection code used.

The purpose of the third set of experiments has been to conduct a fair and objective comparison of the different error detection and correction circuits. To this end, we used a metric based on information theory to provide an attack-independent evaluation of each error detection/correction circuit. Based on this, we report how the number of information bits that are available to the attacker changes as a function of the measurement noise.

Finally, in the fourth set of experiments we analyze for each of the error detection/correction circuits how the success rate of an adversary (exploiting the correlation coefficients and the Hamming weight) varies as a function of the model used during the attack. These experiments show whether the additional information available to the attacker can be beneficial even if the attacker is unaware of the presence of the specific error detection and correction code.

15.4.1 Evaluating the Effects of the Added Check Bits on Kocher's Difference of Means DPA

The goal of these experiments is twofold: determine whether the redundancy added to a cryptographic circuit affects the effectiveness of the classical Kocher's DPA, and find out whether the redundancy bits themselves are susceptible to this attack as are the data bits. We also want to explore whether the choice of a particular error detection code influences the results. We thus performed Kocher's DPA attack on various implementations of the AES S-box, with and without error detection.

For the reference circuit, we mounted Kocher's DPA attacks targeting, one by one, all the output bits of the S-box. In the attacks on implementations that include a fault detection circuit, we distinguished between two cases: in the first case we targeted only the eight output bits of the S-box, mimicking the situation in which the attacker is unaware of the presence of the error detection circuit; in the second case, we included in our hypothesis the redundancy bits, assuming that the attacker knows about the specific error detection check bits that have been added to the S-box. All these attacks were performed directly on the noise-free power traces produced by the SPICE-level simulator described in Sect. 15.3.

Figures 15.4 and 15.5 show the results of a Kocher's DPA attack on one output bit of the AES S-box in the reference circuit and on the same bit of the S-box with an added error detection circuit based on complementary parity bits, respectively. The differential trace corresponding to the correct key is plotted in black, while all the others are in gray. As can be seen, during the computation of the output that corresponds to the initial part of the graph (approximately up to 1,000 ps), the presence of the parity bits seems to make the attack more difficult. However, when the result is stored into the register (at about 1,750 ps) the peak is, in both cases, of approximately the same height. The above situation repeats itself for all the other codes. Since the adversary typically targets the point that yields the highest probability of succeeding with the DPA attack, and since this point is usually the time when the computed value is stored into the register, we can conclude that the presence of an error detection circuit does not substantially affect Kocher's DPA.

The situation is slightly different when the adversary attacks one of the redundant check bits. When the target bit (used as selection function) generates two sets that contain approximately the same number of elements (as in the case of the parity bit, the complementary parity, the Hamming code and the dual parity), the value of the

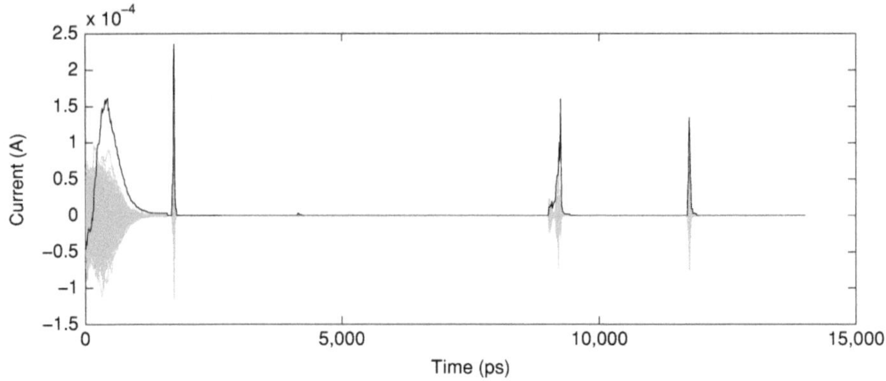

Fig. 15.4 Kocher's DPA attack on the reference implementation of the AES S-box

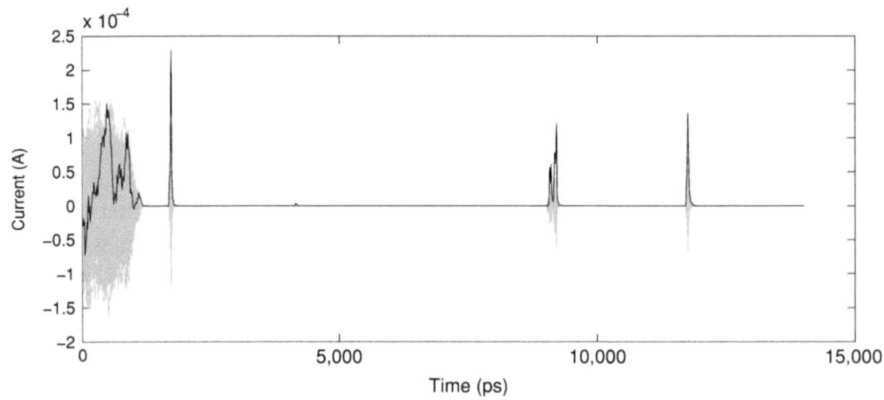

Fig. 15.5 Kocher's DPA attack on the implementation of the AES S-box with added complementary parity

peak for the attacks on a check bit is approximately the same height as that for an attack on an information bit. This shows that an attack on the parity bit is just as efficient as an attack on any other bit. In contrast, when the selection bit does not split the set evenly (as in the case of residue modulo 3 or 7), the strength of Kocher's DPA attack is significantly reduced and it may even fail sometimes.

In summary, the experiments described above show that Kocher's DPA attacks based on the difference of means and performed on noise-free traces are not substantially affected by the presence of an error detection circuit. Additionally, very often the check bits can also be exploited by the attacker. Consequently, as a general rule, when adding countermeasures against power analysis to an implementation that includes an error detection circuit, the check bits should be protected as well.

15.4.2 Evaluating the Effects of Check Bits on Pearson's Correlation-Based DPA

The second set of experiments focused on Pearson's correlation coefficient DPA, which uses Hamming weight with the goal of finding out whether one of the circuits is easier to attack than the others. Another objective is exploring whether knowledge of the presence of an error detection circuit can be exploited by the attacker. We performed a series of correlation-based DPA attacks against all the considered implementations of the AES S-box. It is important to notice that since we reset the circuit after each computation, the Hamming weight and the Hamming distance models would yield exactly the same results.

In the case of AES, the attacker usually hypothesizes all the eight output bits of the S-box. However, when the S-box is extended to provide error detection or correction, the number of output bits is higher than 8. Since the correlation value changes depending on the number of bits included in the hypothesis, it is worthwhile exploring the effects of the added check bits in two cases: when the adversary hypothesizes only the eight output bits of the AES S-box (i.e., the attacker is unaware of the presence of an error detection circuit) and when the adversary hypothesizes all the output bits, including the check bits (i.e., the attacker is aware of the particular error detection code used).

As in the previous series of experiments, we performed our evaluation on the noise-free traces produced by the SPICE-level simulator and we considered the circuit depicted in Fig. 15.1. Furthermore, as in the case of Kocher's DPA, since the traces obtained by simulation are noise-free and the size of the S-box is eight bits, we can fully characterize the device by simulating all the 256 input plaintexts for each of the 256 possible keys.

The first series of attacks was performed including in the attack hypothesis all the bits of the target register. This corresponds to the situation where the adversary is aware of the particular error code used. Then, we included in the attack hypothesis only the eight output bits of the S-box, i.e., we assumed that the adversary is unaware of the presence of an error detection circuit.

Figures 15.6 and 15.7 show the results of a Pearson's correlation coefficient DPA attack on the output of the AES S-box in the reference circuit and on the output of the S-box with an added error detection circuit based on double parity, respectively, when the presence of the error detection code is unknown to the attacker. The figures show the time period during which the outputs of the S-box and, when present, the check bits, are computed (approximately up to 2,500 ps), and the time interval in which the results and the check bits are stored into the register (approximately from 2,500 to 3,000 ps).

As can be seen, this is the best situation for the attacker, and this is confirmed by the high values of the correlation coefficients for the time intervals which correspond to the register write operations. In fact, in the attack mounted when the presence of the error correction code was known to the attacker, all the circuits showed a correlation value approximately equal to 1. Additionally, when the presence of the

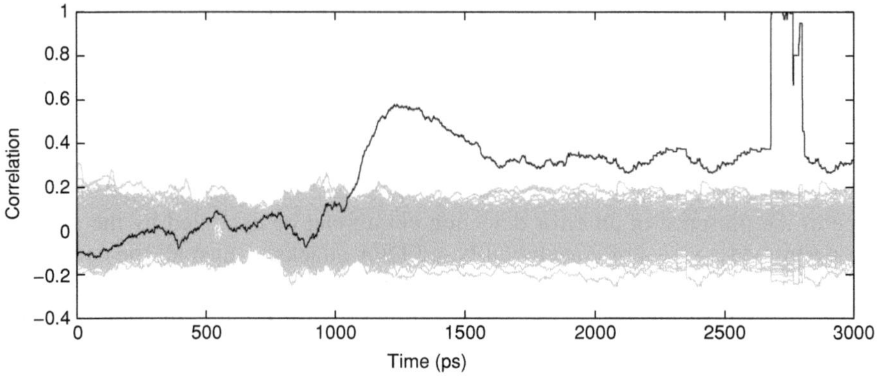

Fig. 15.6 Pearson's correlation coefficients DPA attack on the reference implementation of the AES S-box

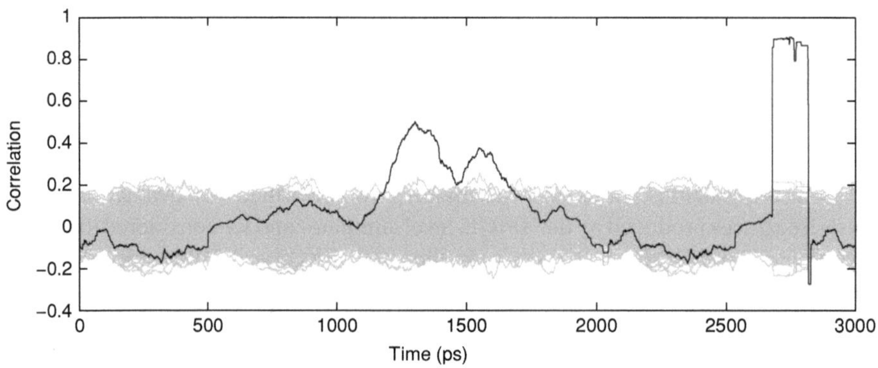

Fig. 15.7 Pearson's correlation coefficients DPA attack on the AES S-box with added double parity with the attacker being unaware of the error detection circuit

error detection code was unknown to the attacker, for all the considered circuits the correlation coefficients were slightly decreased, but still very high (never below 0.81), thus sufficient for easily extracting the secret key.

We can also see that during the computation, from the beginning of the trace to approximately 2,500 ps, the difference between the correlation coefficients of the wrong key guesses and those of the correct key guess decreases, mainly because of the reduction in the correlation coefficient of the correct key guess. However, this does not represent a significant problem for the adversary since the easiest attack point is still the register store operation, and there the correlation, as shown before, is not significantly affected by the presence of the error detection/correction circuit.

These results show that when the adversary has noise-free traces or, alternatively, a sufficient number of traces to completely filter out the noise, the correlation-based DPA which uses the Hamming weight model is always successful, independently of the particular error detection and correction code used and of the attacker's knowledge of the specific circuit implemented.

15.4.3 Evaluating the Effects of the Check Bits Using Information Theory

As previously shown, when the power consumption traces are noise-free or include very little noise, the presence of any one of the considered error detection/correction circuits does not significantly impact the resistance against the two most common power analysis attacks.

In order to obtain a more fair comparison among the different error detection and correction schemes and to quantify the effect that the redundancy may have on the resistance to power analysis attacks, it is necessary to analyze the behavior of each circuit in the presence of noise, independently of the particular attack hypothesis and scenario.

To this end, we use the metric based on information theory proposed by Standaert et al. [384], which was developed to allow the evaluation of side-channel information leakage.

Intuitively, this information-theoretic metric measures the resistance against the strongest possible type of side-channel attack and allows an evaluation that is independent of a particular attack scenario. Practically, the metric measures how much uncertainty about the secret key remains after the attacker takes advantage of the given information leakage.

More formally, let K be a random variable representing the key that the adversary wants to recover in the side-channel attack. Let X be a random variable representing the known plaintext entering the target operations (in our case the S-box with or without an error detection/correction circuit) and let L be a random variable representing the power consumption traces generated by a computation with input X and key K. L is obtained by adding a certain amount of normally distributed random noise R to the power trace T produced by a SPICE-level simulation, i.e., $L = T + R$. The conditional entropy H between the key K and its corresponding leakage L is defined as follows:

$$\mathrm{H}[K|L] = -\sum_{k} \Pr[k] \cdot \sum_{x} \Pr[x] \int \Pr[l|k,x] \cdot \log_2 \Pr[k|l,x] dl \tag{15.1}$$

where $\Pr[k]$ is the probability of the key k, $\Pr[x]$ is the probability of the plaintext x, $\Pr[l|k,x]$ is the probability of the leakage l given the key-plaintext pair (k,x), and $\Pr[k|l,x]$ is the probability of the key k given the leakage-plaintext pair (l,x). The probability density function of L is assumed to be approximated by the normal distribution $\mathcal{N}(\mu_{k,x}, \sigma^2)$, where $\mu_{k,x}$ is the noise-free leakage measured during the computation of the (k,x) pair and σ is the standard deviation of the noise. As a result, the conditional entropy of K given L, can be rewritten as

$$H[K|L] = -\sum_k Pr[k] \cdot \sum_x Pr[x] \cdot \int_{-\infty}^{\infty} \mathcal{N}_l(\mu_{k,x}, \sigma^2) \cdot \log_2 \frac{\mathcal{N}_l(\mu_{k,x}, \sigma^2)}{\sum_{k^*} \mathcal{N}_l(\mu_{k^*,x}, \sigma^2)} \, dl \, .$$

There are different factors that influence this conditional entropy. The first one is obviously the simulated power trace. The second one is the standard deviation of the noise in the leakage. The size of the power traces is also important: simulated traces are typically composed of several thousands of samples. Hence, directly applying multivariate statistics on these large dimensionality variables is hardly tractable.

In order to reduce the dimensionality of the power traces, some compression techniques such as the Principal Component Analysis (PCA) [259] and integration over the full trace were proposed in the past. For the experiment carried out in this work, we used the latter, namely we first integrated the noise-free trace to reduce its dimensionality and then we evaluated the entropy. Therefore, the mutual information is extracted from a trace that was first compressed to one, single sample.

Given the conditional entropy one can calculate the so-called mutual information [384] (which intuitively quantifies what the adversary knows about the secret key K assuming that he has the knowledge of the leakage L) as follows:

$$\mathrm{I}[K, L] = \mathrm{H}[K] - \mathrm{H}[K|L]$$

where $\mathrm{H}[K]$ is the entropy. Since all key values are equally probable, $\mathrm{H}[K]$ is equal to n which is the number of bits of the key K.

Figure 15.8 depicts the value of the mutual information for each considered error detection/correction circuit as a function of the standard deviation of the noise. Intuitively, the higher the number of bits available to the attacker, the lower the resistance against the power analysis attack. In the left part of the graph, where the noise level is under a certain threshold, all the circuits have information leakage as high as 8. This means that the attack is not affected by the presence of the particular circuit used, since it will be successful in any case. This confirms the results presented in Sects. 15.4.1 and 15.4.2, which have shown that when noise is completely absent, the effectiveness of both Kocher's and Pearson's DPAs was not affected by the presence of error detection/correction circuits. Figure 15.8 also depicts the dual situation, which corresponds to the case where the noise level is higher than a certain threshold; here too, the attack is not affected by the particular circuit used. However, in this case the mutual information is always 0, and the adversary will not be able to retrieve the secret key in any case.

When the standard deviation of the noise is in the middle interval, it is possible to quantify the effect that each of the error detection circuits can have on the strongest possible power analysis attack. The reference S-box (the one without any error detection code) is characterized by the smallest number of bits leaked, followed by the parity scheme. The two worst circuits are the ones implementing the residue codes modulo 3 or 7. The graph thus confirms the intuition that, except for the case

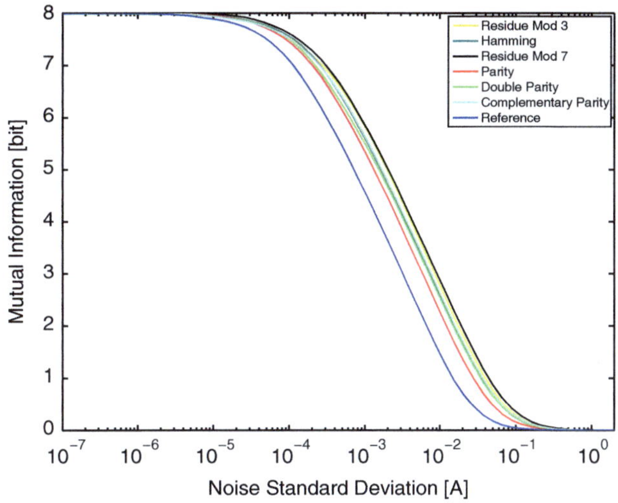

Fig. 15.8 Mutual information as a function of the noise's standard deviation for each of the considered error detection and correction codes

of the Hamming error correction code, there is a direct relation between the number of check bits and the amount of information leakage. It is important to emphasize that the ranking obtained using the information theory metric shows the number of bits available to the strongest possible attacker; it is possible that in a specific attack scenario in which the adversary is not able to exploit all the information present in the power trace, the ranking of the circuits will be different. Still, this metric is the most objective one since it does not depend on the specific attack hypothesis. In the next section we show how the information stored in the power traces may be exploited by an attacker who uses the correlation-based DPA and the Hamming weight model.

15.4.4 Evaluating the Effect of Check Bits on the Success Rate

The goal of the last set of experiments was to evaluate how the success rate of an attacker who uses the correlation-based DPA varies for each different circuit as a function of the power model used. We discuss in this section whether one of the considered circuits is easier to attack than the others when the same attack is used.

Intuitively, the better the power model is, the easier it is for an adversary to recover a key. One important result which can be obtained from these experiments is whether the additional information leakage generated by the error detection/correction circuit can be exploited by the attacker even if he is unaware of its existence.

To this end, we performed a set of correlation-based DPA attacks against all the considered implementations of the AES S-box, using different attack hypotheses and increasing at each run the number of traces. The attacks were performed using

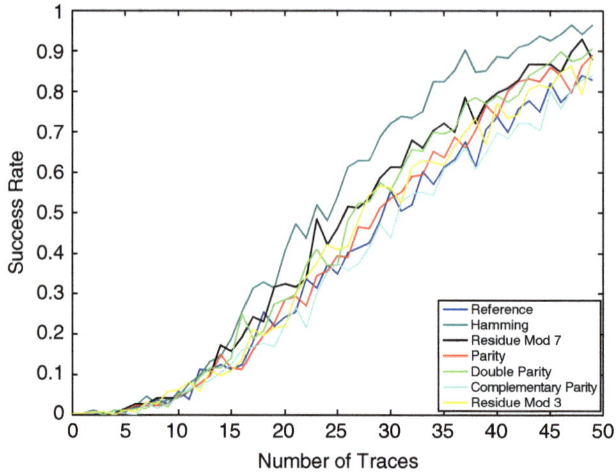

Fig. 15.9 Success rate of the correlation-based DPA attack versus the number of traces using attack hypothesis of the full size of the output register (the adversary is aware of the error detection and correction code used)

all the 256 possible keys and randomly selecting the input plaintext. For the first set of experiments, the correlation between the Hamming weight of the eight-bit S-box output and the power traces was calculated for all the considered circuits. The underlying assumption was that the adversary is unaware of the presence of the error detection/correction circuit and can therefore construct only an approximated power model. In the second set of experiments we computed the correlation between the power traces and the Hamming weight of the full output of the S-box (including the check bits), thus assuming that the attacker is aware of the particular redundancy added to the S-box circuit.

Using the results of all the attacks that were mounted, we can compute the first order success rate [384], defined below. Given an adversary who attacks a secret key K and generates n key guesses $g = [g_1, g_2, \ldots, g_n]$ that are sorted according to the attack result (correlation-based DPA in our case), we define a function f that returns 1 if the correct key is g_1 and 0 otherwise. The first-order success rate of the attack against the secret key K is defined as

$$\mathrm{Succ}^K_{\mathrm{attack}} = \Pr[f = 1]\,. \tag{15.2}$$

We concentrate on one of the intermediate situations discussed in Sect. 15.4.3, where the noise is present but is not sufficiently high to completely overshadow the signal and cause the attack to fail. To simulate the noise conditions, we added white noise to the traces generated by the transistor-level simulator.

Figures 15.9 and 15.10 show the success rate of the correlation-based DPA as a function of the number of traces for a fixed value of the noise standard deviation (equal to 5×10^{-4}) using two different attack hypotheses. Figure 15.9 shows the

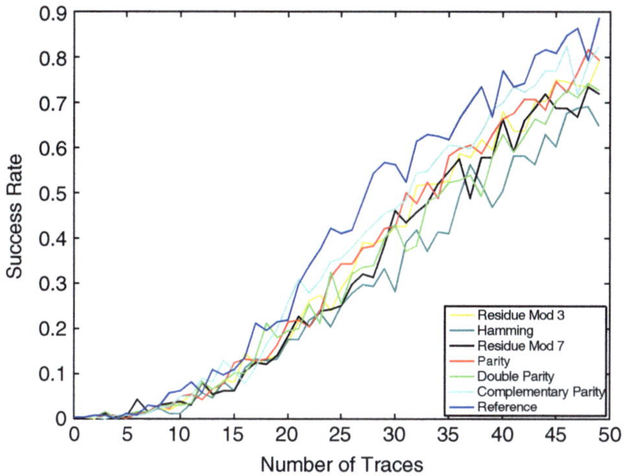

Fig. 15.10 Success rate of the correlation-based DPA attack versus the number of traces using attack hypothesis of 8 bits only (the adversary is unaware of the error detection and correction code used)

success rate of the correlation-based DPA when the attacker is aware of the presence of the code, thus targeting all the bits of the output register. Figure 15.10 shows the success rate of the DPA when the attack hypothesis is based on the Hamming weight of only the eight output bits of the S-box, i.e., the attacker is unaware of the code used. In both figures, the faster the curve approaches 1, the easier it is for the attacker to recover the secret key.

As can be seen from Fig. 15.9, the error detection/correction circuit that yields the worst resistance against the correlation-based DPA attacks is the one that uses the Hamming correcting code. The only code that seems to be slightly more resistant than the reference S-Box is the one based on complementary parity. We can therefore state that when the adversary mounting a correlation-based DPA attack is aware of the particular error correction code used, the added redundancy significantly helps the adversary. Note that the ranking of the circuits in Fig. 15.9 is in agreement with that reported in Sect. 15.4.3. The information theory metric was computed using all the points of the trace and assumes a strong adversary; thus it incorporates also information which can not be exploited by the attack considered in this case.

In contrast, as shown in Fig. 15.10, when the presence of the particular error detection/correction circuit is not known, the success rate of the reference S-box is higher than all the others. This, however, does not mean that the presence of the check bits can never help the attacker when he is unaware of them. The implementations that include check bits still generate a large amount of information leakage, as indicated by the information theory metric (and shown in Sect. 15.4.3), and it is possible that different technological libraries or more sophisticated attacks can show a significant improvement even when the presence of the code is unknown.

15.5 Conclusion

We have presented in this chapter an evaluation of the effect that an error detection circuit may have on the resistance to power analysis attacks on hardware implementations of cryptographic S-Boxes. The evaluation was carried out using the STMicroelectronics 90 nm CMOS technology library and specific synthesis and place and route options to prevent the tool from removing the redundancy that is present due to the added error checking circuitry.

Our results show that the presence of the error detection/correction circuit increases the amount of information available to the attacker. We also show that, depending on the particular attack hypothesis, the adversary may take advantage of this additional information.

It is important, however, to mention that our conclusions regarding the impact of the different error detection and correction codes on the vulnerability to power analysis attacks may be different for other design environments. As is typical with hardware designs, even the same high-level description of a module may lead to a quite different VLSI circuit if the design options and technology library are changed.

Nevertheless, when incorporating fault detection or correction circuits into hardware implementations of cryptographic algorithms, it is of crucial importance to be aware of the possible effects that the added redundancy may have on the robustness against power analysis attacks.

The experiments presented in this chapter provide an example of how to perform an evaluation of the vulnerability of the designed circuit to power analysis attacks prior to manufacturing.

Part V
Implementing Fault Attacks

Chapter 16
Injection Technologies for Fault Attacks on Microprocessors

Alessandro Barenghi, Guido M. Bertoni, Luca Breveglieri, Mauro Pelliccioli and Gerardo Pelosi

Abstract The dependability of computing systems running cryptographic primitives is a critical factor for evaluating the practical security of any cryptographic scheme. Indeed, the observation of erroneous results produced by a computing device after the artificial injection of transient faults is one of the most effective side-channel attacks. This chapter reviews the (semi-)invasive fault injection techniques that have been successfully used to recover the secret parameters of a cryptographic component. Subsequently, a complete characterization of the fault model derived from the constant underfeeding of a general-purpose microprocessor is described, in order to infer how the faulty behavior causes exploitable software errors.

16.1 Introduction

Attacks on cryptographic components, both hardware accelerators and software routines, based on the injection of intentional faults, are the most recent trend in side-channel attacks. It is sufficiently proved by the existing case studies [64, 65, 66, 69] that at least some devices can be very successfully attacked in this way, and, therefore, the topic has gained considerable academic and industrial interest in the last decade.

These attacks are based on two factors: First of all, the existence of a deterministic or statistical procedure to exploit the erroneous results (typically ciphertexts) of the faulty cryptographic computations in order to infer the whole key, or the plaintext, or, more generally, of pieces of information that help the attacker infer the unknown parameters of the cryptoscheme; second, the availability of suitable technologies

A. Barenghi (✉) · L. Breveglieri · M. Pelliccioli · G. Pelosi
Dipartimento di Elettronica e Informazione (DEI), Politecnico di Milano, Milan, Italy
e-mail: barenghi@elet.polimi.it

G. M. Bertoni
STMicroelectronics, Milan, Italy

M. Joye and M. Tunstall (eds.), *Fault Analysis in Cryptography*,
Information Security and Cryptography, DOI: 10.1007/978-3-642-29656-7_16,

to inject faults, in a reliable and repeatable fashion, into the computation of the target cryptographic primitive (implemented in either hardware or software). The first factor has received a number of responses and every widely employed cryptographic algorithm has a number of possible attack procedures based on appropriate fault and error types [64–66, 69]. Sometimes however, such procedures are based on very peculiar or well-targeted fault models and errors, which are either practically infeasible or widely believed to be physically realizable with great difficulty. This is especially true when the fault model mandates a radical change in the circuit design, more than just an inducible misbehavior of the hardware which comes with some unavoidable degree of randomness and imprecision, despite the accuracy of the injection methodology. The second factor is more critical as it relies on the feasibility of inducing an alteration in the working environment, which is determined by the technological advances in the employed equipment. It is not easy at all to inject an exploitable fault into a computing device with a sensible degree of reproducibility and accuracy, regardless of the fact that the targeted device is a dedicated hardware accelerator or a general purpose microprocessor executing a software routine.

It is possible to split the injection techniques into two categories, depending on the cost of the required equipment. Among the most viable techniques, i.e. those whose cost may be a casual but motivated attacker's budget, the most common ones rely on tampering with one of the input lines of the computing device, either the clock signal or the power supply line [12, 23, 24, 359, 369]. Among the techniques requiring a significant budget to be practically carried out, the best documented one is the induction of faults in a computing circuit through irradiation with a coherent light beam (i.e., a laser) [377]. Since the most viable and applied techniques to inject faults are those needing a low budget, we will propose, as a case study, the injection of faults during the execution of a software primitive on a general-purpose microprocessor operating under the a condition of power underfeeding. The effects of supply voltage reductions on an hardware cryptographic accelerator are described in Sect. 17.3.

16.2 Fault Injection Technologies

The techniques employed to alter the correct functioning of a computing device are basically an in vitro reproduction of the causes of natural hazards. These causes are selected from the whole range of working environment alterations, such as changes in the power supply, irregularities in the generation of the clock for synchronous circuits, anomalous irradiation, electromagnetic disturbances or physical alteration of the circuit due to nanoscale wear. Since the range of technological options to induce these alterations is large and getting wider, in order to provide a clearer overview, the description of the techniques will follow a classification, which divides them according to the required budget. We will point out the degree of technical skill and knowledge of the implementation required to employ the injection methods, and we will characterize the model of the achievable faults with respect to their effectiveness and temporal spatial accuracy. The combination of these criteria enables a chip designer to recognize the possible threats to his secure implementation, depending on the assumed skill set and budget of the attackers she will be put against.

16.2.1 High-Cost Fault Injection Techniques

A class of possible threats that cannot be ignored if the potential attackers have access to a large budget (roughly above 3,000 euros and in general, up to millions of euros) is the one represented by the fault injection methodologies that target the secure chip by gaining direct access to the silicon die and by targeting the perturbation action in a very precise fashion. These techniques, despite their leaving evident traces of tampering, are very powerful due to the high spatial precision of the injected faults. Another notable aspect is the fact that the attacker is able to gain a great deal of knowledge regarding the implementation strategies of the secure chip under attack, since he will be able to directly inspect it, and thus to localize sensitive parts, despite any possible hardware obfuscation techniques.

The simplest of these techniques relies on strong and very precisely focused light pulses to induce alterations in the behavior of one or more logic gates of a circuit. In particular, the strong irradiation of a transistor may induce the formation of a temporary conductive channel in the dielectric, akin to the one which is regularly present which the transistor is correctly polarized. This, in turn, may lead a logic circuit to switch state in a very precise and controlled manner. For instance it is possible to flip-up or down at will an SRAM cell, by targeting one of its transistors as reported in [377]. In order to obtain very precisely focused light pulses, the light emitted from a camera flash is concentrated with the aid of a precision optical microscope by applying it to the eyepiece after the device under attack has been carefully placed on the slide holder. In order to avoid over-irradiation of the device, which might lead to permanent damage to the circuit, care must be taken in selecting an appropriate magnification level for the microscope lens. This requires only moderate technical skill, since what is needed is to properly operate an optical microscope and to design a synchronizing circuit to trigger the flashes. The main limitations of this method are represented by the nonpolarized nature of the white light emitted by the camera flash, which in turn implies that the light beam is not perfectly coherent. The intrinsic scattering the light undergoes when it is focused through imperfect lenses further spoils the focus of the beam, thus lowering the accuracy of this technique. Moreover, it is no longer possible to hit a single SRAM cell with the current etching technologies, since the width of the gate dielectric is now more than ten times smaller than the shortest wavelength of the visible light. This technique has moderate time accuracy, whose limit is represented by the synchronization of the flashlight with the computing device. Since common flashlights are unable to emit a sequence of flashes within a very short time frame, this technique cannot induce multiple faults in a single run of an algorithm (see Table 16.1).

The most straightforward refinement of this technique is to employ a laser beam instead of a simple flashlight. The fault model induced in the chip is substantially

Table 16.1 Summary of high-cost fault injection techniques: all methods require a detailed knowledge of specific implementation aspects

Technique	Accuracy [space]	Accuracy [time]	Technical skill	Permanent modifications
Light pulses	Moderate	Moderate	Moderate	Possibly
Laser beams	High	High	High	Possibly
Focused ion beams	Complete	Complete	Very high	Yes

the same as the one obtained through a concentrated light beam [379], except for the fact that the laser beam turns out to be more efficient in inducing one (or more) faults per irradiation, with higher accuracy in space. Laser beam attacks allow also the irradiation of the back of the silicon die (i.e. the side where the chip is not etched). This can be achieved using near infra-red (NIR) lasers, since pure silicon has a transmittance of around 50 % for wavelengths between 1 μm and 5 μm. This technique allows the attacker to successfully hit also parts of the circuit on the bottom layers, although with less precision since the silicon substrate scatters the beam while it is traveling through it. However, the irradiation energy must be carefully calibrated since it is quite likely that an improper setting will permanently burn the chip. The commercial fault injection workstations [344] are composed of a laser emitter together with a proper focusing lens and a placement pad endowed with stepper motors in order to achieve very precise targeting of the beam. The technical skills required to properly operate a laser injection workbench are very specific and specific training must be followed by anyone willing to attempt such attacks. The limitation of this fault injection technique is the fact that it is impossible to achieve sub-wavelength precision in the irradiated zone. This limits the minimum number of gates hit by the irradiation depending on the etching technology and the laser wavelength. The time accuracy is higher than the one based on flashes, since the laser emitter needs less time to recharge; it is thus possible to inject multiple faults within the same execution of a cryptographic algorithm.

The most accurate and powerful fault injection technique employs Focused Ion Beam (FIB) microsurgery. A FIB enables an attacker to arbitrarily modify the structure of a circuit, in order to reconstruct missing buses, cut existing wires, mill through layers and rebuild them. Usually FIB workstations are employed to debug and patch chip prototypes, or to reverse engineer an unknown design by adding probing wires to parts of the circuit that are not commonly accessible. For instance, the successful reconstruction of a whole read bus for a portion of a chip Flash memory containing a cryptographic key, without any damage to its contents [398]. Current FIBs are able to operate with precision up to 2.5 nm, i.e. less than a tenth of the gate width of the smallest transistor that can currently be etched. FIB workstations are very expensive to run and require a strong technical and scientific background in order to be properly operated. Usually their use is limited to extremely well-equipped reverse engineering laboratories. The only limitation of the FIB technology is represented by the diameter of the atoms whose ions are used as a scalpel. Currently, due to physical properties (in particular due to its low melting point), the most common choice for circuit etching is gallium, which sets the lower bound for the etching accuracy to roughly 0.135 nm. It is also possible to perform the deposition of both insulating layer and conducting material, although the cost of the FIB station equipped with multi-material ion sources is even higher.

In Table 16.1 we recap the mentioned techniques regarding high-budget attacks, providing a summary of their features.

16.2.2 Low-Cost Fault Injection Techniques

We can consider "low cost" injection technologies as all the technical options available to an attacker which require less than 3,000 euros of equipment in order to set up the workbench at the present time. This budget is well within the financial means of a single motivated attacker, and thus these fault injection techniques should be considered as a serious threat to the implementations of a secure chip, since they can successfully inject faults. Low-cost techniques are not able to achieve the precision, in time or space, reachable with more expensive techniques. This in turn implies that, instead of injecting a small number of very precise faults, the attacker collects a massive number of faulty computations and selects which ones actually yield useful information for retrieving the secret parameters, provided the attack technique has a distinguisher for exploitable faults.

The fault injection technique we will analyze in detail in Sect. 16.3, is the constant under-powering of a computing device. By running the chip with a depleted power supply, the attacker is able to insert transient faults starting from single-bit errors and becoming more and more invasive as the supply voltage gets lower. The errors appear as single-bit setup failures in the latches of the circuit, due to the slower transition speed of the most current demanding lines (usually those leaving long combinatorial cones). On the one hand, this leads to high accuracy in the spatial pattern of the error, which tends to repeat with nearly perfect precision. On the other hand, since no timing or synchronization controls are employed, faults cannot be injected a precise time segment of the running algorithm. Since this technique does not involve precise timing for the reduction of the feeding voltage, the faults tend to uniformly happen over the whole computation of the targeted device, thus forcing the attacker to discard the erroneous results which do not fit his attack model for the specific cipher. This methodology is reported to be effective also on large-scale integrated circuits [23, 24] such as the ARM9 described in Sect. 16.3. The underfeeding is achieved by employing a tunable precision power supply unit to feed the device with a different voltage and requires the attacker to be able to tap into the power supply line (which is usually exposed on the PCB). This requires only basic skill and can be easily achieved in practice without leaving evidence of tampering on the device. Moreover, no knowledge of the implementation details of the platform is needed, since it is possible to profile the behavior of the device when it is operating in fault inducing conditions, and to infer the injected fault model. The technological limit of this technique is represented by the fact that low-power devices exhibit higher sensitivity to voltage that is too low and thus require the attacker to be able to tune the voltage drop with fine steps. However, in [24] it is reported that the attack technique is hindered neither by the different working clock frequency of the device, nor by the etching technology employed to build it (except for the lower power consumption it might result in).

A refinement of the aforementioned technique is represented by the injection of well-timed power spikes or temporary brownouts on the supply line of a circuit. It is possible to skip the execution of a single instruction of microprocessor code, by carefully reducing the feeding voltage for the duration of a single clock cycle. The possible causes for this misbehavior are related either to the lapse in power supply, causing a failure in the storage of the results of an operation in the target register during the write-back phase, or to a misinterpretation of the instruction during the decode stage. The authors of [359] report a successful application of this technique to eight-bit microprocessor, while a description of successful over-voltage spikes that have been applied to depackaged Radio Frequency Identification tags in [186]. In order to inject a timed voltage lapse, the attacker needs a tunable precision power supply and a custom circuit able to drop the feeding voltage under a certain threshold by temporarily grounding the feeding line. This circuit is also supplied with the same clock that drives the microprocessor and is thus able to correctly time the injection of the glitch. The temporal precision of the fault injection is directly dependent on the accuracy of the voltage drop in terms of both duration and synchronization with the target device. This technique is more difficult to apply when the working clock rate of the attacked circuit is higher, since dropping suddenly the feeding voltage below the working threshold (and ideally to zero) is more difficult due to the mutual induction of the feeding line.

Another viable option for an attacker is to tamper with the clock signal driving the computing device. In particular, it is possible to shorten the length of a single cycle by forcing a premature toggling of the clock signal. The reduction in the single clock cycle length, according to the authors of [12], causes multiple errors involving the corruption of a stored byte or multiple bytes during the computation.

The errors are transient and the device does not incur any damage; thus it is possible to induce faults at will without leaving any evidence of tampering. In order to be able to alter the clock duty cycle, the attacker needs to have direct control over the clock line, which is the typical case when smart cards are targeted [12]. It is also possible to cut the clock line and to connect a new clock line supplied by the attacker, but no reports of this kind of attack are known at the moment. It is impossible to attack chips that employ an on-die independent oscillator to generate their own clock signal, since the attacker is not able to disconnect the clock line from the circuit. The employed workbench in the attack mentioned in [12] involves a modified smart card reader, which is able to shorten the duration of a specific clock cycle by either anticipating the raising edge or delaying the falling edge, depending on the kind of smart card. The modification is not trivial and involves a careful choice of the delaying equipment, but it can be performed without any special tools, except for the extra electronic components required for the modification and a proper soldering station. Clock alteration techniques are hindered by the need to supply a regular clock within the working range of the device, while retaining the ability to alter a single clock edge. This implies that the equipment inducing the alteration must be working at a clock frequency higher than that of the attacked device, and this is intrinsically more difficult as the target device working frequency increases.

Another way to induce nondestructive faults in a digital circuit is to steadily overclock the device under attack; as shown in [172], the number of induced faults increases gradually when the computations are performed at a frequency higher than the nominal one. In particular, it is possible to exploit a clock frequency interval where the appearance of the fault is very limited in order to obtain localized alterations in the computation which can be exploited successfully.

Another possibility for an attacker willing to inject controlled faults into a computing device is to alter the environmental conditions around the device under attack, for instance, causing the temperature to rise. A global rise of the temperature around common desktop DRAM modules has been reported to cause multiple multi-bit errors in the memory content [169]. The authors report a thermal fault induction attack against the DRAM memory chips of a common desktop PC. The reported number of flipped bits is around ten per 32-bit word, when the working temperature of the DRAM is brought up to 100 °C. The number of faulty words is also reported to be in the range of tenths of the RAM cells. In order to raise the temperature to a fault inducing level, a common 50 W light bulb was employed to illuminate the chips and the level of heating was tuned by modifying the distance between the light bulb and the chips. The temperature level was monitored with a simple commercial thermometer, by placing the temperature probe on the DRAM. The workbench setup requires minimal technical knowledge and the equipment is readily available to anyone. The limitations of this fault induction technique are due to its very coarse grain, which tends to easily corrupt large amounts of data. Another downside of this technique is the fact that it is actually possible to destroy parts of the circuit through excessive heating, since some components have a very narrow tolerance range with respect to the working temperature.

A practical way to induce faults without needing to physically tap into the working device is to cause strong electromagnetic (EM) disturbances near it. The eddy currents induced by strong EM pulses into the lines of the circuit cause temporary alterations at the level of the signals, which in turn may be recorded by a latch, thus causing an erroneous value to be employed in a computation. Since the EM pulse affects uniformly the whole attacked device, it is necessary to shield the components that should not be subject to faults with either a metal plate or a mesh, connected to a proper grounding point. This technique has been proved effective against eight-bit micro-controllers by [360], employing as a source of EM disturbances a spark generator and placing the two endpoints between which the spark is produced very close to the chip package. The authors also demonstrated that better results in terms of more efficient fault injection may be obtained by removing the plastic package of the chip before the spark generator is placed. The spark generator employed in the reported article consists of a simple piezoelectric gas lighter, whose wires have been extended to obtain an easily placeable spark gap, which is held directly above or below the device under attack. All the parts of the circuit that did not need to

Table 16.2 Summary of low-cost fault injection techniques: all methods require only basic or moderate technical skills

Technique	Accuracy [space]	Accuracy [time]	Permanent modifications	HTA[a]	KPIA[b]
Underfeeding	High	None	No	No	No
Voltage spikes	Low	Moderate	No	No	Partial
Clock glitches	Low	High	No	Yes	Yes
Heat	Low	None	Possibly	Partial	Yes
EM pulses	Low	Moderate	Possibly	No	No
Light irradiation	Low	Low	Yes	Yes	No

[a]Hindered by technological advances [b]KPIA: requires knowledge of physical implementation aspects

be disturbed (e.g., the board on which the chip under attack was mounted) were properly shielded by aluminum plates connected to a common ground. The spark generator can be triggered by a circuit, related to the clock signal of the device under attack, in order to achieve moderate timing accuracy for the fault injection. The only technological limitation of this technique is represented by the chips that employ grounded metal packaging (usually for heat sinking purposes), which act as an EM shield. The decapsulation of the chip is mandatory to successfully lead the attack, and this adds a step that requires uncommon technical skill. Nonetheless, the decapsulation procedure may be performed with low-cost equipment (nitric acid and common glassware from a chemistry lab), without noticeably raising the cost of the attack.

Assuming the attacker is able to successfully decapsulate a chip, it is possible to perform fault injection attacks by illuminating the die with a high energy light source such as an UV lamp or a camera flash. The strong irradiation on the silicon surface can cause the erasure of EEPROM and Flash memory cells, where constants important for an algorithm's execution are kept. Depending on the duration of the irradiation process, the authors of [361] report a progressive blanking of all the memory cells, together with the resetting of the internal protection fuses of the microprocessor targeted during the attack. Schmidt et al. [361] provide a survey of the speeds at which the memories of different types of microprocessors get erased, and conclude that it is possible also to selectively wipe out only a part of the stored data by exposing only a part of the die to UV irradiation. The employed workbench to blank part of the microprocessor memory cells is very simple and involves only a UV lamp, which is shone on the exposed silicon die of the microprocessor from very close. In order to shield the parts of the circuit that do not need to be exposed, it is possible to cover them with UV-resistant, dye which can be bought in hardware stores. This technique is limited by the possible placing of the memory cells on layers buried under the top one on a multi-layer chip. Since the manufacturing process of a multi-layer die needs to have a flat surface before the next layer is etched, the memory cells laid on bottom layers are covered with metal tiles, which act as a shield from irradiation.

Table 16.2 provides a summary of the fault injection techniques that rely on a low budget, making a comparison among them. From this summary it emerges that low cost fault injection techniques are unable to obtain high precision of the injected faults, and that it is less common to achieve spatial than temporal accuracy.

16.3 Low-Voltage Faults on General-Purpose Processors

This section provides a complete characterization of the faults happening when a general-purpose microprocessor is constantly underfed, and subsequently characterizes the errors induced on the computed outputs in terms of position, shape and timing, together with the methodology followed to obtain such faults. A complete description of the working environment is provided to properly outline the workflow we follow in order to coalesce the new fault and error model.

16.3.1 CPU Architecture and Experimental Settings

The processor architecture taken into account in this study is the ARMv5TE, in particular the version implemented by the ARM9E microprocessor. Our choice was driven by the wide deployment of this CPU, which is nowadays the dominant choice for smartphones, network appliances, portable gaming platforms and low power computers, and is thus quite likely to be used also to compute cryptographic primitives when the possession of a possible attacker.

Our target chip is an ARM926EJ-S [17]; a 32-bit RISC Harvard architecture CPU with 16 general purpose registers and a five stage pipeline. The ARM processor has a full MMU and separate data and instruction caches, each 16 KB wide, coupled with a 16-entry write buffer which avoids stalls in the CPU when memory writebacks are performed. In particular the ARM926EJ-S is also endowed with a hardware Java bytecode interpreter able to directly run Java bytecode. The richness of the available features justifies the vast popularity achieved by this model in the consumer mobile devices. The CPU is embedded in a system on-chip mounted on a development board, specifically a SPEAr Head200 [389] built by STMicroelectronics, which is used as reference board to design ARM-based devices equipped with 64 MB DDR RAM clocked at 133 MHz, 32 MB of on-board Flash storage, two USB Host ports, an RS-232 serial interface and a 100 Mbps Ethernet network card. The system is endowed with a U-Boot [118] embedded bootloader, which is able to load the binary code to be run via TFTP [382] protocol. This allows the board either to run a specific binary, compiled to be independently executed on the ARM9 CPU, or to boot a full-fledged operating system. In the following experiments, raw binaries were employed in order to precisely characterize the fault model of this system. On the other hand, for the sake of practical applicability, all the attacks on cryptosystems were led with a full vanilla Linux 2.6 kernel (DENX distribution) employing an NFS [78] partition as root filesystem. All the binaries were compiled with the GCC 3.4-based development toolchain for ARM9 provided by CodeSourcery [96]. All the fault characterization

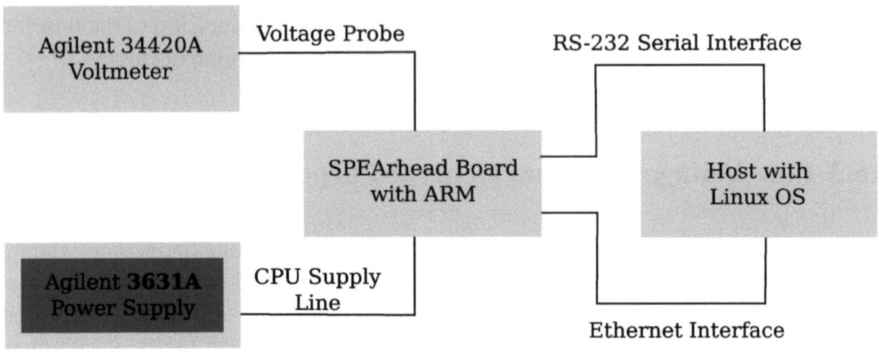

Fig. 16.1 Workbench used in order to accurately characterize the voltage range

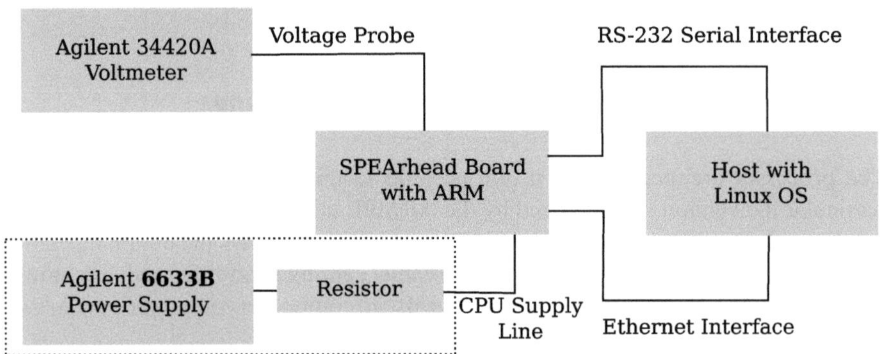

Fig. 16.2 Workbench used to perform the error characterization and the attacks

tests were performed on two or more sample boards, reporting analogous results; for the sake of clarity we present the results on a single board.

Two experimental workbenches were employed during this process: the first one, aimed at producing a precise characterization of the effect of power supply lapses, was endowed with a high-precision power supply; the second one, aimed at carrying the attacks with a lower budget, employed less expensive equipment, without loss in the efficacy of the attacks. The two workbenches are identical except for the change in the Power Supply Unit (PSU).

The first workbench is depicted in Fig. 16.1: the board was at first fed through an Agilent E3631A PSU [4] with a precision of 1 mV, while on the second workbench, depicted in Fig. 16.2, it was fed through an Agilent 6633B [2] power supply with 0.01 V precision. In order to achieve rough sub-centivolt precision, we used a resistive partitor with a common commercial grade resistor; whilst this solution does not provide the same accuracy voltage control as using a high-precision PSU, it proved effective enough to successfully implement all the attacks. All the voltage measures were taken with an Agilent 34420A [3] voltmeter with a 1 nV precision probe, which was already available to us; nonetheless the needed precision was only up to 0.5 mV.

The board was connected to a PC with both a null modem cable and an Ethernet cable: the first provided an interface with the Linux shell running on the ARM chip, while the second was used to provide the network connection needed for both booting the board via TFTP and providing the storage via NFS.

16.3.2 Graceful Degradation of Outputs

The first experiment run on the target chip was aimed at investigating whether the appearance of errors in the system followed the gradual behavior we expected. The ARM9 processor has three separate supply lines, one for the core, one for the I/O buses and one for the memory interface; we chose to interact with the one feeding the computational part, due to its critical importance for the correct execution of the binaries run on the device.

To detect the frequency of the appearance of faults, in order to determine how fast they appear during the execution of a program, we tested the correct functioning of the CPU using a simple probe program, whose core loop is as follows:

```
for (a = i = 0; i < 1000000; i++) {
  a = a + 1;
  if (a != i + 1) {
     printc('?');
     if (a != i + 1) {
        printc('#');
       a = i + 1;       /* fix the fault */
       if (a != i + 1)
         a = i + 1;
     }
  }
}
```

The aforementioned code increments a variable a million times, and checks if a fault has happened exactly after the increment. A redundant check is added to lower the likelihood of a false positive occurring in the detection: we consider an actual fault as such only if both checks confirm it. This program was run multiple times while decreasing the voltage of the power supply of 1 mV at a time: 500×10^3 runs were performed at each voltage level probed and the results output by the code were stored.

Figure 16.3 represents the percentage of correct computations over 5×10^7 runs for each voltage level probed. The errors in the output grow linearly with the lapse in the voltage supply, thus confirming our hypothesis of gradual degradation in the quality of the results. The dashed line in the figure points out the voltage point where the number of faulty computations is the same as that of the correct ones.

After ascertaining that the faulty computations of a program were happening slowly, we moved on to consider the number of faults appearing during each single

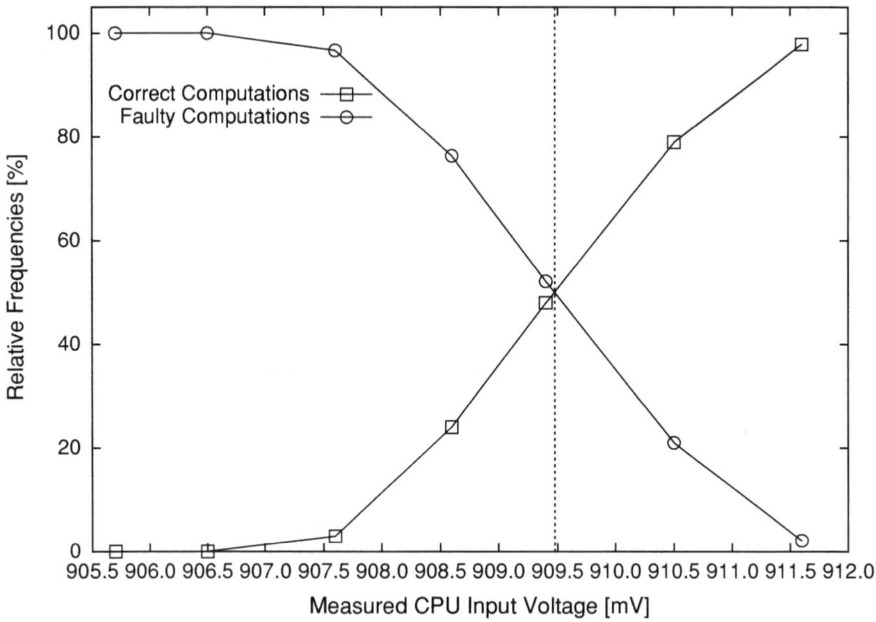

Fig. 16.3 Percentage of correct and wrong computations averaged over 500×10^3 runs

computation. Since our goal was to inject single faults, we were interested in seeing whether the errors in the outputs of the former computations were caused by one or more faults during their executions. Reclassifying the faulted runs from the former experiments according to the number of faults, it is possible to observe, as shown in Fig. 16.4, that there is a 1 mV wide voltage range where only a single fault happens with dominant probability. Moreover, in the adjacent 1 mV range, the probability of having a faulty computation triggered by a single fault ranges from 2 to 40 %. This probability dwarfs the one of having multiple faults contributing to the erroneous result, while in the voltage range where less than half of the computations are faulted. When working under the 50 % faulty computation threshold, the number of possible faults starts growing, and multiple fault scenarios start dominating the fault profile, as can be expected in a degrading environment. After lowering the power supply voltage even more, the board stops outputting data from the RS-232 interface used to communicate, thus preventing the results from being collected. It is therefore clear that, in the voltage region where the correct computations are the majority and faulty ones begin to appear, the faults occurring in the computations are single and not represented by small bursts.

Since the results until now have been obtained while keeping the number of machine instructions per binary fixed for each measurement, and since there is no specific timing for the insertion of the hazard causing faults, it is reasonable to assume that a growth in the executable code size will be complemented by an analogous rise in the probability of a fault appearing during the computation. This is not an obstacle

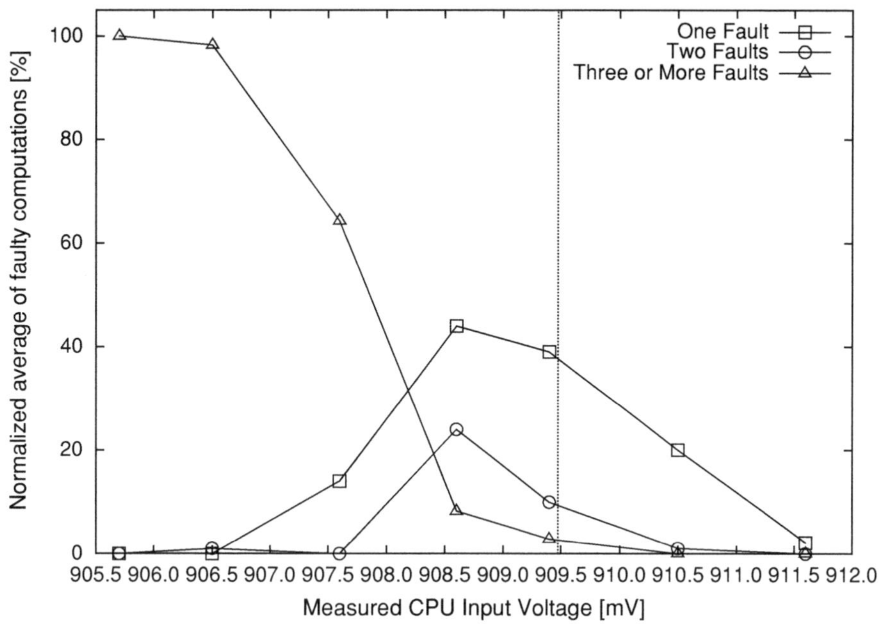

Fig. 16.4 Distribution of injected faults as a function of the voltage

to injecting single faults, as the fault incidence per single execution may be tuned for binaries of different lengths by lowering or raising the voltage accordingly.

16.3.3 Fault Type Characterization

After ascertaining the possibility of injecting a single fault per computation, we moved on to investigate what kind of fault has actually hit the computation by characterizing its effect on the executed code.

By analyzing the binary at the assembly level, all the possible instructions executed by a CPU may be split into three categories based on the architectural units composing the CPU that are used to complete them. The three categories are arithmetic-logical operations, memory reads and writes, and branch instructions. Since memory instructions represent the most expensive operation class in terms of power consumption, they are expected to be the most vulnerable to underfeeding issues.

In order to validate this hypothesis, we recompiled the same probe program, instructing the compiler to keep the variables in the CPU registers during the whole computation, in order to avoid any memory operations while computing the result. Only the instruction cache of the CPU was enabled, leaving only the data loading operations uncached. The execution of the tuned program showed no faults, thus indicating that the wrong values detected by the checks were uniquely to be ascribed to memory operations, while both arithmetic-logical and branch instructions ran correctly despite the voltage drop. All the experiments were conducted by collecting erroneous outputs after the binary had been running for a couple of seconds; this allowed the caching of the whole probe program due to its tiny size.

The low-voltage fault immunity shown by the CPU registers is to be ascribed to the low capacitance design of their implementation, which yields faster switching times than does average logic, partially compensating for the slowdown induced by the lapse in the supply. This feature is mandated by the architectural need to provide fast access to the component, which is critical for designing efficient units and is thus to be expected in all the common CPUs.

Since only the memory operations are affected by faults, the next natural step was to check if all the instructions (i.e. both `loads` and `stores`) were equally affected. In order to determine what kinds of memory instructions are affected by faults, it is possible to use a register-held value as a fault-free reference for computing checks. We set up a probe program that loaded and saved from the memory 0- and 1-filled words, and ran it multiple times while sweeping the whole voltage range of the previous campaign.

The only instructions to report faulty results were the `load` instructions, while all the write instructions were safely performed. This behavior may be sensibly ascribed to the fact that only the memory operations which store values on the underfed part (i.e. the `load` operations which store information in the registers) suffer from the lapse in the power supply. On the other hand the `store` operations place the data on a properly fed part of the architecture. Moreover, the ARM926EJ-S (similarly to all modern CPUs) is endowed with a 16-entry write buffer between the CPU and the memory in order to perform aggressive instruction reordering. The presence of the buffer helps to cut down the capacitive load of the path to the memory and thus helps perform correct writes.

16.3.4 Fault Spread

The experiments run up to now characterize the faults as affecting only `load` instructions and, as far as their number in a single execution goes, depending on the supplied voltage. We are now willing to investigate whether the faulty behavior of the `load` instructions depends on the referenced memory address from which the load is performed. In order to understand this key point, a probe program was designed to overwrite a one million 32-bit word array with 1s, and subsequently to check the values which were loaded back into the registers, while keeping the voltage in the single-fault functioning range. To avoid any possible disturbances, during this test the data cache of the ARM9 processor was disabled, thus forcing the CPU to load each value from the main memory.

Figure 16.5 shows the number of faults occurred while performing 10^6 `load` operations of a 1-filled 32-bit integer from the aforementioned array. In order to analyze the data, the probed memory was clustered into 40 KB-wide zones. We

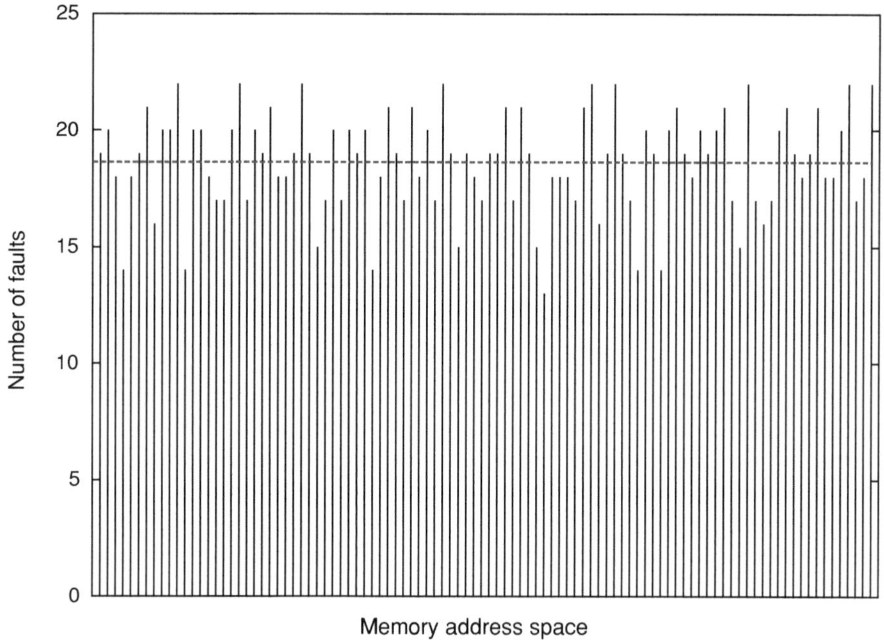

Fig. 16.5 Distribution of the quantity of the injected faults as a function of the position in the address space. The dashed line indicates the expected average value in the hypothesis they are uniformly spread

encountered 1,864 faulty `loads` while running the program; thus we are expecting an average of 18.64 faults per zone if the faults fall uniformly over the address space. The dashed line in Fig. 16.5 indicates the expected number of faults occurring for each zone, assuming a uniform distribution of the faults over the memory.

To confirm the uniform distribution of the faults over the whole address space, we modeled the position hit by the fault as a random variable and we conducted a Pearson χ^2 test to assess the goodness of fit. The results confirmed our hypothesis with confidence over 99.99 %.

16.3.5 Error Characterization and Effects of Frequency Scaling

After fully characterizing the frequency and the occurrence conditions of the faults, our natural target of the investigation becomes characterizing the kinds of errors induced in the computations by the faults. By analyzing the data collected during the last experiment, we were able to notice that all the faulty loads were affected by flip-downs in the bit value loaded. Wanting to ascertain if only flip-downs were

Table 16.3 Normalized frequencies of different error patterns per CPU clock setting

CPU Clock [MHz]	Loaded Pattern	[%]
140	{3}	58.76
	{21}	10.01
	{3, 21}	31.22
224	{21}	100.0
266 (Full)	{10, 16}	38.72
	{10}	53.23
	{9, 10, 11, 12, 15, 16, 21}	2.95
	{ 9, 10, 16}	3.34

possible, we ran the same memory exploration program changing the loaded value to both a zero-filled 32-bit word and some random values. In all the cases only bit flip-downs occurred, and there was never a single instance of a flip-up. When analyzing the position of the bits that are flipped down during the faulty loads, we detected that only a very small number of flip-down patterns were present (namely four) and one of them accounted for more than 50 % of the fault occurrences. When repeating the tests on different boards, the recurring patterns changed, but their number and frequency did not, allowing us to deduce that some bits within a word are more susceptible to flip-downs when the CPU performs a load operation while under-powered. This may be ascribed to the capacitive load of the signal lines of the CPU, and is due to routing issues which may force the I/O lines of a register to have different lengths.

To complete the analysis of the error patterns, we decided to run the same error pattern detection experiment while varying the CPU frequency according to the allowed working range. By piloting the clock generator on the board we were able to scale the frequency of the CPU, mimicking a real-world scenario where the ARM processor is often run at frequencies lower than the maximum allowed, in order to save power. It is possible to choose from among a number of frequency settings which alter globally the working frequency by writing in the Phase Locked Loop (PLL) generator register interface. This causes both the board and the CPU to switch their working frequency: the board is run at the frequency written in the register while the CPU is run at twice the set value. The clock setting is retained until the board is rebooted, but it is possible to customize the deployment model in order to either lock it permanently or leave the frequency scaling to the operating system. We wanted to investigate whether the faulty behavior had any changes while working in different frequency environments; therefore we locked the running frequency in order to collect homogeneous samples of the behaviors. In a real-world scenario this may happen to be the actual working environment, since it is quite common to lock the CPU frequency at a lower value than the maximum allowed in order to save power. The possibility that the frequency choice is left to the operating system does not impair our analysis since the CPU will be running at constant frequency in discrete time slices, thus reporting the same faulty behavior per time slice. Table 16.3 reports the result of the experiments performed and shows how, regardless of the frequency at which we are running, the error patterns are few and characterized by one, which is dominant as far as the occurrence frequency goes.

16.3.6 Effects of the Errors on the Computation

After fully characterizing the errors induced by our fault injection technique, the last part of this study sums up the possible effects on the computation caused by such errors. Albeit they originate from the same cause, i.e. faulty `load` operations, we may distinguish between two different effects of the faults depending on whether the load is related to an instruction fetch or to a data load. In the latter case a data load error occurs, while in the former case an instruction substitution may occur. For the sake of clarity, we will deal separately with the two outcomes in order to distinguish between their possible effects on the computation of cryptographic primitives.

16.3.6.1 Data Load Errors

Data-related errors are representable as transient changes in the value of a t-bit wide variable c during an execution. In particular they are single-bit flip-downs placed in a fixed position within the microprocessor word. The faulty value $\widetilde{c}$ equals the correct one c minus a power of 2, 2^{ε}, where ε is the position of the fault. Possible values of ε are expressed in the form $\varepsilon = k\,w + i$ with w equal to the word length, $i \in \{0, \ldots, w-1\}$ and $k \in \{0, \ldots, \lfloor \frac{t}{w} \rfloor\}$. The value of ε is dependent on the single chip instance under examination, and fixed for each sample. These changes in the loaded value are very precise in the way they cause the alterations and therefore may easily leak sensitive information.

16.3.6.2 Instruction Substitution Errors

Bit flipping during an instruction fetch may alter either the opcode or the arguments of the instructions, depending on which bit is affected by the flip-down. In particular, the affected instructions will be transformed into ones having binary encoding, differing only by a flip-down of the faulty bit. In the case of the ARM architecture, this may result in either a substitution of one kind of instruction with another or in a reversal of the triggering condition of a conditional instruction.

An example of a possible instruction substitution through a single-bit flip-down as follows:

```
AND R1,R1,#0x42 @ Fault Free
EOR R1,R1,#0x42 @ Faulty
```

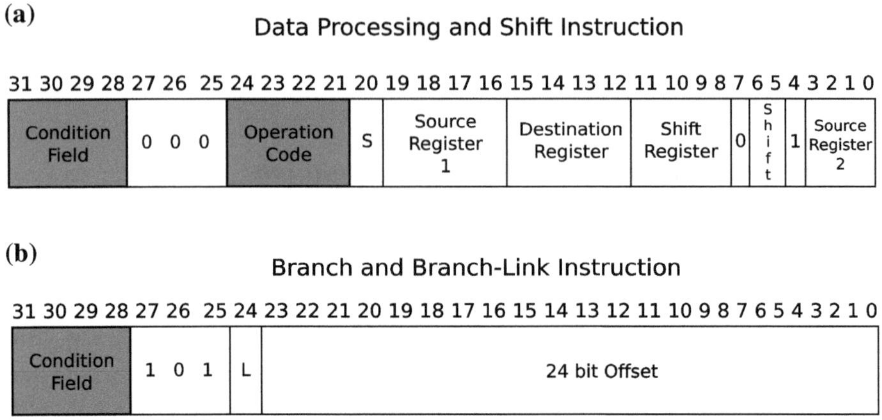

Fig. 16.6 Excerpt from the ARM ISA description [17] depicting branch and data processing instructions. Grayed areas point out the interesting fields during fault injection

Since the AND and the XOR instructions have radically different behavior, it is possible to alter the inner working of the algorithm by exchanging them, thus leading to the possible computation of a weaker version.

Given that the ARM architecture allows the conditional execution of all the arithmetic-logical instructions, and stores the kind of condition in a suffix of the opcode, as Fig. 16.6 depicts, it is possible for the error to actually invert the condition of the predicate instruction.

For instance, in the following code sample, the two instructions share the same opcode except for the zero-condition bit setting:

```
ADDNE R1,R1,#0x42 @ Fault Free
ADDEQ R1,R1,#0x42 @ Faulty
```

This behavior could lead to mis-executions of the algorithm leaking significant content, especially if the conditional instructions are directly related to the key value (e.g. in the common square-and-multiply algorithm used to perform fast exponentiation). Moreover, since also the branch instructions rely on the same condition bits of the common condition, the control flow of the program may be equally altered if the condition bit of a branch is flipped, as in the following sample:

```
BNE LOOP @ Fault Free
BEQ LOOP @ Faulty
```

This kind of alteration may lead to substantial control flow alterations, which can lead to lowering the number of times a loop is executed or to skipping it altogether, thus resulting in substantial reductions in the complexity of a cryptographic primitive computed on the device.

We have been able to reproduce all the aforementioned alterations on our chip samples by running the probing programs without enabling the instruction cache, and thus allowing the code loading operations to be performed directly from memory. Since the alterations are chip-dependent, the exploitation of this kind of fault requires us to precisely know which bit is affected by the fault, and thus to determine which instruction substitutions are performed. Nonetheless, since our methodology of probing does not compromise the computing architecture, it is possible to scan a sample chip in order to understand which of these code mutations are performed and to devise specific attacks.

16.4 Conclusion

This chapter reviews invasive and semi-invasive fault injection techniques, putting forth a classification of the state-of-the-art methodologies in terms of high-cost and low-cost injection equipment and methods. The fault model derived from the constant underfeeding of a general-purpose microprocessor is fully characterized in order to infer how faulty hardware behavior causes software errors. The induced faults have been characterized in both position and corruption patterns, by splitting the effects into two classes: data corruption and instruction substitution. The most appealing features of the model derive from the ease of induction and the absence of future hurdles caused by the continuing evolution of chip building techniques.

Chapter 17
Global Faults on Cryptographic Circuits

Sylvain Guilley and Jean-Luc Danger

Abstract Methods of injecting faults in a laboratory are numerous and varied. We divide the state of the art in methods of injecting faults in electronic circuits into two categories. The first is global attacks, which disturb all the equipotentials of a netlist simultaneously. The second is local attacks, which target a more specific zone of the components' surface, rear or front. Global attacks are a less accurate method of injecting faults but require a much lower budget. This chapter further discusses the specifics of global versus local faults. Then, it provides models for global faults and demonstrates that most theoretical fault attack constructions can be obtained in practice by means of global fault injection. To illustrate this, we provide an extensive characterization of fault models by emulation (FPGA) on application-specific devices (ASICs). Finally, this chapter ends with an exhaustive survey of the experimental means of injecting global faults and their effect as a critical path setup time violation phenomenon.

17.1 Introduction

The behavior of electronic systems can be transiently or permanently modified as soon as they are made to operate out of their specified functional environment. An attacker can take advantage of this to mount active attacks, with a view to bypass or defeat cryptographic mechanisms. Depending on the physical accessibility and the tamper-resistance of the target device, an attacker can opt for either a global or a local attack. A global attack can be induced by many vectors, such as glitches, on global input or bidirectional lines, over-clocking, under-powering or heating. This contrasts with local faults, caused by the exposure of a portion of a decapsulated component to

---

S. Guilley (✉) · J.-L. Danger
Institut TELECOM/TELECOM ParisTech,
Paris, France
e-mail: guilley@enst.fr

M. Joye and M. Tunstall (eds.), *Fault Analysis in Cryptography*,
Information Security and Cryptography, DOI: 10.1007/978-3-642-29656-7_17,

Table 17.1 Comparison of global and local attacks features

Feature	Global	Local
Invasive	no	yes
Cost in equipment	low	high
Required expertise	low	high
Easiness of detectability	yes	no
Controllability in space	no	yes
Controllability in time	yes	yes

something that will induce a fault, such as a laser beam, a particle source, or strong eddy currents.

Table 17.1 compares global and local attacks according to different features. The goal of this table is to define the contexts in which one or the other type of attack best applies.

The advantage of global attacks is that they can be conducted on devices with little preparation. Typically, in the case of a device such as a smart card, all the ports (data, control and power inputs) are under the control of an attacker. This means that the component is ready to be mounted on a test platform without any further customization. Proposals to have smart cards be autonomous in terms of energy (thanks to micro-batteries [373] or photovoltaic coupling [128]) do exist; however, they fail to protect against active attacks. Module-level encapsulation (e.g. Sishell and ACSIP solutions proposed by Axalto) does deny access to the chip surface, but still leaves the input/output ports unprotected. In the case of autonomous components, such as those present in a printed circuit board (PCB), global attacks will require that they be unsoldered so that they can be placed in a testing environment. But in both cases, an attack can be carried out irrespective of any chemical preparation. However, such a delicate operation is mandatory for most local attacks since they require the device to be physically close to the source of perturbation. Using the terminology coined by Skorobogatov, global attacks are referred to as "noninvasive" whereas local attacks are qualified as "semi-invasive" [377]. This latter term signifies that local attacks require a chip to be depackaged but do not require electrical contact with its metal surface. Active probing and circuit editing with a focused ion beam (FIB) tool are two examples of "invasive" attacks.

As discussed in [397, Sect. 1], blasting a device with photons using white light flashes can be considered a global attack, although it requires access to the surface of a microprocessor. Indeed, a flash bulb does not allow an attacker to focus the optical energy precisely, as opposed to a laser beam. However, the article [361] explains that the surface of a chip can be made opaque by covering it with dark ink. Then, by simply scratching the thin ink film off any zones of interest, an attacker can make sure that light enters into a device by only this local opening. Therefore, in the following, we will consider light injections as out of the scope of global attacks.

As global attacks are less sophisticated than local ones, they also involve low-cost equipment. Most of them are indeed commercially available as off-the-shelf tools for functional debugging or conformance testing. The level of expertise required to use such equipment is also low, since the tools are mainstream and delivered with a detailed user manual. On the other hand, local attacks typically require a detailed description of the employed test benches, which must be fully understood in order to exploit their capabilities.

The counterpoint to this is that global attacks are less powerful than local attacks. For instance, they are easier to detect. Indeed, one global sensor for the whole chip can be enough to raise an alarm. For example, a razor approach [115] can be used to monitor in real time (at every clock cycle) the speed of the device: should the clock period be too short, the razor immediately identifies it. Alternatively, the clock can be stabilized internally by dedicated logic [405]; such monitoring makes it possible to detect an abnormal slowdown caused by any kind of global perturbation. Also, sensors can be designed to test a device's internal temperature [258].

Another drawback of global attacks is that the location of fault injection in space is unknown a priori. Critical zones can be targeted with local attacks, albeit with knowledge of the layout or a reverse engineering of the device (e.g. by layout reconstruction via tedious delayering [398] or by OBIC [378]). However, we note that the most powerful fault attacks do not require an attacker to know the exact spatial location of a fault; for instance, the so-called "Bellcore" attack on RSA [56] is able to break an implementation of RSA operating in CRT mode regardless of the location of the fault. The same remark applies to the attacks on the penultimate or antepenultimate rounds of the AES data path [324, 403] or key schedule [229, 394]. Now, if the target is not a hardware but a software implementation, any "place" of the algorithm can be perturbed with a global attack simply by knowing the exact "time" (in terms of clock cycles) it is executed by the processor. Therefore, attacks on software benefit from a very small advantage when compared to local fault injection techniques.

Regarding the accuracy of an attack, it can be tuned to meet the expected behavior of a device. For instance, in both global and local cases, the stress can be permanent or transient. Transient faults are less likely to be detected because of their brevity, but, if coupled with the known timing of an algorithm, they can be applied at critical instants. This would minimize the number of faults that produce results that cannot be exploited.

We invite the reader's attention to the fact that, in some cases, local attacks cannot be performed. One representative situation is where a microprocessor is well shielded and protected by miscellaneous coatings. Another example is multi-chips assembled in packed system-in-package (SiP), which become nonfunctional if separated. The FPGAs are another example, where a focused stress tends to alter the configuration (representing the large ratio in terms of surface usage) rather than the user logic [79], which can be of interest but not in the context of differential fault analysis. In contrast, global attacks selectively perturb the running logic without affecting the steady configuration.

From a scientific point of view, an attractive feature of global faults is that they can be readily modeled in theory and produced with accurate control in both emulation and real hardware. This makes it possible to precisely define their behavior, with the view to better fight them. Also, some local faults can be viewed as global faults on a more confined area. This is notably expected to be so for low-cost electromagnetic

Table 17.2 Table for the rating factors of global and local fault attacks (see classification terminology in [75, p. 21])

	Global faults		Local faults	
Factors	Identification	Exploitation	Identification	Exploitation
Elapsed time	< one hour	< one week	< one month	< one hour
Expertise	Layman	Layman	Expert	Layman
Knowledge of the TOE[†]	Public	Public	Critical	Public
Access to TOE[†]	< 10 samples	< 10 samples	< 100 samples	< 10 samples
Equipment	None	Standard	Specialized	Specialized
Open samples	Public	Public	Public	Public

[†] *TOE is short for "Target Of Evaluation".*

Caption:	Easy	Nominal	Hard

fault injection [332, 404], which induces a local voltage drop, thereby mimicking an under-powering attack.

With regard to the common criteria (CC) international standard for computer security certification [100], a reference for the definition of such attacks has been agreed upon in a supplemental document [75]. It defines the difference between two phases in an attack scenario: the identification and the exploitation. Basically, the difference between global and local attacks is that

- global attacks need no identification phase, but are not always successful in the exploitation phase, whereas
- local attacks require an extensive and methodical identification phase, but enable an extremely fast and focused exploitation phase.

An example of the common criteria definition is given in Table 17.2; it makes it possible to contrast global versus local attack strengths, based on a coarse rating of three categories: easy, nominal, hard.

17.2 Faults Model

The majority of the global faults on clocked circuits are caused by a timing constraint violation. This section starts with a brief review of propagation delays in CMOS logic styles. Then, timing constraint related to usage of synchronous logic are explained, and the effect of constraints violations is analyzed. Finally, these theoretical predictions are transposed to nonCMOS logic styles, typically involved in protected implementations.

17.2.1 Propagation Delays in CMOS Logic Gates

In CMOS logic [419], every gate has a propagation delay, which is inherent in the existence of capacitances (most often unwanted, i.e. parasitic) within and around it. Thus, every stimulus at the input of a logic gate requires some time to produce its effect at the output. The propagation delay depends on many factors.

The most important factor is the function of a given gate. For instance, an AND gate with one input equal to 0 will evaluate to 0 irrespective of the arrival time of the second input. This "early evaluation" would not have happened with the same data configuration on an OR gate [243, 390]. Indeed, an OR gate cannot decide on the final result before being supplied with the value of its latest input if the fastest ones are equal to 0.

The propagation time also depends on the shape of the input signals. An input signal transiting slowly will take longer time to propagate through the gate, since a signal coming from a long line is likely to have a smooth edge, whereas a well-buffered signal that has a short interconnect length will probably have a steep transition slope. Thus the routing also influences the propagation time. The cross-coupling of the wires also has a subtle effect has a signal's propagation speed and transition speed: depending on the neighboring wires, any signal can be either slowed down, sped up or have its waveform modified.

Additionally, some gates can answer in nondeterministic time. One representative example is the XOR gate that is placed in the situation of an arbiter; this occurs when the two inputs change almost simultaneously in antinomic directions, e.g. from $(0, 1)$ to $(1, 0)$. Then, depending on the relative delay between the two inputs, the gate can start to glitch. The answer time is thus dependent on thermal fluctuations.

The CMOS gates can glitch, which means that some transitions can be generated even if they are not the final values. Once generated, the glitches propagate and lead to other glitches. The final value of the computation is taken only after no other changes occur. Indeed, the digital computations can be modeled as the transmission of events on an oriented graph: the signal propositions being causal, every computation in a combinational netlist necessarily converges to a steady state.

To summarize, the propagation time in CMOS logics depends on the data and on the routing. Furthermore, the longest path for the same input data might not be the same, because of the nondeterministic behavior of some gates.

17.2.2 Timing Constraints in Synchronous Logic

In sequential logic, one global signal, called the clock, coordinates all the combinational computation in parallel, at the same pace. One implicit condition is that all the gates are expected to have finished propagating their data when the clock's rising edge arrives. Of particular importance is the longest path, which determines the maximal clock frequency. As discussed in the previous subsection, this critical

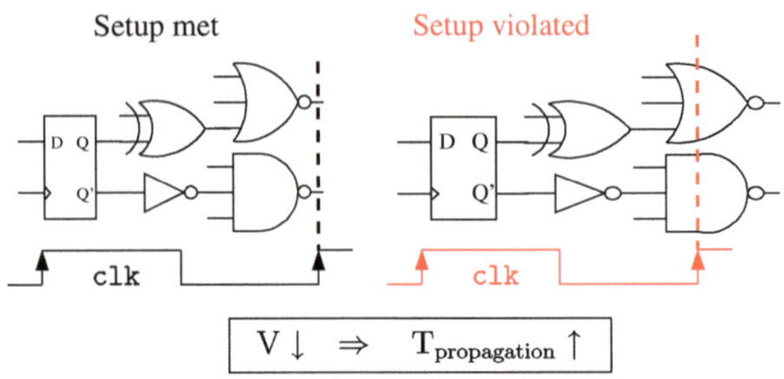

Fig. 17.1 Illustration of setup violations when the clock period is reduced below the setup time

path is also data-dependent. It is called the setup time and corresponds to the clock's smallest admissible period.

If, for whatever reason, the clock period is less than the setup time, faults can occur. This is illustrated in Fig. 17.1: when either the clock accelerates or the data signals decelerate, the setup time can be violated.

The slowdown of the logic gates when their supply voltage is reduced is illustrated on a SPICE [301] simulation in Fig. 17.2. This figure shows the voltage variation of one bit from a combinational circuit that starts evaluating at $t = 1$ ns. The initial value for this bit is "logical 1" and its final result is "logical 0" (0 V, i.e. the ground potential). The transition between those two values is simulated for various supply voltages, from the nominal 1.2 V down to 0.4 V in steps of −0.1 V. This voltage is also present in the electrical value representing "logical 1". We note that the time needed for this bit to take its final value gets larger as the supply voltage is reduced. Even for a supply voltage of 0.4 V, close to the threshold of the transistors, the gate output does not converge within a reasonable amount of time; the validity of the simulation models is questionable under such extraordinary conditions. Nonetheless, the effect of the supply voltage on the speed of evaluation is undebatable. Also, this phenomenon is all the more important as the voltage is reduced.

More precisely, it can happen that the critical path is not activated, resulting in all the signals being steady at the clock's rising edge. Consequently, no deviation from normal behavior can be observed; but if the critical path is indeed violated then two options are likely to be possible. Either the value of the longest path is sampled at the rising edge of a clock as the correct value. In this case, the behavior is faulty, but the result is correct by chance. Or the sampled value is the logical inverse of the correct one, and the computation starts to be erroneous.

Figure 17.3 shows how the propagation delay is affected by the temperature of the circuit. The same test bench used to generate the results in Fig. 17.2 is used, at nominal voltage (1.2 V), but the temperature takes the values −150, −50, +50 and +150 °C. It is clear that the temperature has a smaller impact on the delay; thus, underfeeding is a very effective means of fault injection.

Figures 17.2 and 17.3 show simulation results on a S-box, which is the most critical resource in symmetric encryption algorithms, in terms of area and timing. This S-box is made up of few gates (21 standard cells), which speeds up the computation, but it is representative of any combinational block.

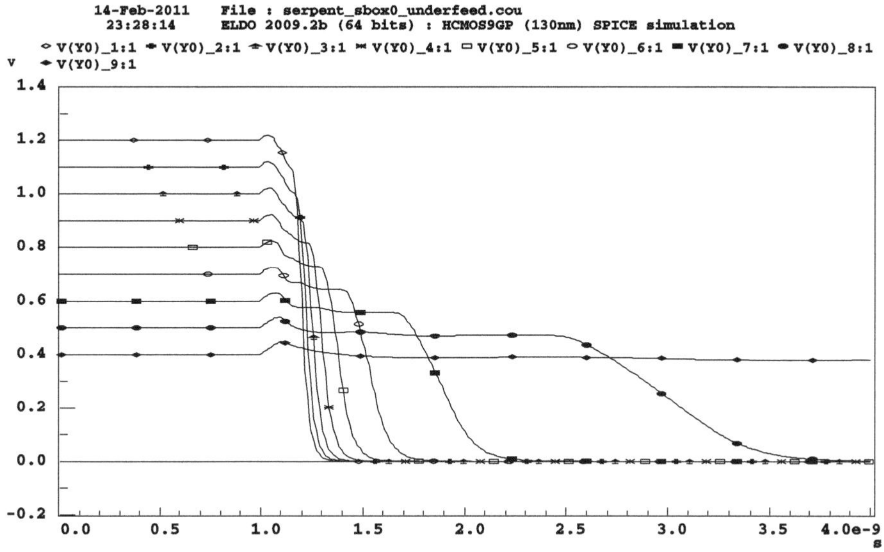

Fig. 17.2 SPICE simulation of the evaluation date of the least significant bit output of the S-box #0 of SERPENT for different voltages (1.2 V is nominal) in low-leakage 0.13 μm CMOS technology from STMicroelectronics

17.2.3 Physical Explanation of the Bit-Flip Fault Model

In hardware, such a violation is compatible with the "bit-flip" model. Indeed, CMOS gates are balanced in timing (rising and falling edges have similar transition durations). Thus evaluating a signal as 1 instead of 0 is as likely as the converse. In software, this accounts for failures such as instruction skips or the setting of the output signals of an XOR to 0, notably observed by Clavier [92, 93]. We can therefore state that software derouting is a mere consequence of a failure of the hardware that executes it.

We assert that in mainstream processors (unprotected CMOS) the timings are data-dependent. Ideally, from an algorithmic optimization standpoint, all the paths in a data path can be critical: it is the logic synthesizer's objective to create a design that complies with the timing's closure constraints. This suggests that faults can occur at various locations in the data path. This is beneficial to the attacker, because she can take advantage of the faults occurring on various bits of the data path to retrieve all the bits of the key.

Various means to inject global faults exist, most of which cause setup time violations; we refer to them as "stress methods". When the stress is low, the perturbation results in one single path being violated. This causes a single bit-flip at the output of a combinatorial path; in turn, this causes a bit-flip in the next state register. As the critical path is definitely data-dependent, the bit-flip can actually occur everywhere, especially when the logic is well balanced. This is true of a block cipher, but not of

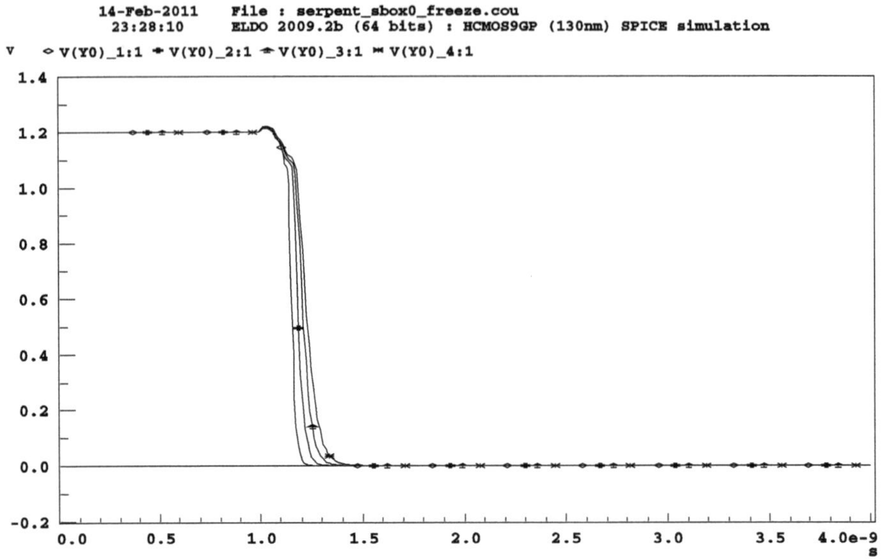

Fig. 17.3 SPICE simulation of the identical setup to Fig. 17.2 at nominal supply voltage, but where the temperature varies in −150, −50, +50, +150 °C

an integer multiplier, for which the most significant bit is much more critical than the least significant bit. We note that the notion of critical path that we use is the dynamic one rather than the static one. CAD tools compute the maximum frequency a circuit can reach by evaluating a static metric; indeed, it is computationally impossible to estimate a dynamic timing because of combinational explosion [380]. Therefore, gates are modeled as nodes with delays and their interconnections are labeled with estimated propagation times. A synthesizer looks for the critical path with an A* like algorithm, which happens to be the longest path in this netlist. However, in practice, this path may not be activated, and other paths can become critical instead (albeit with shortened timings). This accounts for the random repartitioning of global faults in every register. When the stress gets stronger, multiple paths are violated simultaneously, which can cause double, triple or more faults located in one word (e.g. one byte) or in several words.

Finally, we conclude with the suitability of global attacks to system-on-chips (SoCs), even if they embed a wealth of coprocessors, such as control blocks and device drivers, along with any cryptographic coprocessors. In such a complex, compound system, it is not taken for granted that a global fault provokes a failure selectively in one block. However, in practice, cryptographic modules are by far the most computation-intensive, and therefore are very likely to host the critical path.

17.2.4 Global Fault Models on NonCMOS Logics

All logic styles that are built upon transistors feature a propagation time, irrespective of the representation of the data. For those logic styles that use a clock, the effect from a perturbation of the environment naturally leads to setup violation faults. We detail two examples:

- dual rail with precharge logics (DPL [114]) and
- quasi-delay-insensitive logics (QDI [272]).

In DPL with a (0, 0) spacer, the whole circuit is initialized to zero every other clock period. Thus, the only fault compatible with the setup violation are $1 \to 0$, which is an asymmetric model [368].

Asynchronous logic styles [291] are a special case since they are not clocked. Therefore, if the circuit slows down because of a global fault, the functioning is not altered. This makes QDI styles very dependable, reliable, and, in addition, natively immune to global faults.

17.3 Experiments in Emulation in FPGA and in Real Hardware in ASIC

In this section we detail an experimental confirmation of the properties described in the previous section:

- bit-flips or byte-flips with low Hamming weight do exist at the end of the clock rounds (and thus at the end of rounds of iterative block ciphers);
- monobit faults appear first with low stress intensity;
- faults are scattered across the complete data path.

In this respect, an ASIC implementation of AES is studied in terms of faults. In contrast to Sect. 16.3, we describe here faults within a device, and not incurred by delays between two integrated circuits (a processor and its memory). Then, other experimental results discussed in the literature are analyzed.

17.3.1 Statistics on Faults in a Hardware Implementation of AES

Setup-time violation faults are induced by under-powering an AES [142] where AES is a block cipher with three variants: 128-, 192- and 256-bit keys. In the 128-bit version there are ten rounds that we denote by R0, R1, ..., R8 and R9. The rounds R7 and R8 are especially vulnerable to differential fault analysis, especially those that are derived from Piret and Quisquater's attack [324].

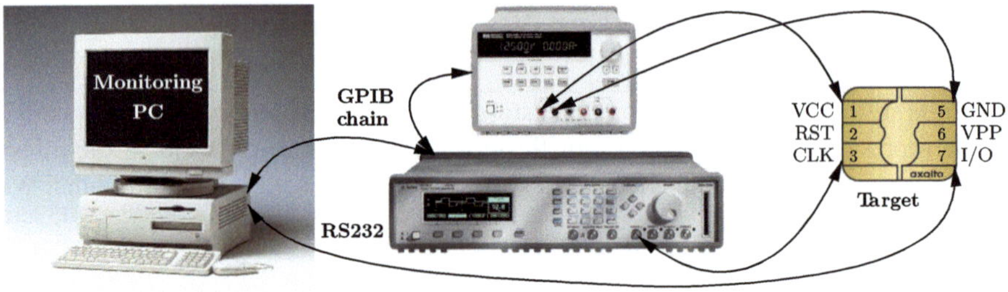

Fig. 17.4 Experimental fault platform for attacks on AES in an ASIC

17.3.1.1 Experimental Setup

In this section, we present the first experimental setup used for an under-powering fault injection attack demonstration [369]. The setup consists of standard communication with a smart card, except that the power generator supplies the card with a nonnominal continuous voltage VCC and that an arbitrary waveform generator is used to supply the clock signal. Figure 17.4 sketches the experimental setup. The power supply and the waveform generator are controllable remotely, in such a way that various values of VCC and various clock shapes can be tested successively.

In our experiments, the smart card was a 130 nm ASIC with an embedded AES coprocessor (amongst others). The nominal voltage of the chip was 1.2 V. We observe that the circuit remains functional for a VCC as low as 700 mV: at this low voltage, the CPU, in charge of controlling the I/Os and of delegating encryption to the AES module, crashes. However, the AES module starts to output erroneous results for a voltage VCC of about 800 mV and less. In the following, we use the fact that the smart card remains functional within the range [775–825] mV to explore the faults for several chosen intermediate voltages. The power supply could deliver a voltage with an accuracy of half a millivolt. Therefore, we conducted the following acquisitions:

- The triples {message, key, ciphertext} were recorded for 20,000 encryptions at each 100 values of VCC in steps of 0.5 mV.
- In a view to simulate an attack, the key was kept at a constant value.
- Conversely, to collect representative results, the input message was varied randomly at each encryption.

As a result, the entire acquisition campaign consisted of two million encryptions.

17.3.1.2 Motivation for This Modus Operandi

In most fault attacks on cryptographic devices the fault model is stringent. An attacker is expected to be able to inject "single" faults into "precise" words or rounds. These constraints can be relaxed in some attack scenarios. For instance, in Piret and Quisquater's differential fault analysis [324] adversary model, an attacker needs to

```
aes_128 aes_128::encrypt(
  reg_128 const& m,
  reg_128 const& k,
  fault_t const& f = ZERO)
{
  aes_128 aes (m, k); // The initial state
  aes.AddRoundKey ();
  aes.KeySchedule (0);
  for (int round = 1; round < 10; ++round)
  {
    // Fault injection
    aes.set_state (aes.get_state () ^ f[round - 1]);
    aes.SubBytes ();
    aes.ShiftRows ();
    aes.MixColumns ();
    aes.AddRoundKey ();
    aes.KeySchedule (round);
  }
  // Fault injection
  aes.set_state (aes.get_state () ^ f[9]);
  aes.SubBytes ();
  aes.ShiftRows ();
  aes.AddRoundKey ();
  return aes; // The faulted encryption (value 'c')
}
```

Fig. 17.5 Adapted encrypt function that computes AES with two arguments and a faulted AES otherwise

know neither where nor when the errors occurred. However, to use this fault model based on a "byte-flip", a surgical fault injection is often carried out [379]. At the opposite end we investigated a low-cost "global" fault injection technique based on under-powering. This method is not surgical but rather "global" in the sense that the whole algorithm is targeted simultaneously and continuously during the encryption process. The voltage reduction allows the attacker to generate the single faults most differential fault attacks are based on. The reason for this is that the stress caused by the insufficient power supply remains gentle; dysfunctions do not appear suddenly and thus complex multiple faults do not show up.

17.3.1.3 Analysis of Faults

In this section, we present an analysis of the two million results obtained from the acquisition campaign described in Sect. 17.3.1.1. In order to analyze the faults, we assume that the message and the key are known to an attacker and that they are not faulted. In other words, the faults concern only the computation of the encryption and thus can affect only the ciphertext. We use a fast register transfer level (RTL) model of the AES ("encrypt" function), adapted to corrupt any byte of the state matrix at any round. The function is sketched in the C++ code snippet in Fig. 17.5.

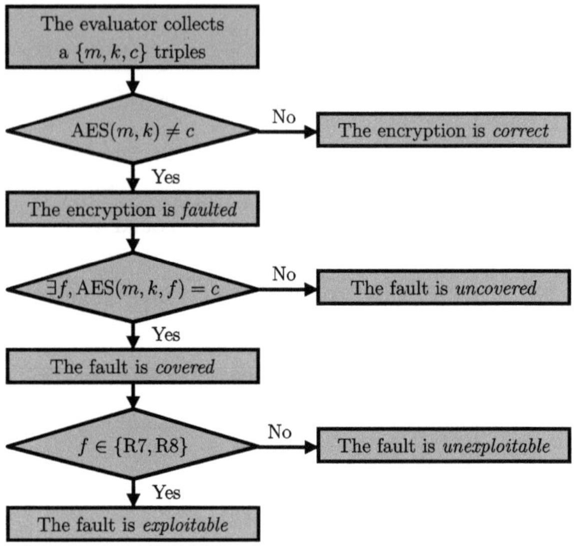

Fig. 17.6 AES-128 faults analysis; in our research, the value of f is zero, but for exactly one byte

In this model we can see that the f value XORed with the message permits any fault injection. The array `fault_t` is defined as:

```
// Mask ready to be applied
// to every round' state
typedef reg_128 fault_t[10];
```

where `reg_128` is typedef'ed for `char[4][4]`. The default value of the fault f, `aes_128::ZERO`, corresponds to 10×16 zeroes. To simulate a single-fault injection, the `fault_t f` is initialized to all-zero, with the exception of a single byte (`char`) within the `reg_128` state for a single round index. The fault analysis consists in calling this function for all the possible fault values, and comparing the output with the ciphertext obtained from the experimental platform. Indeed, the only evidence that a fault has occurred experimentally is a faulted ciphertext. The exhaustive search of single "byte-flip" requires $(2^8 - 1) \times 16 \times 10 = 40{,}800$ calls to the "encrypt" routine. This allowed us to

- find the experimental faulty ciphertext, in this case the fault is called "covered", or
- declare that the fault is "uncovered", if the faulty ciphertext matches none of the "encrypt" function results.

The purpose of this implementation is to identify the different types of faults that occur in a smart card, as shown in Fig. 17.6. In this figure, the AES function with two arguments is the regular implementation of AES-128, whereas the AES function with three arguments is the "encrypt" routine detailed previously.

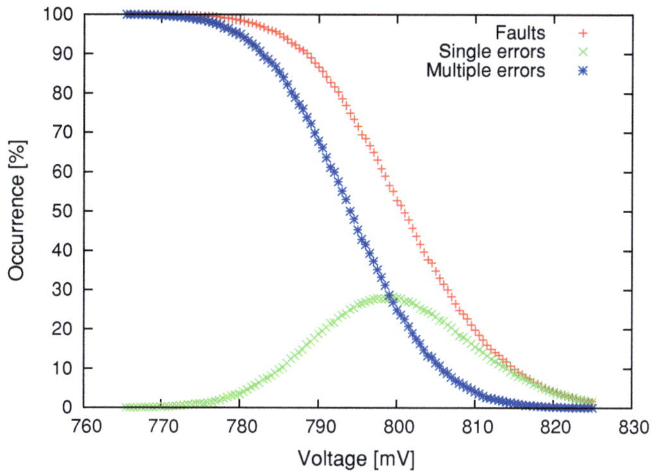

Fig. 17.7 Occurrence of faults

17.3.1.4 Experimental Results: Coverage Estimation Software Tool

In this section, a description of a fault analysis is used to find the occurrence of a single byte fault that affects the state matrix of AES. Attention is focused on the data path, while the key schedule is assumed here to be fault-free. This choice is motivated by our goal to reproduce Piret and Quisquater's differential fault analysis [397] on a smart card. This attack is detailed more thoroughly in Sect. 4.2.2. It is straightforward to adapt the results obtained in this section to other attacks, such as attacks on the key schedule [160]. The purpose of this study is to demonstrate the effect of under-powering the device on the faults generated throughout the encryption process.

Figure 17.7 shows the occurrence of faults for a given under-powering of the device (this kind of characterization is identical to the one already presented Fig. 16.4). It appears that within about 60 mV, the device moves from an error-free state to a fully erroneous behavior. As already explained in Fig. 17.6, faults are partitioned into "single" and "multiple", depending on whether they are covered by the "encrypt" function for a fault f. Single faults have a distribution in a "bell shape", where the maximum is reached at voltage $\approx$800 mV, where 30 % of detected faults are single bits. This behavior is compatible with a fault model where errors are caused by a setup violation on a critical combinatorial path. The lower the power supply, the more likely a critical path is violated, and thus the most frequent faults are single faults. Nevertheless, below the $\approx$800 mV threshold, multiple critical paths are violated. Hence an augmentation of multiple faults and a subsequent diminution of single faults was observed.

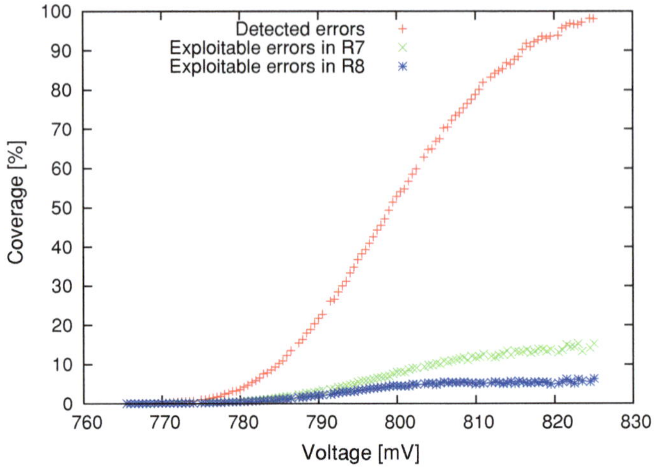

Fig. 17.8 Coverage of single faults, and detail of exploitable faults

17.3.1.5 Spatio-Temporal Characterization of Faults

The "covered" faults are single byte errors detected as per the search algorithm depicted in Fig. 17.6, and they are represented in Fig. 17.8. We continue the analysis in this section in terms of their spatio-temporal locality:

- in which rounds (from R0 to R9) are they more likely? (refer to Fig. 17.9) and
- in which byte of the state are they more likely? (refer to Fig. 17.10).

Figures 17.9 and 17.10 show the empirical PMF (Probability Mass Functions) of the faults.

As already mentioned, we can see in Fig. 17.9 that the first round (R0) is never affected by faults. This observation was indeed predictable, since the first round is special: it consists of the AddRoundKey transformation alone. Therefore, the critical path is not in this round.

It can seem counterintuitive that faults occur at one round and not at the others. In a static timing analysis (STA) of a design, the critical path is the same for every iteration. Therefore, one might expect that if a critical path is violated at one round, then all the rounds will be faulty. However, we observe single errors localized at a given round. The reason could be that the critical path is highly data-dependant.

From the analysis of Figs. 17.9 and 17.10, we show that the faults are not uniformly distributed over time and space. This observation, albeit not general since our setup is very particular, can be valuable for the designers in charge of implementing countermeasures.

We can see that the first round is never affected by faults. Also, it appears clearly that the faults are not uniformly distributed over time and space.

Some hints can be given about possible reasons for the heterogeneous distribution of the faults. First of all, several instances of the S-Boxes operate in parallel in the

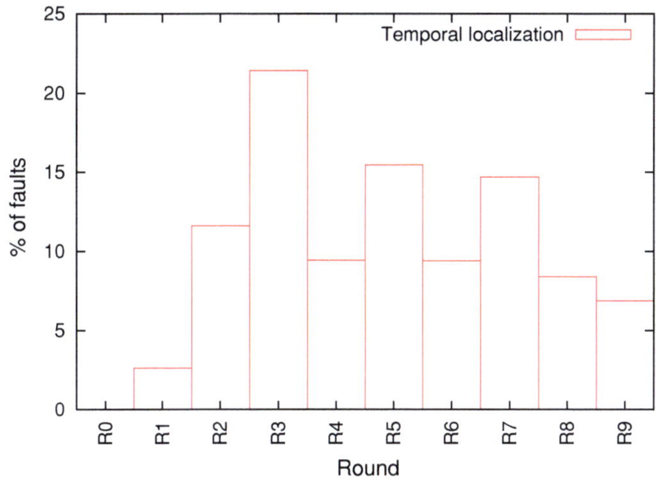

Fig. 17.9 Temporal localization of single faults

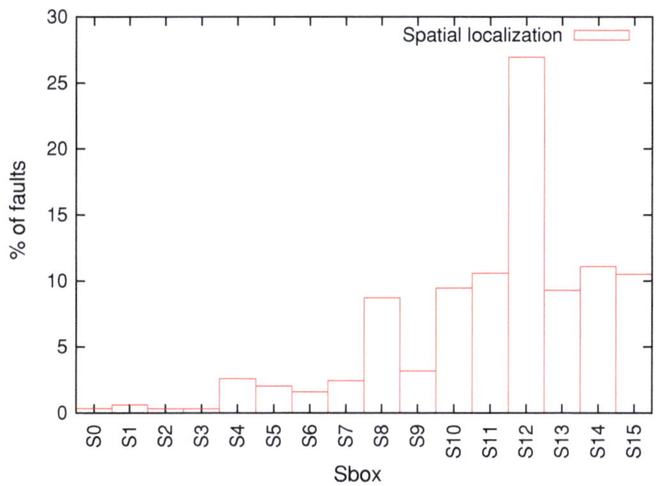

Fig. 17.10 Spatial localization of single faults. The SubBytes S-box $s_{i,j}$ has index $4 \times i + j$ in the PMF histogram

studied architecture. However, there is no reason for those S-Boxes to be synthesized or to be placed and routed in exactly the same manner. This is one cause of spatial heterogeneity (Fig. 17.10). Second, the temporal heterogeneity (Fig. 17.9) could be ascribed to the time between two consecutive faults. For instance, if this average time is close to half the encryption duration, then it is likely that a fault occurring in the first rounds is followed by another fault in the subsequent rounds. Similarly, a single fault in the last rounds is rare since it is highly probable that another fault actually happened

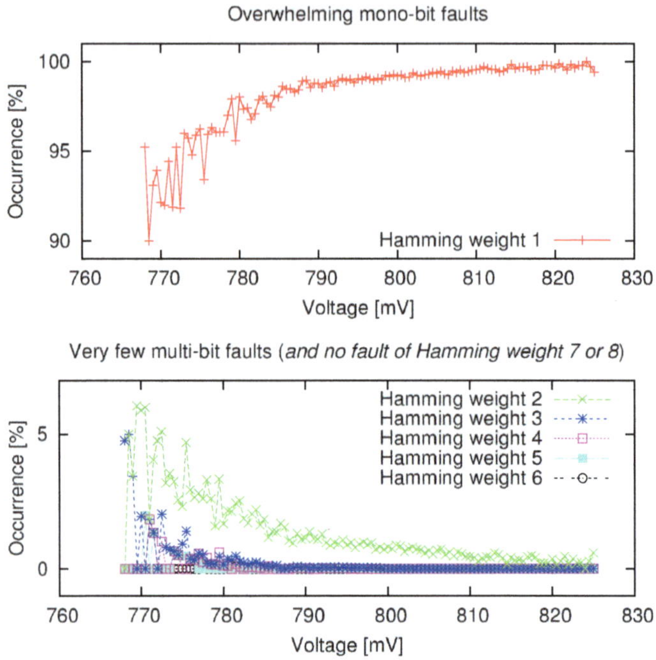

Fig. 17.11 Number of bits flipped amongst the single faults detected during the AES encryption

in the first rounds. Thus the maximum number of faults is expected to take place exactly at the middle of the encryption (around round R5), accounting for the shape of the occurrence curve. The average and variance of the time between two faults depends on the stress level (in our case on the voltage) and on the implementation type, as observed in [43].

The number of bits flipped in a byte is shown in Fig. 17.11. It clearly appears that for a low level of stress, most faults consist of a single bit-flip. This is consistent with the fact that one single critical path gets violated. Conversely, we note that when the stress gets higher single faults mainly consist of multiple erroneous bits.

17.3.2 Fault Statistics in Other Targets

As described in [43, 223], the same manipulations done in Sect. 17.3.1 on an ASIC can be ported on an FPGA. For instance, the faults follow an occurrence law represented in Fig. 17.12. It is qualitatively similar to that of an ASIC (refer to Fig. 17.7); thus FPGAs are suitable platforms for studying global faults.

Now, if the critical component in a SoC is the CPU, then faults can modify either the execution flow or the manipulated data. A complete study is carried out in [22, 24] on a 32-bit CPU of the ARM family running an AES. The fault model is found to be original: faults occur only when the CPU is interacting with its internal RAM.

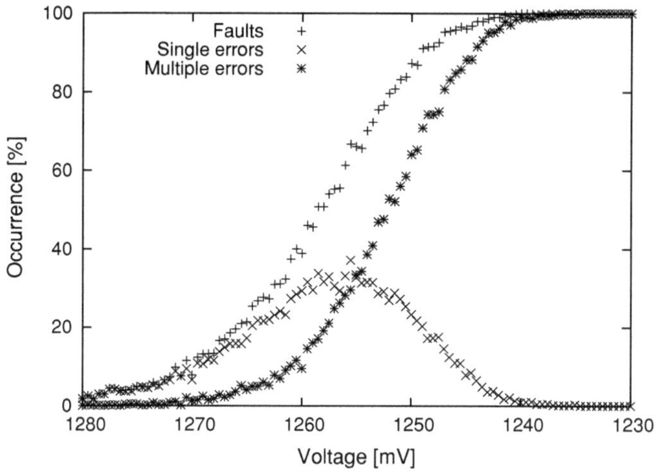

Fig. 17.12 Occurrence of faults in an FPGA where the S-Boxes use the $GF(2^4)$ representation

All the write accesses are correct, but read accesses can be faulty. This asymmetry is caused by the heterogeneity of the CPU architecture.

17.4 Conclusion

This section mentions and details global injection techniques already studied.

- **Glitches on global lines** are described using the clock in [15] and in the power line with the modified CLIO board produced by Gemplus Card International [21]. Glitches on the clock obviously contribute to violating the critical path, whereas glitches on the power line increase the propagation time through the gate, thus causing faults when the accumulated propagation time gets larger than the clock period.
- **Overclocking** is simulated in [136, 114] and realized experimentally with an FPGA [5, 148] and with an ASIC [353]. Overclocking is the continuous counterpart of glitches in the clock.
- **Under-powering** is described in [23]. It also causes a slowdown of the combinational logic, thus permitting setup violations.
- **Heating** can be a means to inject global faults, as discussed in [73, 169, 408]. As exemplified by the CPU burn bench [336], errors caused by excess of heat can be caused by an internally generated abnormally high activity.

In conclusion, all of the methods described in this section come down to the violation of a critical path. Thus, the attacker has a wide array of methods to induce faults, and all these threats should be considered, since a skilled attacker will naturally use the easiest method of attack available.

Chapter 18
Fault Injection and Key Retrieval Experiments on an Evaluation Board

Junko Takahashi, Toshinori Fukunaga, Shigeto Gomisawa, Yang Li, Kazuo Sakiyama and Kazuo Ohta

Abstract This chapter presents fault injection experiments using a side-channel evaluation board called SASEBO, which was developed to unify testing environments for side-channel analysis. We describe experiments where faults were injected into a cryptographic LSI mounted on a SASEBO board using a clock glitch. In this experiment, the faults can be induced at any desired point in time during the computation of an algorithm. We show the results of injecting faults into block cipher and public key modules implemented on the LSI. We also show the key retrieval from standard ciphers using the faulty outputs obtained in these experiments. This work contributes to the study of how a fault is injected into a target device, such as an LSI mounted on an evaluation board, and verifies various theoretical fault analyses using an experimental environment.

18.1 Introduction

Fault analyses against various ciphers have been proposed [49, 55, 84, 127, 160, 296, 322, 324, 394, 395] and most of these studies are theoretical; therefore, the attack assumptions are based on the capability of a likely attacker and the characteristics of cryptographic devices. A fault model, in particular, in which fault properties such as fault injection area and variation, is important to the discussion of the practicality of these attacks. In order to evaluate the possibility of a fault attack in these theoretical studies, some practical fault attacks against actual hardware were proposed [148, 223, 226, 353, 369]. As an example, Selmane et al. reported the results of practical experiments of fault analysis on smart cards with an AES co-processor

J. Takahashi (✉) · T. Fukunaga
NTT Information Sharing Platform Laboratories, Tokyo, Japan
e-mail: takahashi.junko@lab.ntt.co.jp

J. Takahashi · S. Gomisawa · Y. Li · K. Sakiyama · K. Ohta
The University of Electro-Communications, Chofu, Japan

M. Joye and M. Tunstall (eds.), *Fault Analysis in Cryptography*,
Information Security and Cryptography, DOI: 10.1007/978-3-642-29656-7_18,

[223, 369] where faults were induced by under-powering the circuit, and showed that approximately 16 % of the corrected ciphertexts were suitable for application to Piret and Quisquater's attack [324] (see Sect. 4.2.2. for a thorough description), which needs two faulty ciphertexts caused by faults that affect one byte. They succeeded in recovering an AES key by applying Piret and Quisquater's attack with the faulty ciphertexts obtained in their experiments. Another method of fault injection and key retrieval that has been introduced is global faults, detailed in Sect. 17.3.

There are many experimental results of fault analysis. However, it is difficult to standardize the evaluation of fault analysis and describe an experimental environment that anyone can construct. In order to unify the testing environment and obtain uniform experimental results, the Side-channel Attack Standard Evaluation Board (SASEBO) was developed by the Research Center for Information Security (RCIS) of Advanced Industrial Science and Technology (AIST) in Japan in 2007 [342]. The evaluation boards are distributed to research institutes with detailed documentation as a common experimental platform. There are currently five types of SASEBOs, four of which (SASEBO, SASEBO-G, SASEBO-B, SASEBO-GII) are FPGA versions and one (SASEBO-R) is an evaluation board on which LSI can be mounted for the evaluation of ASIC LSIs. Three kinds of LSIs which are compliant with SASEBO-R were developed [340]. Recently, the evaluations of various side-channel analyses including the fault analysis using SASEBO have been reported [148, 254, 353].

In this chapter, we present the experimental results of fault injection into LSIs mounted on a SASEBO-R board. We describe an experimental environment that supplies a clock signal with a clock glitch to the target device. In these experiments, faults could be injected into any desired point in time during the computation of a cipher. We also demonstrate the key retrieval of the de facto standard ciphers, AES and RSA, as an example, using the faulty outputs obtained from the experiments. We succeed in retrieving the AES or RSA key by applying the relevant theoretical fault analysis [56, 324, 429].

The rest of this chapter is organized as follows. We describe the fault injection mechanism in Sect. 18.2. We describe a way to construct an experimental environment using SASEBO-R in Sect. 18.3 and present the experimental results in Sect. 18.4. We show the key retrieval using the actual faulty outputs in Sect. 18.5. Finally, we conclude this chapter in Sect. 18.6.

18.2 Fault Injection Mechanism

Various fault injection techniques have been proposed, such as supplying under-power voltage [223, 369], supplying an irregular clock frequency [148, 353] to the target devices and irradiating the target device with a laser beam [361, 378, 379], to induce a wrong calculation in order to apply fault analysis. In our case, we use an irregular clock frequency, referred to as a clock glitch, in order to induce faults because the sets of equipment are inexpensive and we can easily experiment with fault injection. Using a clock glitch, faults can be induced precisely during the computation

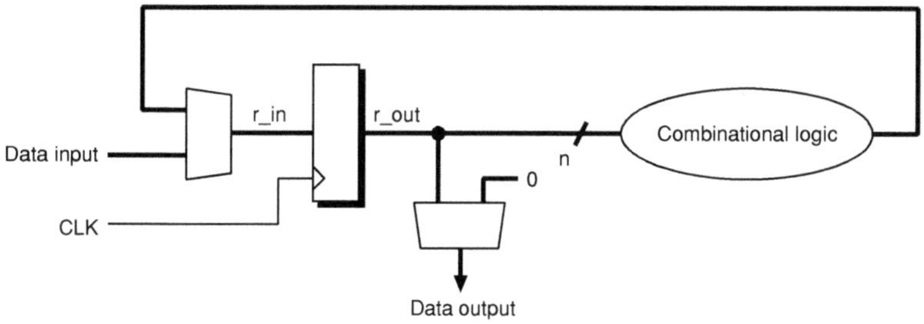

Fig. 18.1 Basic loop architecture

of an algorithm, for example in one byte during a particular round of an instantiation of block cipher, which is commonly done in theoretical fault analysis.

A schematic with the basic loop architecture is shown in Fig. 18.1. The CLK signal denotes the clock signal. The r_in signal represents the input of the data register and equals the output of the combination logic used to compute the cipher. The r_out signal represents the output of the data register and is the value stored in the register. The fault injection mechanism for inserting a clock glitch is described in Fig. 18.2 for a normal clock with a glitch. During normal operation, as shown in Fig. 18.2a, all the intervals of the positive edges for each clock are longer than the maximum delay of the circuit and are correctly stored in an intermediate state in the data register. If a glitch occurs in the clock signal during the computation and the shortened interval is shorter than the maximum delay of the circuit, as shown in Fig. 18.2b, the value stored in the data register r_in is likely to become corrupted due to a timing violation. Therefore, we can inject a fault into the desired intermediate state during the computation of an algorithm in the targeted device if we can generate a glitch at the appropriate time.

18.3 Experimental Environments for Fault Injection

In this section, we explain the methods for generating a clock glitch and targeting devices and algorithms with fault injection.

18.3.1 Developed Experimental Environment

Precise control of the timing of a clock glitch is important for investigating the relationship between the shortened interval and the injected fault. To achieve this, we developed an experimental environment based on the following concepts.

- Use two clock sources for which we control their phases in order to be able to generate a clock glitch with precise timing.

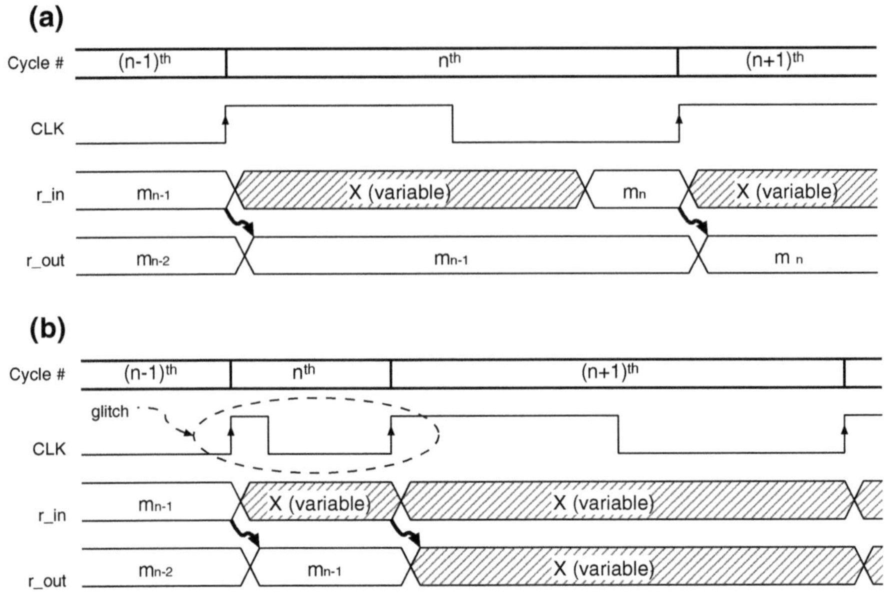

Fig. 18.2 Example waveforms of input and output from data resistor: **a** Normal operation **b** Glitch occurs in clock signal during n-th cycle calculation

- Control the switching timing by detecting the LSI execution using an oscilloscope to change the time that a fault is injected, such as a particular round for block ciphers.
- Measure the actual shortened interval using an oscilloscope in order to collect precise results.

We give the list of the equipment used in the experiments in Table 18.1 and an overview of the developed experimental environment based on the above concepts in Fig. 18.3. A PC controls the cryptographic LSI and collects the results of faulty calculations. An oscilloscope records the shortened interval of the clock for analysis. The cryptographic LSI is mounted on a SASEBO-R board and controlled from a PC via an FPGA on the evaluation board.

We developed an LSI controller in the FPGA written in Verilog HDL. This supplies the LSI with a clock signal with a glitch, which is a combination of two external clocks generated by a two-channel pulse generator. These two external clocks have the same frequency but have different phases. The glitch is generated by changing the selection of these clock sources at the appropriate point in time. To obtain the appropriate timing, the glitch controller uses the Trigger_out signal of the oscilloscope, which outputs a signal when it detects the EXEC signal of the LSI, which in turn outputs the EXEC signal during the execution of the crypt calculation. Instead of using the EXEC signal in the experiment, we can use a power consumption trace of the LSI by setting a trigger using the characteristic form of a trace which appears in the execution

Table 18.1 Equipments used in experiments

Equipment	Product name and model number
Oscilloscope	LeCroy WaveMaster 8600A
Pulse generator	Agilent 81150A
DC power supply	Agilent E3646A
Target device	Cryptographic LSI (I) or (II) mounted on SASEBO-R shown in Table 18.2

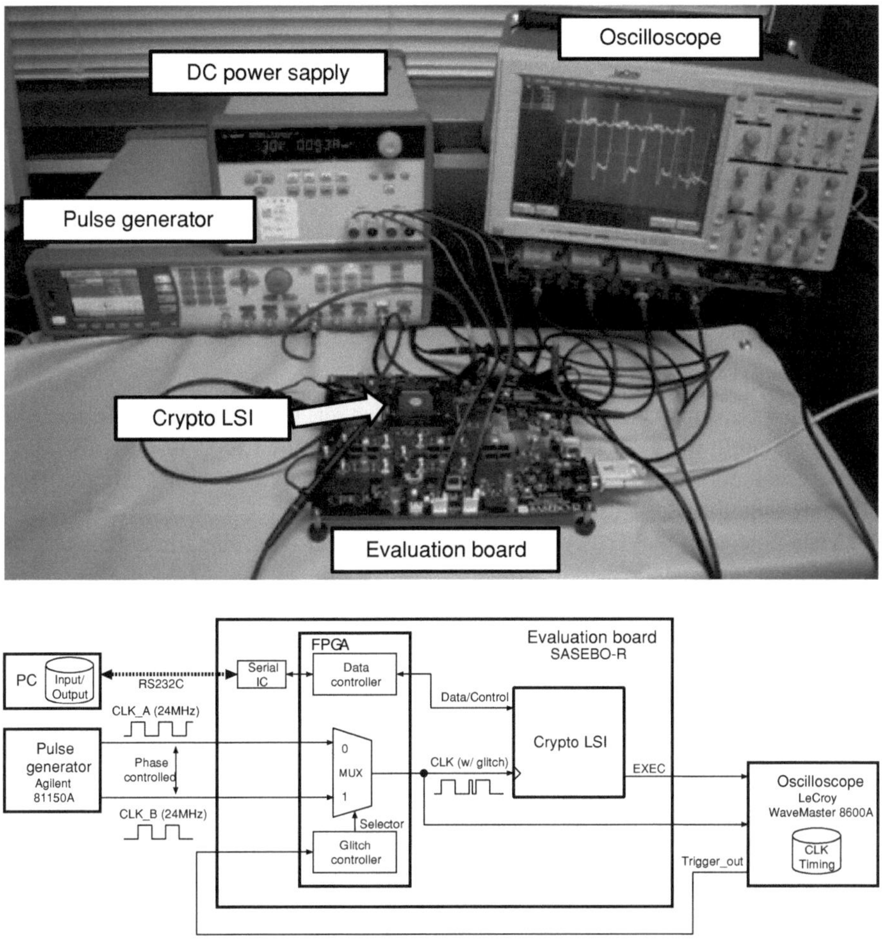

Fig. 18.3 Overview of experimental environment

of the crypt calculation. This experience is useful if we evaluate another LSI that does not output any signal related to the execution of a cryptographic algorithm.

We show a schematic for generating a clock with the glitch in Fig. 18.4. This graph shows the method for generating a clock with the glitch from two clock sources

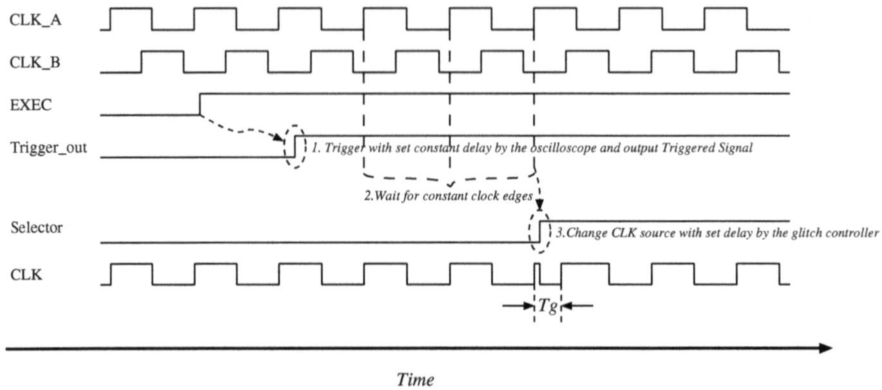

Fig. 18.4 Schematic for generating a clock glitch

CLK_A and CLK_B. In these experiments, we set both CLK_A and CLK_B to 24 MHz (a period of approximately 41.7 ns). At the start of the calculation, the following steps are performed.

1. The oscilloscope detects the positive edge of the EXEC signal, and outputs the Trigger_out signal with a constant delay that we set. The timing of the Trigger_out signal output does not need to be synchronized with CLK_A or CLK_B. Then, the glitch controller module detects it. We can set any length of delay using the delay function of the oscilloscope.
2. After detecting the Trigger_out signal, the glitch controller module waits for constant clock edges, which we set.
3. Then, the glitch controller module changes the CLK source from CLK_A to CLK_B with a short constant delay, which we set.

In this example, the glitch controller module waits for three constant positive clock edges, and then causes the next CLK positive edge to rise at a different time. One clock with the glitch is generated; however, continuous clocks with the glitch can be generated by setting a high-frequency for CLK_B and switching the two clocks at the appropriate time.

The experimental jitter of the Trigger_out signal from the oscilloscope is 19.2 ns, which accounts for 80 % of the 24 MHz LSI clock frequency. However, the construction of the glitch controller, i.e., switching two clock sources with the clock edge timing, enables us to limit the jitter to 35 ps. This is defined by the precision of the phase difference of the two clock sources of the pulse generator. Thus, we can inject a fault into intermediate values using a timing violation as mentioned in Sect. 18.2.

Table 18.2 Summary of cryptographic LSIs

Version no.	Process rule (nm)	LSI version	Year	Supported cryptography
(I)	130	GAIA	2008	ISO/IEC 18,033 block ciphers, RSA
(II)	130	MASH	2009	ISO/IEC 18,033 block ciphers, RSA (six kinds of implementation), ECC
(III)	90	CHAR	2009	ISO/IEC 18,033 block ciphers, RSA (six kinds of implementation), ECC

18.3.2 Target Cryptographic LSI

A prototype ASIC called the 'Standard Cryptographic LSI' is available for studying side-channel analysis [340]. This cryptographic LSI is manufactured using a Taiwan Semiconductor Manufacturing 130 nm or 90 nm CMOS processor and packaged in a 160-pin QFP. This LSI is designed for side-channel analysis and its IP core, which is written in Verilog HDL, is publicly available. The fact that the IP core is open is very useful because we can inspect the source code and use this information to study how the implementation style affects the results of the fault injection. Three kinds of LSIs have been developed, version (I), version (II), and version (III). Table 18.2 shows the summary of the three kinds of LSIs. We can place all three kinds of LSIs on the SASEBO-R board. In this study, we evaluate the block cipher modules implemented on version (I) LSI and public cryptography implemented on version (II) LSI for the experiments. The evaluated ciphers for version (I) or version (II) LSI and their maximum path delay T_d, written in the specification of the cryptographic LSI [341], are given in Tables 18.3 or 18.4. All modules implemented on the LSI are directly driven by an external clock supplied via a clock pin, a buffer, and a clock control circuit.

18.4 Results of Fault Injection Experiments

This section shows the experimental results that can be obtained using the experimental environment of the fault injection developed.

18.4.1 Experimental Results for Block Ciphers

This subsection presents the interrelation between the fault injection and the clock glitch width for block ciphers.

Table 18.3 Evaluated cipher modules of version (I) LSI and their maximum path delay T_d

Module name	T_d (ns)[a]	Description
AES_TBL	7.409	Direct mapping S-Box with case syntax
AES_PPRM3	9.407	Three-stage positive prime Reed-Muller (PPRM) based S-Box
AES_Comp_ENC_top	10.567	Encryption part of AES_Comp
AES_S	10.612	Composite field-based S-Box. Back annotated netlist is used in order to confirm delay to FPGA implementation
AES_Comp	13.788	Composite field-based S-Box
AES_PPRM1	15.407	One-stage positive prime Reed-Muller (PPRM)-based S-Box
AES_SSS1	30.884	Composite field-based S-Box. Random switching logic (RSL) is applied for side-channel countermeasure
DES	6.499	Single-DES, 64-bit key, 16 rounds
Camellia	10.612	Camellia, 128-bit block, 18 rounds
CAST128	20.263	CAST-128, 64-bit block, 128-bit key, 16 rounds
SEED	23.098	SEED, 64-bit block, 128-bit key, 16 rounds
MISTY1	23.620	MISTY1, 64-bit block, 128-bit key, eight rounds

[a] @25 °C, 1.20 V

Table 18.4 Evaluated cipher modules of version (II) LSI and their maximum path delay T_d

Module name	T_d (ns)[a]	Description
RSA	19.486	RSA using the six kinds of modular multiplications and each has two modes, 512-bit data and key. Mode: CRT mode (CRT/nonCRT) Functions: (0) Left-to-right binary method, (1) Right-to-left binary method, (2) Left-to-right binary method with dummy multiplication, (3) Right-to-left binary method with dummy multiplication, (4) Montgomery powering ladder, (5) Square-multiply exponentiation method
ECC	9.841	Elliptic curve multiplication with Montgomery powering algorithm over GF(2^{61}), 61-bit data, 64-bit key

[a] @25 °C, 1.20 V

18.4.1.1 Interrelation Among Number of Error Bytes, Glitch Round and Glitch Width T_g

Using the environments developed, we can investigate how the glitch affects the number of error bytes in the encryption output with its injected round and its width T_g. The reason we target the number of bytes is that the execution of many block ciphers involves many bytewise operations, and many theoretical fault attacks use these bytewise operation characteristics. As an example, a glitch could be injected into the last or penultimate round of the encryption of each module because these two rounds are convenient for investigating the effect of injected faults and are frequently used in theoretical fault attacks. The fault injection round can be changed by setting the delay of the Trigger_out signal of the oscilloscope appropriately.

The results for all AES modules are shown in Fig. 18.5, and for other block cipher modules in Fig. 18.6. Each line in both graphs represents the average of 5,000 samples in steps of 0.2 ns. Each sample is encrypted with a random plaintext and key.

The results support the idea that the cause of the fault injection is a set-up time violation as explained in Sect. 18.2. In particular, both the number of error bytes and the error starting time show this characteristic. When we gradually shorten T_g, the number of error bytes increases as T_g decreases.

The error starting time for each module is somewhat shorter than each of their respective maximum path delays, T_d, and the order of modules sorted by their error starting time is the same as that sorted by T_d. For example, the results of the AES_SSS1 module, which has the longest T_d of 30.884 ns of all evaluated AES modules, indicates that the error starting time is approximately 26 ns, which is the longest in all the evaluated AES modules. The results of the AES_TBL module, which has the shortest T_d of 7.409 ns of all evaluated AES modules, indicates that the error starting time is approximately 3.5 ns in the last round. As is shown in Fig. 18.5b, AES_SSS1 module has a very steep slope; then, compared to other AES modules, it is difficult to set T_g in the experiments as only one byte of the intermediate states is corrupted. As shown in Fig. 18.6, the results of the MISTY1 module, which has the longest T_d of the evaluated modules, indicate that the error byte starting time is approximately 13.5 ns, which is the largest in the evaluated modules. The results for the DES module, which has the shortest T_d indicate that the error byte starting time is approximately 3.0 ns which is the shortest in the evaluated modules. In Fig. 18.6, when the faults are injected into the penultimate round (or last round), the number of error bytes is all (or half) the bytes of the register, caused by the characteristics of the Feistel structure when we use the shortest T_g.

The results in Fig. 18.7 show the probability of the number of error bytes for the AES_PPRM1 module as an example when the fault is injected into the last round. Each color of bar in the graph indicates the number of the error bytes. As T_g is decreased, the probability of the number of bytes affected increases and the expected number of bytes affected increases from one to 16. This figure shows that the one-byte error starts to appear at approximately $T_g = 8.5$ ns, and all 16 bytes are corrupted at approximately $T_g = 2.5$ ns. This result also shows that we can inject a one-byte fault when we set T_g to the absolute edge, i.e., 8.5 ns in this example. The fact that we can inject a one-byte fault is useful because some theoretical fault analyses use this as a fault model, that is, a model where one byte of an intermediate state is randomly corrupted. The results also indicate that the position of the error byte is the same when we change the fault injection round while maintaining the same T_g. Therefore, we can also inject a one-byte fault in an arbitrary round in the same way as we do in the last round.

From the above fact, a one-byte fault can be injected into only one round at any time of the encryption. Even when the target device does not output the intermediate state, we can calculate the value of the faults injected into the intermediate state from the correct and faulty ciphertexts by computing the inverse operations of AES to determine how the fault was propagated, given that we know the correct and faulty ciphertexts and the secret key we set. Specifically, we compute the difference

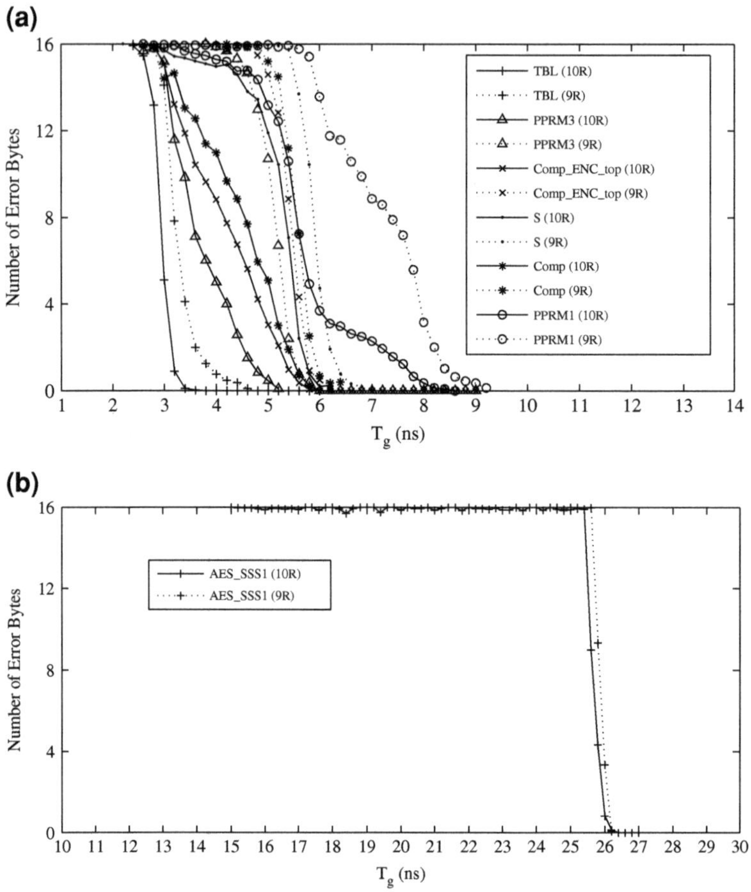

Fig. 18.5 Average number of error bytes for AES modules. **a** Results from AES module; **b** Results from AES-SSS1 module

between the two intermediate states obtained by the correct and faulty ciphertexts using the secret key we set with a PC. Table 18.5 compiles an example of the fault propagation result, the values of the output of each round calculated from the correct-faulty ciphertexts and secret key by computing the inverse operations of AES for the AES_SSS1 module. This result shows that a one-byte fault can be injected into the output of round 7, and a one-byte fault injected there is propagated to four bytes in the next round's output by the MixColumns transformation.[1] The key retrieval procedure will be discussed in Sect. 18.5.1.

[1] In order to inject faults into the output of round 7, we need to shorten T_g in round 8 in the case of the AES_SSS1 module.

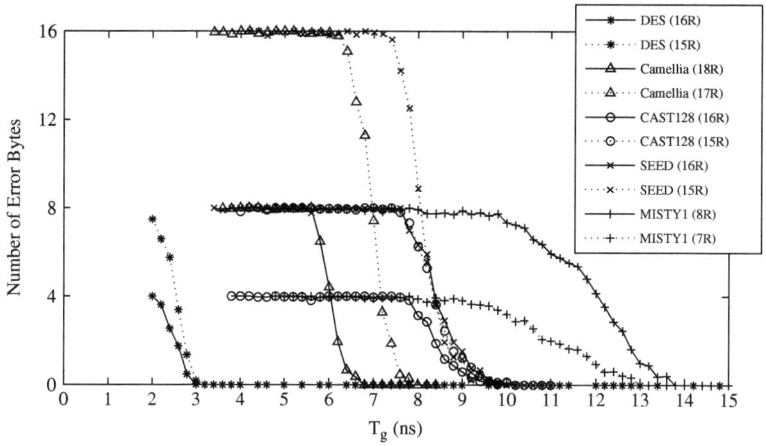

Fig. 18.6 Average number of error bytes for other modules

Table 18.5 Calculated faulty intermediate values from correct and faulty outputs for the AES_SSS1 module

Round	Correct output	Faulty output	Differences correct and faulty outputs
Round 7	08e44dcc40a133bc\ 111388c26d693a71	0a2242d42d5bace5\ f3b491a524d52ffd	0000000000000000\ 0000007800000000
Round 8	47c65ea7b9a44b10\ c01bb154051bf7b7	9b474f7630f607e4\ 3c102793f9c058dd	0000000000000000\ 000000005454fca8
Round 9	4043e107f8740366\ 73c9286d5fff1517	06f7df66a358f18a\ 19cc5a54f91fe3af	5a5aeeb47e82fc7e\ 84f87c7c06030305
Round 10	209b52ed447d30fa\ 1a8d0241ac46e8fe	163ce8530873138d\ e4265a286f9d5984	d3c828782bd7445e\ 3e12a2b786637821

Key 000102030405060708c430b7323ba122

18.4.1.2 Interrelation Among the Position of Error Bits, Glitch Round, and Glitch Width T_g

In this section, we describe how the characteristics of a glitch change the number of bits affected by its being injected into a given round with interval T_g. For example, the results of the AES_PPRM1 module are shown in Fig. 18.8a, b and c when the fault injected into round 10 (the last round) and round 9. The points in these figures represent the error occurrences in the output of round 9 or round 10 (ciphertext), and the output bit in the bit position is different from that of the correct output. Each graph is a result of 5,000 encryptions with various values for T_g. In these examples, we randomly set the plaintexts and keys.

In these results, we can see many fragmented streaks of error bits. We believe that the length of each streak largely corresponds to its path delay. In Fig. 18.8a and b, we can also see a similar trend in the fragmented streaks of error bit positions at the

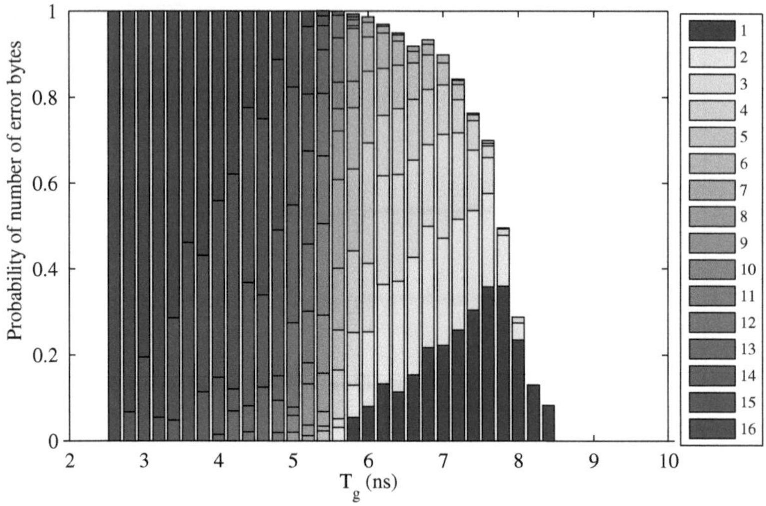

Fig. 18.7 Probability of number of error bytes for AES_PPRM1 module when the faults are injected into the last round

area when the fault starts to be injected. This means that we can inject a fault into almost the same bits between rounds 9 and 10 when we set T_g to the same length as that of the starting point of the fault injection. Therefore, this figure shows that a one-byte fault injected into any bit position of round 10 can also be injected into the same position of the earlier round. Compared to the results of Fig. 18.8a, the length of the fragmented streaks is short and the number of errors of Fig. 18.8b increases when $T_g < 8$ ns when a clock glitch is generated in round 9. We conjecture that this is because of the fault diffusion from the MixColumns operation in the AES block cipher, which is not executed in the last round.

18.4.2 Experimental Results for Public Key Cryptography

This subsection presents the interrelation between the success probabilities of the fault injection and the glitch width T_g for public key cryptography.

In these experiments, for RSA modules, a continuous 60-cycle clock glitch is injected at a specific time during the computation of an RSA signature. For an ECC module, a continuous ten-cycle clock glitch is injected at a specific time during an ECC scalar multiplication. The detailed relationship between the success probabilities of the fault injection and the glitch width T_g is shown in Fig. 18.9. Each success probability shown in Fig. 18.9 is produced from 1000 measurements. For each RSA or ECC module, the success probability has been divided into 11 sections and shown in different colors. For example, with $T_g = 8.5$ ns, no fault is injected for LR binary with a dummy RSA module and the fault can be injected with probability of 0.1 for the LR binary RSA module.

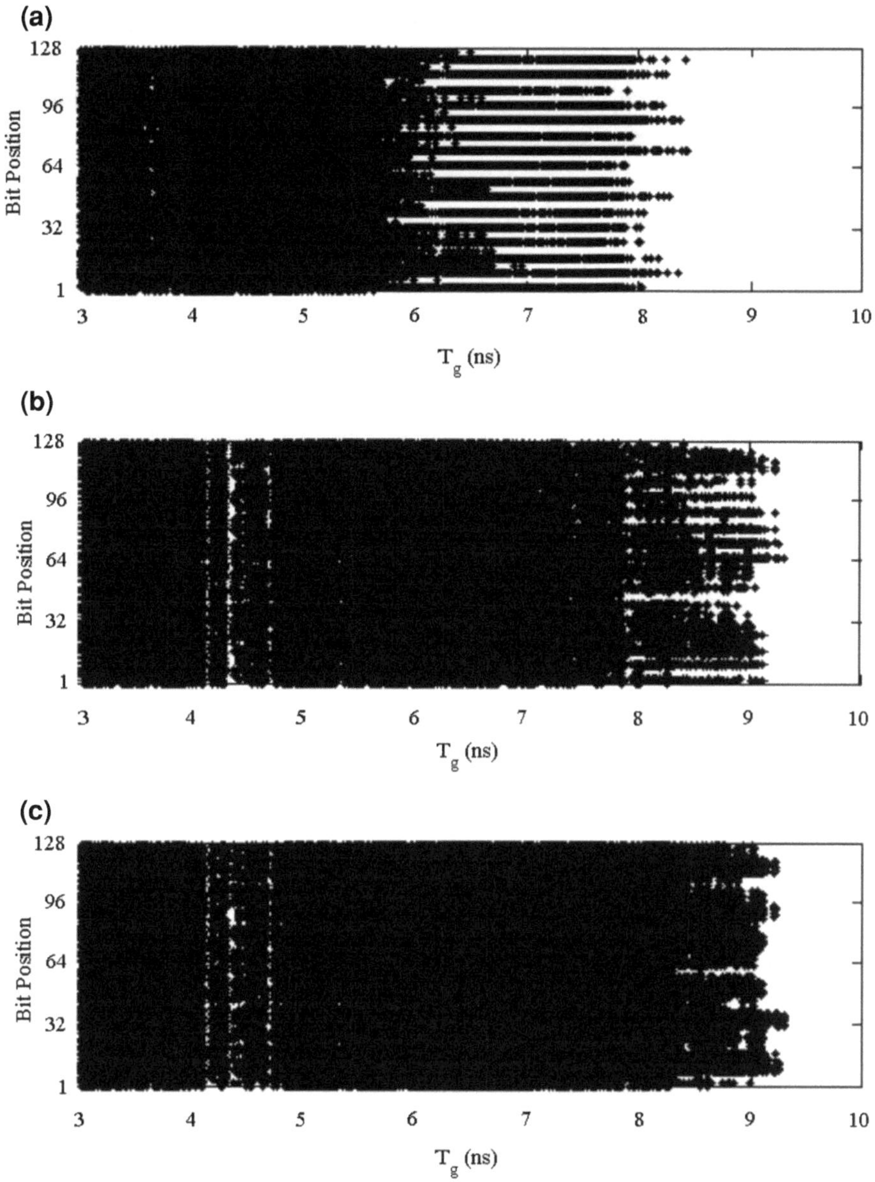

Fig. 18.8 Error bit position of AES_PPRM1 module. **a** The difference between correct and faulty ciphertexts when the fault is injected into round ten, **b** The difference between correct and faulty outputs of round nine when the fault is injected into round nine, **c** The difference between correct and faulty ciphertexts when the fault is injected into round nine

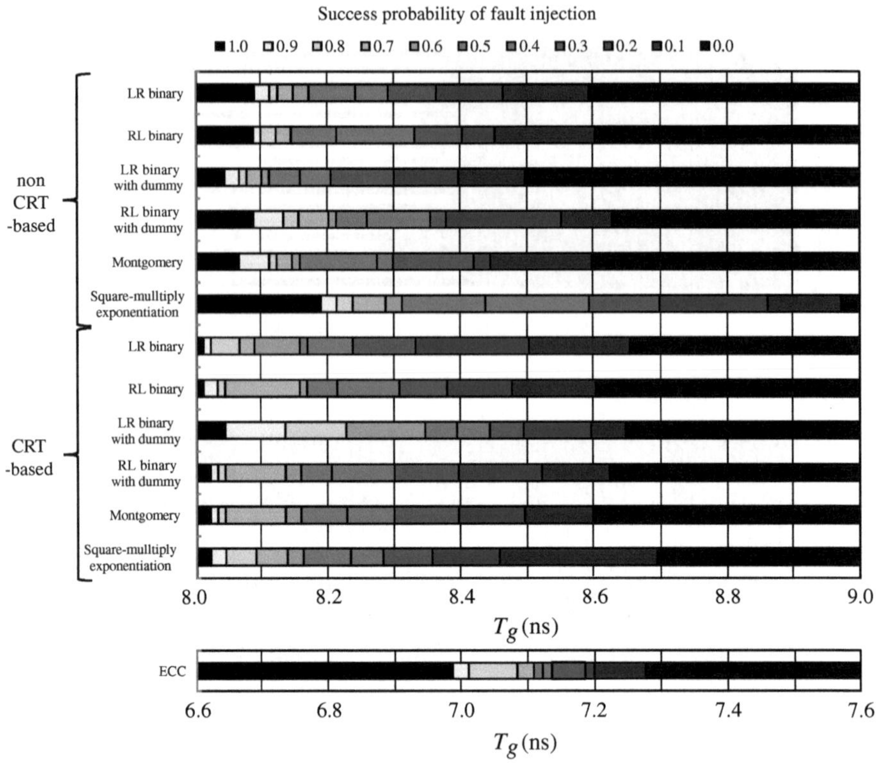

Fig. 18.9 Probability of fault injection for public cryptography. LR (RL) binary refers to the left-to-right (right-to-left) binary method

The important thing to note is that the fault can be injected with probability 1.0 when the glitch width T_g is small enough, e.g. $T_g < 8.0$ ns for both the CRT-based and the nonCRT-based RSA modules and $T_g < 7.0$ ns for the ECC module. The key retrieval procedure will be discussed in Sects. 18.5.2 and 18.5.3, respectively, for all the CRT-based RSA modules and the dummy-operation-based nonCRT-based RSA models.

18.5 Key Retrievals Using Faulty Ciphertexts

In order to verify the various attack methods of the theoretical fault analysis, we perform the key retrieval from a faulty ciphertext obtained in our experiments. As examples, we show the results of key retrievals using well-known fault analyses on AES and RSA.

18.5.1 Differential Fault Analysis Attack Against AES

In this section, we describe an implementation of the well-known fault attack against AES proposed by Piret and Quisquater [324], which uses a random fault model. An attacker can recover a 128-bit key using one pair of correct and faulty ciphertexts with a brute-force search when a one-byte random error occurs in the intermediate state of the output of round 7.

The experimental results of the fault injection in Sect. 18.4.1.1 in Fig. 18.5 show that we can inject a one-byte error into a part of the intermediate state with a success rate of almost 100 % when T_g is set appropriately. Then, we assume that we can inject a one-byte fault into the intermediate state at the end of round 7 with the same glitch characteristics. We attempted retrieving the secret key using Piret and Quisquater's attack with one pair of correct and faulty ciphertexts. If the faults are appropriately injected into the intermediate state, we can retrieve the secret key using a brute-force search with an expected 2^{32} candidate secret key values. The details of the attempt of the key retrievals are given hereafter.

1. We execute normal encryption using a randomly selected plaintext and record the correct ciphertext.
2. We generate a clock glitch into the last round (round 10) with the plaintext selected above. Then, we maximize T_g in the region in which the fault is injected into the ciphertext by controlling the phase difference of the two clock sources. This cause corruption in one byte in the ciphertext. We note the position of the error byte, which can easily be obtained from the faulty ciphertext.
3. We change the glitch position to round 7 while maintaining T_g as set above.[2] This can be easily done by changing the trigger delay of the oscilloscope. We record the faulty ciphertext.
4. We reduce the key candidate space using Piret and Quisquater's attack with one pair of correct and faulty ciphertexts recorded above. Subsequently, we execute a brute-force attack on the reduced key space. These attacks were implemented in C code and executed on a Core2 Duo 3.0 GHz PC.

We succeeded in retrieving the key set in the LSI from all AES modules (the retrieved key is the same as the key set in the LSI). The results agree fairly well with the theoretical results that the average reduced key candidate space is 2^{32}, and show, with high probability, that in a practical application the one-byte error position is the same, regardless of the round in which the clock glitch was generated, when we maintain the same T_g.

[2] In the case of the AES_SSS1 module, we change the glitch position to round 8 to inject the fault into the output of round 8 because of the characteristics of implementation with the side-channel countermeasure.

18.5.2 Boneh et al.'s Attack Against CRT-Based RSA

Before introducing Boneh et al.'s attack against CRT-based RSA, we review the security of RSA against side-channel attacks. First of all, all the CRT-based RSA modules are potentially vulnerable to Boneh et al.'s attack with one fault injection. The attacker can easily distinguish the CRT-based modules from others by conducting a simple power analysis (SPA) on a power consumption trace (as described in Chap. 2). An attacker can also distinguish whether or not the LR or RL binary RSA module is used using SPA. In the case of the LR or RL binary RSA module, the attacker can retrieve a private key bit by bit using SPA. If the target device is not the LR or RL binary RSA module, the attacker can conduct a safe-error attack to retrieve the key if the dummy operations are involved in the exponentiation. We first introduce Boneh et al.'s attack in this subsection, and then we describe safe-error attacks in the next subsection.

Boneh et al. proposed a fault analysis attack against the CRT-based RSA implementations [56]. This attack can recover a private key using one pair of correct and faulty outputs. Without loss of generality, suppose a fault occurs only during the computation of one of two modular exponentiations using the CRT, but no fault occurs during the other computation of the modular exponentiation. Then, the public modulus can be easily factored by calculating the greatest common divisor of the difference between the correct and faulty outputs and the public modulus.

As mentioned above, SPA based on a power trace can help an attacker to identify whether or not the CRT is used. This indicates that an attacker can find an appropriate timing of fault injection. For example, Fig. 18.10 shows the power consumption of a CRT-based LR binary RSA module with dummy multiplication. The power trace shown in Fig. 18.10 can be divided into two parts, denoted by periods A and B. We can guess that periods A and B of the power consumption correspond to the two exponentiations required to compute the CRT algorithm, i.e. $C_p = y^d \mod p$ and $C_q = y^d \mod q$, where y is output, p and q are the prime numbers, and d is the secret key.

By following the experimental results of the fault injection in Sect. 18.4.2, we inject a fault at any timing during the computation corresponding to the period A shown in Fig. 18.10. We then try to factor the public modulus and retrieve two primes and the secret key with one pair of correct and faulty outputs. For example, Table 18.6 shows the actual values in the experiments and the secret values calculated using the above attack method in the case of a CRT-based LR binary method with the dummy multiplication module. In the same way, we can retrieve the private key from CRT-based RSA modules implemented on the LSI.

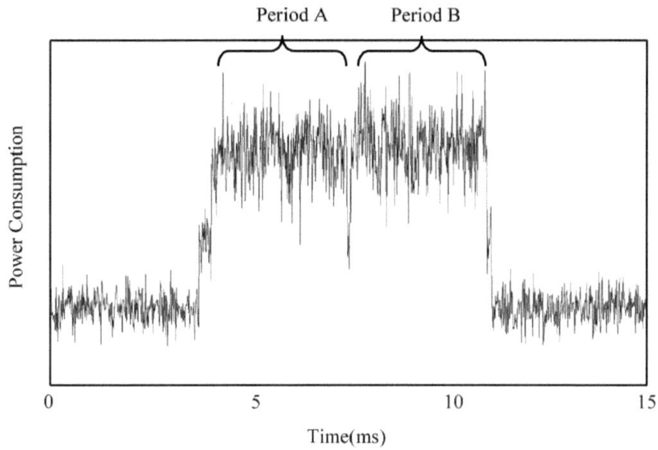

Fig. 18.10 Power consumption of a CRT-based LR binary RSA with dummy multiplication

Table 18.6 Actual values based on experiments and secret values using RSA attack

Actual values	
Correct output	008f535623122751c9c780c2b16c7dca52d0c595026eb46fd8e25296996e45c8\
	ffca0e9e4080a2169379cd41ce1f336c07b25bf43dba5f27c95bf575552bcb8c
Faulty output	010625188f795efd6d05a008eff5b1914b76ead9f2af98529280f47867770ab0\
	acdc5c2b9479c5f53f076deba24d1c6b4188c4ca9ee9cdd9692417b9a48fa556
Public modulus	00958f36f47ee7f4cb62db37c420e55ea18c4726f5d94b66d9dcb4d9df784767\
	c5a1870fbfd39fd6ab48ef1bb9cf97cb829086c8b72f037e031422b191d07d5f
Secret values	
Prime p	3a909b053b310fe1e26d72d317108622d9d4934e9834b7a0ef3b9881ca2779b7
Prime q	028dc1e58b8f557e51774d93224d24e5e3df06f31546942ae7a2f19786d93999
Secret key	002d833545560f9dfeae40d1d621b4ce23c91ac406de399f32f52b81e403d006\
	6282048a2ad1b3c1168ff402b16bdc6f15652be337a8071360256367e048db8f

18.5.3 Yen et al.'s Attack Against RSA with Dummy Operation (Safe-Error Attack)

Yen et al. proposed a computational safe-error attack, the so-called C safe-error attack, against a classical exponentiation algorithm with countermeasures for side-channel attacks, such as the square-and-multiply-always exponentiation algorithm using a dummy operation [429]. The C safe-error attack is performed by inducing transient random computational faults during a potentially dummy multiplication. This attack uses the fact that when the current exponent bit of a private key is equal to 0, the fault injection has no effect since the multiplication is a dummy operation. Therefore, the exponent bit can be obtained by examining the correctness of the output.

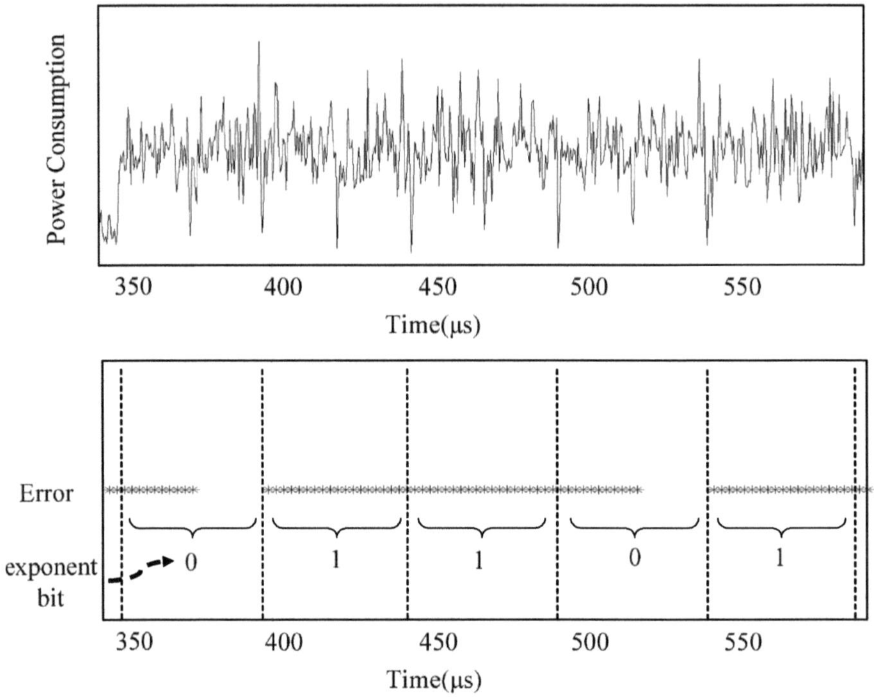

Fig. 18.11 Power consumption and result of C safe-error attack against LR binary using dummy operations

When we inject faults into the RSA calculation, we shift the timing of the fault injection that consists of a continuous clock of 60 cycles with the glitch. We try to retrieve the exponent bit of the secret key, for the RSA modules with the dummy operation using the C safe error attack. As an example, Fig. 18.11 shows the power consumption and the result of the C safe-error attack against the left to right binary method with dummy operations. The points in this figure show the error occurrence in the output. In this figure, we cannot recognize which operation is the dummy from the power consumption. However, by executing the C safe-error attack, we can easily find that the calculation with no error is a dummy operation, and the exponent bits can be obtained by examining whether or not the output is correct. In the same way, we also succeed in retrieving the private key from other RSA modules with dummy operations.

18.6 Conclusion

In this chapter, we describe an experimental environment for injecting faults into an LSI mounted on a SASEBO board that uses a clock glitch to allow us to study how a fault is injected and propagated in an LSI. We describe an experimental methodology for fault injection using two clocks and show how the clock glitch affects the error occurrence in the cryptographic modules. In order to verify the theoretical attack methods, we perform the key retrieval of AES and RSA as examples, using the faulty ciphertexts obtained from the experiments. The results in this work contribute to the future study of fault analyses using actual devices.

References

1. Adleman, L.M.: On breaking generalized knapsack public key cryptosystems. In: 15th Annual ACM Symposium on Theory of Computing, pp. 402–412. ACM Press, Boston (1983)
2. Agilent Technologies: 6633B 100 Watt System Power Supply Datasheet (2006) http://www.home.agilent.com/upload/cmc_upload/All/663xseries_datasheet_Jan06.pdf
3. Agilent Technologies: 34420A NanoVolt, Micro-Ohm Meter Datasheet (2009). http://cp.literature.agilent.com/litweb/pdf/5968-9726EN.pdf
4. Agilent Technologies: E3631A 80W Triple Output Power Supply Datasheet (2009). http://cp.literature.agilent.com/litweb/pdf/5968-9726EN.pdf
5. Agoyan, M., Dutertre, J.M., Naccache, D., Robisson, B., Tria, A.: When clocks fail: On critical paths and clock faults. In: D. Gollmann, J.L. Lanet, J. Iguchi-Cartigny (eds.) Smart Card Research and Advanced Applications (CARDIS 2010). Lecture Notes in Computer Science, vol. 6035, pp. 182–193. Springer (2010)
6. Ajtai, M.: Random lattices and a conjectured 0–1 law about their polynomial time computable properties. In: 43rd Symposium on Foundations of Computer Science (FOCS 2002), pp. 733–742. IEEE Comput. Soc. (2002)
7. Ajtai, M.: Generating random lattices according to the invariant distribution (2006). Draft
8. Ajtai, M., Kumar, R., Sivakumar, D.: A sieve algorithm for the shortest lattice vector problem. In: 33rd Annual ACM Symposium on Theory of Computing, pp. 601–610. ACM Press (2001)
9. Akdemir, K.D., Karakoyunlu, D., Sunar, B.: Nonlinear error detection for elliptic curve cryptosystems (2010). Preprint
10. Akdemir, K.D., Sunar, B.: Generic approach for hardening state machines against strong adversaries. IET Comput. Digit. Tech. **4**(6), 458–470 (2010)
11. Akkar, M.L., Bevan, R., Goubin, L.: Two power analysis attacks against one-mask methods. In: Roy, B.K.., Meier, W. (eds.) Fast Software Encryption (FSE 2004). Lecture Notes in Computer Science, vol. 3017, pp. 332–347. Springer (2004)
12. Amiel, F., Clavier, C., Tunstall, M.: Fault analysis of DPA-resistant algorithms. In: Breveglieri, L., et al. vol. 69, pp. 223–236
13. Anderson, R., Biham, E., Knudsen, L.: Serpent: A proposal for the Advanced Encryption Standard (1998). http://www.cl.cam.ac.uk/rja14/Papers/serpent.pdf
14. Anderson, R.J., Kuhn, M.G.: Tamper resistance—a cautionary note. In: 2nd USENIX Workshop on Electronic Commerce, pp. 1–11. USENIX Association (1996)
15. Anderson, R.J., Kuhn, M.G.: Low cost attacks on tamper resistant devices. In: B. Christianson, B. Crispo, T.M.A. Lomas, M. Roe (eds.) Security Protocols, *Lecture Notes in Computer Science*, vol. 1361, pp. 125–136. Springer (1997)

M. Joye and M. Tunstall (eds.), *Fault Analysis in Cryptography*,
Information Security and Cryptography, DOI: 10.1007/978-3-642-29656-7,

16. Antipa, A., Brown, D.R.L., Menezes, A., Struik, R., Vanstone, S.A.: Validation of elliptic curve public keys. In: Y. Desmedt (ed.) Public Key Cryptography – PKC 2003. Lecture Notes in Computer Science, vol. 2567, pp. 211–223. Springer (2003)
17. ARM Limited: ARM9 family of general-purpose microprocessors. ARM926EJ-S Technical Reference Manual (2008). http://infocenter.arm.com/help/topic/com.arm.doc.ddi0198e/DDI0198E_arm926ejs_r0p5_trm.pdf
18. Aumüller, C., Bier, P., Fischer, W., Hofreiter, P., Seifert, J.P.: Fault attacks on RSA with CRT: Concrete results and practical countermeasures. In: Kaliski Jr.., et al. vol. 208, pp. 260–275
19. Baek, Y.J., Vasyltsov, I.: How to prevent DPA and fault attack in a unified way for ECC scalar multiplication: Ring extension method. In: E. Dawson, D.S. Wong (eds.) Information Security Practice and Experience (ISPEC 2007). Lecture Notes in Computer Science, vol. 4464, pp. 225–237. Springer (2007)
20. Bao, F., Deng, R.H., Han, Y., Jeng, A.B., Narasimhalu, A.D., Ngair, T.H.: Breaking public key cryptosystems on tamper resistant devices in the presence of transient faults. In: B. Christianson, B. Crispo, T.M.A. Lomas, M. Roe (eds.) Security Protocols. Lecture Notes in Computer Science, vol. 1361, pp. 115–124. Springer (1998)
21. Bar-El, H., Choukri, H., Naccache, D., Tunstall, M., Whelan, C.: The sorcerer's apprentice guide to fault attacks. Proceedings the IEEE **94**(2), 370–382 (2006)
22. Barenghi, A., Bertoni, G., Breveglieri, L., Parrinello, E., Pellicioli, M.: Fault attacks against AES 256 (2010). Early Symmetric Crypto (ESC) seminar, Remich, Luxembourg
23. Barenghi, A., Bertoni, G., Parrinello, E., Pelosi, G.: Low voltage fault attacks on the RSA cryptosystem. In: Breveglieri, L., et al. vol. 66, pp. 23–31
24. Barenghi, A., Bertoni, G.M., Breveglieri, L., Pellicioli, M., Pelosi, G.: Low voltage fault attacks to AES. In: Tehranipoor, M., Plusquellic, J. (eds.) IEEE International Workshop on Hardware-Oriented Security and Trust (HOST 2010), pp. 7–12. IEEE Comput. Soc. (2010)
25. Barreto, P.S., Kim, H.Y., Lynn, B., Scott, M.: Efficient algorithms for pairing-based cryptosystems. In: Yung, M. (ed.) Advances in Cryptology—CRYPTO 2002. Lecture Notes in Computer Science, vol. 2442, pp. 354–368. Springer, Heidelberg (2002)
26. Barreto, P.S.L.M., Galbraith, S.D., O'Eigeartaigh, C., Scott, M.: Efficient pairing computation on supersingular abelian varieties. Des. Codes Cryptogr. **42**(3), 239–271 (2007)
27. Barrett, P.D.: Implementing the Rivest Shamir and Adleman public key encryption algorithm on a standard digital signal processor. In: Odlyzko, A.M. (ed.) Advances in Cryptology—CRYPTO '86. Lecture Notes in Computer Science, vol. 263, pp. 311–323. Springer (1987)
28. Bayat-Sarmadi, S., Hasan, M.A.: Concurrent error detection in finite field arithmetic operations using pipelined and systolic architectures. IEEE Trans. Comput. **58**(11), 1553–1567 (2009)
29. Bellare, M., Rogaway, P.: Random oracles are practical: A paradigm for designing efficient protocols. In: 1st ACM Conference on Computer and Communications Security (CCS '93), pp. 62–73. ACM Press (1993)
30. Bellare, M., Rogaway, P.: Optimal asymmetric encryption. In: De Santis, A. (ed.) Advances in Cryptology— EUROCRYPT '94. Lecture Notes in Computer Science, vol. 950, pp. 92–111. Springer, Berlin (1995)
31. Bellare, M., Rogaway, P.: The exact security of digital signatures: How to sign with RSA and Rabin. In: Advances in Cryptology—EUROCRYPT '96. Lecture Notes in Computer Science, vol. 1070, pp. 399–416. Springer (1996)
32. Bellcore: New threat model breaks crypto codes. Press release (1996)
33. Bernstein, D.J., Lange, T.: Faster addition and doubling on elliptic curves. In: Kurosawa, K. (ed.) Advances in Cryptology—ASIACRYPT 2007. Lecture Notes in Computer Science, vol. 4833, pp. 29–50. Springer (2007)

34. Bertoni, G., Breveglieri, L., Koren, I., Maistri, P., Piuri, V.: Error analysis and detection procedures for a hardware implementation of the Advanced Encryption Standard. IEEE Trans. Comput. **52**(4), 492–505 (2003)
35. Bertoni, G., Breveglieri, L., Koren, I., Piuri, V.: Fault detection in the Advanced Encryption Standard. In: Proc. of the International Conference on Massively Parallel Computing Systems (MPCS 2002), pp. 92–97. Ischia, Italy (2002)
36. Berzati, A., Canovas, C., Castagnos, G., Debraize, B., Goubin, L., Gouget, A., Paillier, P., Salgado, S.: Fault analysis of GRAIN-128. In: Tehranipoor, M., Plusquellic, J. (eds.) IEEE International Workshop on Hardware-Oriented Security and Trust (HOST 2009), pp. 7–14. IEEE Comput. Soc. (2009)
37. Berzati, A., Canovas, C., Dumas, J.G., Goubin, L.: Fault attacks on RSA public keys: Left-to-right implementations are also vulnerable. In: Fischlin, M. (ed.) Topics in Cryptology—CT-RSA 2009. Lecture Notes in Computer Science, vol. 5473, pp. 414–428. Springer (2009)
38. Berzati, A., Canovas, C., Goubin, L.: In (security) against fault injection attacks for CRT-RSA implementations. In: Breveglieri, L., et al., vol. 65, pp. 101–107
39. Berzati, A., Canovas, C., Goubin, L.: Perturbating RSA public keys: An improved attack. In: Oswald and Rohatgi, vol. 314, pp. 380–395
40. Berzati, A., Canovas-Dumas, C., Goubin, L.: Fault analysis of Rabbit: Toward a secret key leakage. In: Roy, B.K., Sendrier, N. (eds.) Progress in Cryptology – INDOCRYPT 2009. Lecture Notes in Computer Science, vol. 5922, pp. 72–87. Springer (2009)
41. Berzati, A., Canovas-Dumas, C., Goubin, L.: Public key perturbation of randomized RSA implementations. In: Mangard and Standaert, vol. 271, pp. 306–319
42. Beth, T., Jungnickel, D., Lenz, H.: Design Theory, vol 1. Cambridge University Press, Cambridge (1999)
43. Bhasin, S., Danger, J.L., Guilley, S., Selmane, N.: Security evaluation of different AES implementations against practical setup time violation attacks in FPGAs. In: Tehranipoor, M., Plusquellic, J. (eds.) IEEE International Workshop on Hardware-Oriented Security and Trust (HOST 2009), pp. 15–21. IEEE Comput. Soc. (2009)
44. Biehl, I., Meyer, B., Müller, V.: Differential fault analysis of elliptic curve cryptosystems. In: Bellare, M. (ed.) Advances in Cryptology – CRYPTO 2000. Lecture Notes in Computer Science, vol. 1880, pp. 131–146. Springer (2000)
45. Biham, E., Granboulan, L., Nguyen, P.Q.: Impossible fault analysis of RC4 and differential fault analysis of RC4. In: Gilbert, H., Handschuh, H. (eds.) Fast Software Encryption (FSE 2005). Lecture Notes in Computer Science, vol. 3557, pp. 359–367. Springer (2005)
46. Biham, E., Shamir, A.: Differential cryptanalysis of DES-like cryptosystems. In: Menezes, A.J., Vanstone S.A. (eds.) Advances in Cryptology – CRYPTO '90. Lecture Notes in Computer Science, vol. 537, pp. 2–21. Springer (1991)
47. Biham, E., Shamir, A.: Differential cryptanalysis of DES-like cryptosystems. J. Cryptol. **4**(1), 3–72 (1991)
48. Biham, E., Shamir, A.: A new cryptanalytic attack on DES. Manuscript (1996)
49. Biham, E., Shamir, A.: Differential fault analysis of secret key cryptosystems. In: Kaliski Jr., B.S. (ed.) Advances in Cryptology – CRYPTO '97. Lecture Notes in Computer Science, vol. 1294, pp. 513–525. Springer (1997)
50. Biryukov, A.: Block and stream ciphers and the creatures in between. In: Biham, E., Handschuh, H., Lucks, S., Rijmen, V. (eds.) Symmetric Cryptography, no. 0721 in Dagstuhl Seminar Proceedings. Schloss Dagstuhl—Leibniz-Zentrum für Informatik, Germany (2007)
51. Blömer, J., Krummel, V.: Fault based collision attacks on AES. In: Breveglieri, L., et al. (eds.) vol. 69, pp. 106–120
52. Blömer, J., Otto, M.: Wagner's attack on a secure CRT-RSA algorithm reconsidered. In: Breveglieri, L., et al. vol. 69, pp. 13–23

53. Blömer, J., Otto, M., Seifert, J.P.: A new CRT-RSA algorithm secure against Bellcore attacks. In: Jajodia, S., Atluri, V., Jaeger, T. (eds.) 10th ACM Conference on Computer and Communications Security (CCS 2003), pp. 311–320. ACM Press (2003)
54. Blömer, J., Otto, M., Seifert, J.P.: Sign change fault attacks on elliptic curve cryptosystems. In: Breveglieri, L., et al. vol. 69, pp. 36–52
55. Blömer, J., Seifert, J.P.: Fault based cryptanalysis of the Advanced Encryption Standard (AES). In: Wright, R.N. (ed.) Financial Cryptography (FC 2003). Lecture Notes in Computer Science, vol. 2742, pp. 162–181. Springer, Heidelberg (2003)
56. Boneh, D., DeMillo, R.A., Lipton, R.J.: On the importance of eliminating errors in cryptographic computations. J. Cryptol. **14**(2), 101–119 (2001). Earlier version published in EUROCRYPT '97
57. Boneh, D., Franklin, M.K.: Identity-based encryption from the Weil pairing. SIAM J. Comput. **32**(3), 586–615 (2003)
58. Boneh, D., Venkatesan, R.: Hardness of computing the most significant bits of secret keys in Diffie-Hellman and related schemes. In: N. Koblitz (ed.) Advances in Cryptology—CRYPTO '96. Lecture Notes in Computer Science, vol. 1109, pp. 129–142. Springer, Berlin (1996)
59. Bos, J.N.E.: Practical privacy. Ph.D. thesis, Eindhoven University of Technology (1992)
60. Boscher, A., Handschuh, H.: Masking does not protect against differential fault attacks. In: Breveglieri, L., et al. 65, pp. 35–40
61. Boscher, A., Naciri, R., Prouff, E.: CRT RSA algorithm protected against fault attacks. In: Sauveron, D., Markantonakis, C., Bilas, A., Quisquater, J.J. (eds.) Information Security Theory and Practices (WISTP 2007). Lecture Notes in Computer Science, vol. 4462, pp. 229–243. Springer, Heidelberg (2007)
62. Bosio, A., Di Natale, G.: LIFTING: A flexible open-source fault simulator. In: 17th Asian Test Symposium (ATS 2008), pp. 35–40. IEEE Comput. Soc. (2008)
63. Bousselam, K., Di Natale, G., Flottes, M.L., Rouzeyre, B.: Evaluation of concurrent error detection techniques on the Advanced Encryption Standard. In: 16th IEEE International On-Line Testing Symposium (IOLTS 2010), pp. 223–228. IEEE Comput. Soc. (2010)
64. Breveglieri, L., Gueron, S., Koren, I., Naccache, D., Seifert, J.P. (eds.): Fault Diagnosis and Tolerance in Cryptography—FDTC 2007. IEEE Comput. Soc. (2007)
65. Breveglieri, L., Gueron, S., Koren, I., Naccache, D., Seifert, J.P. (eds.): Fault Diagnosis and Tolerance in Cryptography—FDTC 2008. IEEE Comput. Soc. (2008)
66. Breveglieri, L., Gueron, S., Koren, I., Naccache, D., Seifert, J.P. (eds.): Fault Diagnosis and Tolerance in Cryptography—FDTC 2009. IEEE Compu. Soc. (2009)
67. Breveglieri, L., Joye, M., Koren, I., Naccache, D., Verbauwhede, I. (eds.): Fault Diagnosis and Tolerance in Cryptography—FDTC 2010. IEEE Comput. Soc. (2010)
68. Breveglieri, L., Koren, I. (eds.): 2nd Workshop on Fault Diagnosis and Tolerance in Cryptography FDTC 2005, Edinburgh (2005)
69. Breveglieri, L., Koren, I., Naccache, D., Seifert, J.P. (eds.): Fault Diagnosis and Tolerance in Cryptography—FDTC 2006. Lecture Notes in Computer Science, vol. 4236. Springer, Berlin (2006)
70. Brier, É., Chevallier-Mames, B., Ciet, M., Clavier, C.: Why one should also secure RSA public key elements. In: Goubin and Matsui, vol. 167, pp. 324–338
71. Brier, É., Clavier, C., Coron, J.S., Naccache, D.: Cryptanalysis of RSA signatures with fixed-pattern padding. In: Kilian, J. (ed.) Advances in Cryptology—CRYPTO 2001, Lecture Notes in Computer Science, vol. 2139, pp. 433–439. Springer, Berlin (2001)
72. Brier, É., Clavier, C., Olivier, F.: Correlation power analysis with a leakage model. In: Joye and Quisquater, 202, pp. 16–29
73. Brouchier, J., Kean, T., Marsh, C., Naccache, D.: Temperature attacks. IEEE Secur. Priv. **7**(2), 79–82 (2009)
74. Brumley, D., Boneh, D.: Remote timing attacks are practical. Comput. Netw. **48**(5), 701–716 (2005)

75. BSI: Application of attack potential to smartcards. Supporting Document, version 2.7, revision 1 (2009)
76. Burwick, C., Coppersmith, D., D'Avignon, E., Gennaro, R., Halevi, S., Jutla, C., Matyas Jr., S.M., O'Connor, L., Peyravian, M., Safford, D., Zunic, N.: MARS—a candidate cipher for AES (1999). http://www.research.ibm.com/security/mars.pdf
77. Cadence Corporation: SoC encounter. http://www.cadence.com/products/di/soc_encounter/pages/default.aspx
78. Callaghan, B., Pawlowski, B., Staubach, P.: NFS version 3 protocol specification. RFC 1813 (1995). http://tools.ietf.org/html/rfc1813
79. Canivet, G., Clédière, J., Ferron, J.B., Valette, F., Renaudin, M., Leveugle, R.: Detailed analyses of single laser shot effects in the configuration of a Virtex-II FPGA. In: 14th IEEE International On-Line Testing Symposium (IOLTS 2008), pp. 289–294. IEEE Comput. Soc. (2008)
80. Carlet, C., Ding, C.: Highly nonlinear mappings. J. Complex. **20**(2/3), 205–244 (2004)
81. Chari, S., Jutla, C.S., Rao, J.R., Rohatgi, P.: Towards approaches to counteract power-analysis attacks. In: M.J. Wiener (ed.) Advances in Cryptology—CRYPTO '99. Lecture Notes in Computer Science, vol. 1666, pp. 398–412. Springer, Berlin (1999)
82. Chari, S., Rao, J.R., Rohatgi, P.: Template attacks. In: Kaliski Jr., et al. 208, pp. 13–28
83. Chelton, W.N., Benaissa, M.: Concurrent error detection in $GF(2^m)$ multiplication and its application in elliptic curve cryptography. IET Circuits, Devices and Syst. **2**(3), 289–297 (2008)
84. Chen, C.N., Yen, S.M.: Differential fault analysis on AES key schedule and some countermeasures. In: R. Safavi-Naini, J. Seberry (eds.) Information Security and Privacy (ACISP 2003), Lecture Notes in Computer Science, vol. 2727, pp. 118–129. Springer, Heidelberg (2003)
85. Chevallier-Mames, B., Ciet, M., Joye, M.: Low-cost solutions for preventing simple side-channel analysis: Side-channel atomicity. IEEE Transac. Comput. **53**(6), 760–768 (2004)
86. Chiou, C.W., Chang, C.C., Lee, C.Y., Hou, T.W., Lin, J.M.: Concurrent error detection and correction in Gaussian normal basis multiplier over $GF(2^m)$. IEEE Transac. Comput. **58**(6), 851–857 (2009)
87. Chiou, C.W., Lee, C.Y., Deng, A.W., Lin, J.M.: Concurrent error detection in Montgomery multiplication over $GF(2^m)$. IEICE Transac. Fundam. Electron. Commun. Comput. Sci. **89**(A(2), 566–579 (2006)
88. Chodowiec, P., Gaj, K.: Very compact FPGA implementation of the AES algorithm. In: Walter., et al., vol. 411, pp. 319–333
89. Choukri, H., Tunstall, M.: Round reduction using faults. In: Breveglieri and Koren, I. , vol. 68, pp. 13–24
90. Ciet, M., Joye, M.: Elliptic curve cryptosystems in the presence of permanent and transient faults. Des. Codes and Cryptogr. **36**(1), 33–43 (2005)
91. Ciet, M., Joye, M.: Practical fault countermeasures for Chinese Remaindering based RSA. In: Breveglieri and Koren, I. , vol. 68, pp. 124–132
92. Clavier, C.: De la sécurité des cryptosystèmes embarqués. Ph.D. thesis, Université de Versailles Saint-Quentin-en-Yvelines (2007)
93. Clavier, C.: Secret external encodings do not prevent transient fault analysis. In: Paillier and Verbauwhede vol. 319, pp. 181–194
94. Clavier, C., Feix, B., Gagnerot, G., Roussellet, M.: Passive and active combined attacks on AES. In: Breveglieri, L., et al., vol. 67, pp. 10–19
95. Clavier, C., Gaj, K. (eds.): Cryptographic Hardware and Embedded Systems—CHES 2009, Lecture Notes in Computer Science, vol. 5747. Springer, Berlin (2009)
96. Codesourcery: GNU toolchain for ARM processors (2009). http://www.codesourcery.com/sgpp/lite/arm
97. Cohen, H.: A Course in Computational Algebraic Number Theory, Graduate Texts in Mathematics, vol. 138. Springer, Berlin (1993)

98. Cohen, H., Frey, G. (eds.): Handbook of Elliptic and Hyperelliptic Curve Cryptography, Discrete Mathematics and Its Applications, vol. 34. Chapman & Hall/CRC (2005)
99. Cohen, H., Miyaji, A., Ono, T.: Efficient elliptic curve exponentiation using mixed coordinates. In: K. Ohta, D. Pei (eds.) Advances in Cryptology—ASIACRYPT '98, Lecture Notes in Computer Science, vol. 1514, pp. 51–65. Springer, Berlin (1998)
100. Common Criteria: http://www.commoncriteriaportal.org/
101. Coppersmith, D.: Small solutions to polynomial equations, and low exponent RSA vulnerabilities. J. Cryptol. **10**(4), 233–260 (1997)
102. Coron, J.S.: Resistance against differential power analysis for elliptic curve cryptosystems. In: Koç and Paar, vol. 237, pp.292–302
103. Coron, J.S., Giraud, C., Morin, N., Piret, G., Vigilant, D.: Fault attacks and countermeasures on Vigilant's RSA-CRT algorithm. In: Breveglieri, L., et al., vol.67, pp. 88–96
104. Coron, J.S., Joux, A., Kizhvatov, I., Naccache, D., Paillier, P.: Fault attacks on RSA signatures with partially unknown messages. In: Clavier and Gaj, vol. 95, pp. 444–456
105. Coron, J.S., Kizhvatov, I.: An efficient method for random delay generation in embedded software. In: Clavier and Gaj, vol. 95, pp. 156–170
106. Coron, J.S., Mandal, A.: PSS is secure against random fault attacks. In: Matsui, M. (ed.) Advances in Cryptology—ASIACRYPT 2009, Lecture Notes in Computer Science, vol. 5912, pp. 653–666. Springer (2009)
107. Coron, J.S., Naccache, D., Stern, J.P.: On the security of RSA padding. In: Wiener, M.J., (ed.) Advances in Cryptology—CRYPTO '99, Lecture Notes in Computer Science, vol. 1666, pp. 1–18. Springer Heidelberg (1999)
108. Coron, J.S., Naccache, D., Tibouchi, M.: Fault attacks against EMV signatures. In: Pieprzyk, J. (ed.) Topics in Cryptology—CT-RSA 2010, Lecture Notes in Computer Science, vol. 5985, pp. 208–220. Springer, Heidelberg (2010)
109. Coron, J.S., Naccache, D., Tibouchi, M., Weinmann, R.P.: Practical cryptanalysis of ISO/ IEC 9796-2 and EMV signatures. In: Halevi, S. (ed.) Advances in Cryptology—CRYPTO 2009, Lecture Notes in Computer Science, vol. 5677, pp. 428–444. Springer, Heidelberg (2009)
110. Coxeter, H.S.M.: The Real Projective Plane, 3rd edn. Springer, New York (1995)
111. Daemen, J., Knudsen, L.R., Rijmen, V.: The block cipher Square. In: Biham, E. (ed.) Fast Software Encryption (FSE '97), Lecture Notes in Computer Science, vol. 1267, pp. 149–165. Springer, Berlin (1997)
112. Daemen, J., Rijmen, V.: AES proposal: Rijndael (1999)
113. Daemen, J., Rijmen, V.: The Design of Rijndael. Springer, Berlin (2002)
114. Danger, J.L., Guilley, S., Bhasin, S., Nassar, M.: Overview of dual rail with precharge logic styles to thwart implementation-level attacks on hardware cryptoprocessors. In: 3rd International Conference on Signals, Circuits and Systems (SCS 2009), pp. 1–8. IEEE Press (2009)
115. Das, S., Roberts, D., Lee, S., Pant, S., Blaauw, D., Austin, T., Flautner, K., Mudge, T.: A self-tuning DVS processor using delay-error detection and correction. IEEE J. Solid-State Circuits **41**(4), 792–804 (2006)
116. De Cannière, C., Preneel, B.: Trivium. In: Robshaw, M., Billet, O. (eds.) New Stream Cipher Designs, Lecture Notes in Computer Science, vol. 4986, pp. 244–266. Springer, Berlin (2008)
117. Debraize, B., Corbella, I.M.: Fault Analysis of the Stream Cipher Snow 3G. In: Breveglieri, L., et al., vol. 66, pp. 103–110
118. Denk, W.: Das U-boot bootloader (2009). http://www.denx.de/wiki/U-Boot
119. Di Natale, G., Flottes, M.L., Rouzeyre, B.: A novel parity bit scheme for Sbox in AES circuits. In: P. Girard, A. Krasniewski, E. Gramatová, A. Pawlak, T. Garbolino (eds.) 10th IEEE Workshop on Design & Diagnostics of Electronic Circuits & Systems (DDECS 2007), pp. 267–271. IEEE Comput. Soc. (2007)

120. Di Natale, G., Flottes, M.L., Rouzeyre, B.: An on-line fault detection scheme for Sboxes in secure circuits. In: 13th IEEE International On-Line Testing Symposium (IOLTS 2007), pp. 57–62. IEEE Comput. Soc. (2007)
121. Diffie, W., Hellman, M.E.: New directions in cryptography. IEEE Transac. Info. Theory **22**(6), 644–654 (1976)
122. Domínguez-Oviedo, A.: On fault-based attacks and countermeasures for elliptic curve cryptosystems. Ph.D. thesis, University of Waterloo (2008)
123. Domínguez-Oviedo, A., Hasan, M.A.: Error detection and fault tolerance in ECSM using input randomization. IEEE Trans. Dependable Secur. Comput. **6**(3), 175–187 (2009)
124. Domínguez-Oviedo, A., Hasan, M.A.: Algorithm-level error detection for Montgomery ladder-based ECSM. J. Cryptogr. Eng. **1**(1), 57–69 (2011)
125. Dottax, E., Giraud, C., Rivain, M., Sierra, Y.: On second-order fault analysis resistance for CRT-RSA implementations. In: Markowitch, O., Bilas, A., Hoepman, J.H., Mitchell, C.J., Quisquater, J.J. (eds.) Information Security Theory and Practices (WISTP 2009), Lecture Notes in Computer Science, vol. 5746, pp. 68–83. Springer, Berlin (2009)
126. Doulcier, M., Di Natale, G., Flottes, M.L., Rouzeyre, B.: Self-test techniques for crypto-devices. IEEE Transac. VLSI **18**(2), 329–333 (2010)
127. Dusart, P., Letourneux, G., Vivolo, O.: Differential fault analysis on A.E.S. In: J. Zhou, M. Yung, Y. Han (eds.) Applied Cryptography and Network Security (ACNS 2003), Lecture Notes in Computer Science, vol. 2846, pp. 293–306. Springer, Berlin (2003)
128. Dutertre, J.M., Mirbaha, A.P., Naccache, D., Tria, A.: Very close to perfect solutions against power attacks. Presented at the rump session of EUROCRYPT 2010 (2010)
129. Duursma, I.M., Lee, H.S.: Tate pairing implementation for hyperelliptic curves $y^2 = x^p - x + d$. In: C.S. Laih (ed.) Advances in Cryptology—ASIACRYPT 2003, Lecture Notes in Computer Science, vol. 2894, pp. 111–123. Springer, Berlin (2003)
130. Eisenbarth, T., Kasper, T., Moradi, A., Paar, C., Salmasizadeh, M., Shalmani, M.T.M.: On the power of power analysis in the real world: A complete break of the KeeLoq code hopping scheme. In: Wagner, D. (ed.) Advances in Cryptology— CRYPTO 2008, Lecture Notes in Computer Science, vol. 5157, pp. 203–220. Springer, Berlin (2008)
131. El Mrabet, N.: What about vulnerability to a fault attack of the Miller's algorithm during an identity based protocol. In: Park, J.H., Chen, H.H., Atiquzzaman, M., Lee, C., Kim, T.H., Yeo, S.S. (eds.) Advances in Information Security and Assurance (ISA 2009), Lecture Notes in Computer Science, vol. 5576, pp. 122–134. Springer (2009)
132. EMV: Integrated circuit card specifications for payment systems. Book 2. Security and key management. Version 4.2 (2008). http://www.emvco.com/
133. The eSTREAM project. http://www.ecrypt.eu.org/stream/
134. fail0verflow: Console hacking 2010. Presented at the 27th Annual Chaos Communication Conference (2010)
135. Farhan, S.M., Khan, S.A., Jamal, H.: Mapping of high-bit algorithm to low-bit for optimized hardware implementation. In: 16th International Conference on Microelectronics (ICM 2004), pp. 148–151. IEEE Press (2004)
136. Faurax, O., Tria, A., Freund, L., Bancel, F.: Robustness of circuits under delay-induced faults: test of AES with the PAFI tool. In: 13th IEEE International On-Line Testing Symposium (IOLTS 2007), pp. 185–186. IEEE Comput. Soc. (2007)
137. Feldhofer, M., Dominikus, S., Wolkerstorfer, J.: Strong authentication for RFID systems using the AES algorithm. In: Joye, M., Quisquater, J.-J. (eds.) vol. 202, pp. 357–370
138. Feldhofer, M., Wolkerstorfer, J., Rijmen, V.: AES implementation on a grain of sand. IEE Proc. Info. Secur. **152**(1), 13–20 (2005)
139. Fenn, S.T.J., Gössel, M., Benaissa, M., Taylor, D.: On-line error detection for bit-serial multipliers in $GF(2^m)$. J. Electr. Test. Theory and Appl. **13**(1), 29–40 (1998)
140. Finney, H.: An RC4 cycle that can't happen. Posting to sci.crypt (1994)

141. FIPS PUB 186-3: Digital signature standard (DSS). Federal Information Processing Standards Publication 186-3, National Institute of Standards and Technology (NIST), Gaithersburg (2009)
142. FIPS PUB 197: Advanced Encryption Standard (AES). Federal Information Processing Standards Publication 197, National Institute of Standards and Technology (NIST), Gaithersburg (2001)
143. Fouque, P.A., Lercier, R., Réal, D., Valette, F.: Fault attack on elliptic curve Montgomery ladder implementation. In: Breveglieri, L., et al., vol. 65, pp. 92–98
144. Francq, J., Faurax, O.: Security of several AES implementations against delay faults. In: 12th Nordic Workshop on Secure IT Systems (NordSec 2007), pp. 61–72 (2007)
145. Frey, G., Müller, M., Rück, H.G.: The Tate pairing and the discrete logarithm applied to elliptic curve cryptosystems. IEEE Transac. Info. Theory **45**(5), 1717–1719 (1999)
146. Frey, G., Rück, H.G.: A remark concerning m-divisibility and the discrete logarithm in the divisor class group of curves. Math. Comput. **62**(204), 865–874 (1994)
147. Frieze, A.M., Håstad, J., Kannan, R., Lagarias, J.C., Shamir, A.: Reconstructing truncated integer variables satisfying linear congruences. SIAM J. Comput. **17**(2), 262–280 (1988)
148. Fukunaga, T., Takahashi, J.: Practical fault attack on a cryptographic LSI with ISO/IEC 18033-3 block ciphers. In: Breveglieri, L., et al. vol. 66, pp. 84–92
149. Fumaroli, G., Vigilant, D.: Blinded fault resistant exponentiation. In: Breveglieri, L., et al. vol. 69, pp. 62–70
150. Galbraith, S.D., Hess, F., Vercauteren, F.: Aspects of pairing inversion. IEEE Transac. Info. Theory **54**(12), 5119–5128 (2008)
151. Galbraith, S.D., O'hEigeartaigh, C., Sheedy, C.: Simplified pairing computation and security implications. J. Math. Cryptol. **1**(3), 267–282 (2007)
152. Gama, N., Nguyen, P.Q.: Predicting lattice reduction. In: N.P. Smart (ed.) Advances in Cryptology—EUROCRYPT 2008, Lecture Notes in Computer Science, vol. 4965, pp. 31–51. Springer (2008)
153. Gandolfi, K., Mourtel, C., Olivier, F.: Electromagnetic analysis: Concrete results. In: Koç, C.K. et al., vol. 236, pp. 251–261
154. Garner, H.L.: The residue number system. IRE Transac. Electron. Comput. **8**(6), 140–147 (1959)
155. Gaubatz, G., Sunar, B.: Robust finite field arithmetic for fault-tolerant public-key cryptography. In: Breveglieri. L., et al. vol. 69, pp. 196–210
156. Gaubatz, G., Sunar, B., Karpovsky, M.G.: Non-linear residue codes for robust public-key arithmetic. In: Breveglieri. L., et al. vol. 69, pp. 173–184
157. Gaubatz, G., Sunar, B., Savas, E.: Sequential circuit design for embedded cryptographic applications resilient to adversarial faults. IEEE Transac. Comput **57**(1), 126–138 (2008)
133. Genelle, L., Giraud, C., Prouff, E.: Securing AES implementation against fault attacks. In: Breveglieri, L., et al. vol. 66, pp. 51–64
159. Ghosh, S., Mukhopadhyay, D., Chowdhury, D.R.: Fault attack and countermeasures on pairing based cryptography. Int. J. Netw. Secur. (IJNS)**12**(1), 26–33 (2011)
160. Giraud, C.: DFA on AES. In: Dobbertin, H., Rijmen, V., Sowa, A. (eds.) Advanced Encryption Standard—AES (AES 2004), Lecture Notes in Computer Science, vol. 3373, pp. 27–41. Springer, Heidelberg (2005)
161. Giraud, C.: Fault resistant RSA implementation. In: Breveglieri, L., Koren, I. , I. , vol. 68, pp. 142–151
162. Giraud, C.: An RSA implementation resistant to fault attacks and to simple power analysis. IEEE Transac. Comput. **55**(9), 1116–1120 (2006). Extended abstract in [161]
163. Giraud, C., Thiebeauld, H.: A survey on fault attacks. In: Quisquater, J.J. Paradinas, P. Deswarte, Y. El Kalam, A.A. (eds.) Smart Card Research and Advanced Applications VI (CARDIS 2004), pp. 159–176. Kluwer Academic Publishers, Toulouse (2004)
164. Giraud, C., Thillard, A.: Piret and Quisquater's DFA on AES revisited. Cryptology ePrint Archive, Report 2010/440 (2010)

165. Goldwasser, S., Micali, S.: Probabilistic encryption. J. Comput. Syst. Sci. **28**(2), 270–299 (1984)
166. Good, T., Benaissa, M.: AES on FPGA from the fastest to the smallest. In: Rao, J.R., Sunar, B., vol. 334, pp. 427–440
167. Goubin, L., Matsui, M. (eds.): Cryptographic Hardware and Embedded Systems—CHES 2006, Lecture Notes in Computer Science, vol. 4249. Springer, Berlin (2006)
168. Goubin, L., Patarin, J.: DES and differential power analysis (The "duplication" method). In: Koç, C., Paar, C., vol. 237, pp. 158–172, Springer, Berlin (1999)
169. Govindavajhala, S., Appel, A.W.: Using memory errors to attack a virtual machine. In: 2003 IEEE Symposium on Security and Privacy (S and P 2003), pp. 154–165. IEEE Comput. Soc. (2003)
170. Granger, R., Hess, F., Oyono, R., Thériault, N., Vercauteren, F.: Ate pairing on hyperelliptic curves. In: Naor, M. (ed.) Advances in Cryptology—EUROCRYPT 2007, Lecture Notes in Computer Science, vol. 4515, pp. 430–447. Springer, Berlin (2007)
171. Guilley, S., Sauvage, L., Danger, J.L., Selmane, N.: Fault injection resilience. In: Breveglieri, L., et al., vol. 67, pp. 51–65
172. Guilley, S., Sauvage, L., Danger, J.L., Selmane, N., Pacalet, R.: Silicon-level solutions to counteract passive and active attacks. In: Breveglieri. L., et al., vol. 65, pp. 3–17
173. Habing, D.H.: The use of lasers to simulate radiation-induced transients in semiconductor devices and circuits. IEEE Transac. Nucl. Sci. **12**(5), 91–100 (1965)
174. Hämäläinen, P., Alho, T., Hännikäinen, M., Hämäläinen, T.D.: Design and implementation of low-area and low-power AES encryption hardware core. In: 9th EUROMICRO Conference on Digital System Design (DSD 2006), pp. 577–583. IEEE Press (2006)
175. Hämäläinen, P., Hännikäinen, M., Hämäläinen, T.D.: Efficient hardware implementation of security processing for IEEE 802.15.4 wireless networks. In: 48th Midwest Symposium on Circuits and Systems, pp. 484–487. IEEE Press (2005)
176. Hankerson, D., Menezes, A.J., Vanstone, S.: Guide to Elliptic Curve Cryptography. Springer, Heidelberg (2004)
177. Hariri, A., Reyhani-Masoleh, A.: Fault detection structures for the Montgomery multiplication over binary extension fields. In: Breveglieri, L., et al., vol. 64, pp. 37–46
178. Hemme, L.: A differential fault attack against early rounds of (Triple-)DES. In: Joye, M., Quisquater, J.-J. (eds.), vol. 202, pp. 254–267
179. Herrmann, M., May, A.: Solving linear equations modulo divisors: On factoring given any bits. In: Pieprzyk, J. (ed.) Advances in Cryptology—ASIACRYPT 2008, Lecture Notes in Computer Science, vol. 5350, pp. 406–424. Springer (2008)
180. Hess, F.: Pairing lattices. In: Galbraith, S.D., Paterson, K.G. (eds.) Pairing-Based Cryptography—Pairing 2008, Lecture Notes in Computer Science, vol. 5209, pp. 18–38. Springer (2008)
181. Hess, F., Smart, N.P., Vercauteren, F.: The Eta pairing revisited. IEEE Transac.Info. Theory **52**(10), 4595–4602 (2006)
182. Hoch, J.J., Shamir, A.: Fault analysis of stream ciphers. In: Joye and Quisquater [202], pp. 240–253
183. Hojsík, M., Rudolf, B.: Differential fault analysis of Trivium. In: Nyberg, K. (ed.) Fast Software Encryption (FSE 2008), Lecture Notes in Computer Science, vol. 5086, pp. 158–172. Springer (2008)
184. Hojsík, M., Rudolf, B.: Floating fault analysis of Trivium. In: D.R. Chowdhury, V. Rijmen, A. Das (eds.) Progress in Cryptology—INDOCRYPT 2008, Lecture Notes in Computer Science, vol. 5365, pp. 239–250. Springer (2008)
185. Howgrave-Graham, N., Smart, N.P.: Lattice attacks on digital signature schemes. Des. Codes and Cryptogr. **23**(3), 283–290 (2001)
186. Hutter, M., Plos, T., Schmidt, J.M.: Contact-based fault injections and power analysis on RFID tags. In: 19th IEEE European Conference on Circuit Theory and Design (ECCTD 2009), pp. 409–412. IEEE Press (2009)

187. IEEE Std 1363-2000: IEEE standard specifications for public-key cryptography. Institute of Electrical and Electronics Engineers (IEEE), Piscataway (2000)
188. Ishai, Y., Sahai, A., Wagner, D.: Private circuits: Securing hardware against probing attacks. In: Boneh, D. (ed.) Advances in Cryptology—CRYPTO 2003, Lecture Notes in Computer Science, vol. 2729, pp. 463–481. Springer, Berlin (2003)
189. Ishai, Y., Sahai, A., Wagner, D.: Private circuits ii: Keeping secrets in tamperable circuits. In: Vaudenay, S. (ed.) Advances in Cryptology EUROCRYPT 2006, Lecture Notes in Computer Science, vol. 5479, pp. 308–327. Springer, Heidelberg (2006)
190. ISO/IEC 9796-2: Information technology—Security techniques—Digital signature scheme giving message recovery, Part 2: Mechanisms using a hash-function. International Organization for Standardization (ISO) and International Electrotechnical Commission (IEC), Geneva (1997)
191. ISO/IEC 9796-2:2002: Information technology—Security techniques—Digital signature scheme giving message recovery, Part 2: Integer factorization based mechanisms. International Organization for Standardization (ISO) and International Electrotechnical Commission (IEC), Geneva, Switzerland (2002)
192. http://www.itrs.net/
193. Izu, T., Möller, B., Takagi, T.: Improved elliptic curve multiplication methods resistant against side-channel attacks. In: Menezes, A., Sarkar, P. (eds.) Progress in Cryptology—INDOCRYPT 2002, Lecture Notes in Computer Science, vol. 2551, pp. 296–313. Springer, Heidelberg (2002)
194. Joshi, N., Wu, K., Karri, R.: Concurrent error detection schemes for involution ciphers. In: Joye, M., Quisquater, J.-J., vol. 202, pp. 400–241
195. Joye, M.: Highly regular right-to-left algorithms for scalar multiplication. In: Paillier, P., Verbauwhede, I., vol. 319, pp. 135–147
196. Joye, M.: On the security of a unified countermeasure. In: Breveglieri.L., et al. vol. 65, pp. 87–91
197. Joye, M.: RSA moduli with a predetermined portion: Techniques and applications. In: Chen, L., Mu, Y., Susilo, W. (eds.) Information Security Practice and Experience (ISPEC 2008), Lecture Notes in Computer Science, vol. 4991, pp. 116–130. Springer, Heidelberg (2008)
198. Joye, M.: Protecting RSA against fault attacks: The embedding method. In: Breveglieri, L., et al. vol. 66, pp. 41–45
199. Joye, M., Lenstra, A.K., Quisquater, J.J.: Chinese remaindering based cryptosystems in the presence of faults. J. Cryptol. **12**(4), 241–245 (1999)
200. Joye, M., Manet, P., Rigaud, J.B.: Strengthening hardware AES implementations against fault attacks. IET Info. Secur. **1**(3), 106–110 (2007)
201. Joye, M., Paillier, P., Yen, S.M.: Secure evaluation of modular functions. In: Hwang, R.J., Wu, C.K. (eds.) Procedings of the 2001 International Workshop on Cryptology and Network Security, pp. 227–229. Taipei (2001)
202. Joye, M., Quisquater, J.J. (eds.): Cryptographic Hardware and Embedded Systems—CHES 2004, Lecture Notes in Computer Science, vol. 3156. Springer, Heidelberg (2004)
203. Joye, M., Quisquater, J.J., Bao, F., Deng, R.H.: RSA-type signatures in the presence of transient faults. In: Darnell, M. (ed.) Cryptography and Coding, Lecture Notes in Computer Science, vol. 1355, pp. 155–160. Springer, Heidelberg (1997)
204. Joye, M., Quisquater, J.J., Yen, S.M., Yung, M.: Observability analysis – Detecting when improved cryptosystems fail. In: B. Preneel (ed.) Topics in Cryptology—CT-RSA 2002, Lecture Notes in Computer Science, vol. 2271, pp. 17–29. Springer, Heidelberg (2002)
205. Joye, M., Yen, S.M.: The Montgomery powering ladder. In: Kaliski Jr.., et al. vol. 208, pp. 291–302
206. Jungnickel, D., Pott, A.: Difference sets: An introduction. In: Pott, A., Kumar, P.V., Helleseth, T., Jungnickel, D. (eds.) Difference Sets, Sequences and Their Correlation Properties, NATO Science Series, vol. 542. Kluwer Academic Publishers, Amsterdam (1999)

207. Kahn, D.: The Codebreakers: The Comprehensive History of Secret Communication from Ancient Times to the Internet, 2nd edn. Simon and Schuster Inc. (1997)
208. Kaliski Jr., B.S., Koç, Ç.K., Paar, C. (eds.): Cryptographic Hardware and Embedded Systems CHES 2002, Lecture Notes in Computer Science, vol. 2523. Springer, Heidelberg (2002)
209. Kammer, R.: Advanced Encryption Standard announcement (2000). http://www.nist.gov/director/speeches/rk-aes.cfm
210. Kannan, R.: Improved algorithms for integer programming and related lattice problems. In: 15th Annual ACM Symposium on Theory of Computing, pp. 193–206. ACM Press (1983)
211. Karpovsky, M.G., Kulikowski, K.J., Taubin, A.: Differential fault analysis attack resistant architectures for the Advanced Encryption Standard. In: Quisquater, J.J., Paradinas, P., Deswarte, Y., El Kalam, A.A. (eds.) Smart Card Research and Advanced Applications VI (CARDIS 2004), pp. 177–192. Kluwer Academic Publishers, Amsterdam (2004)
212. Karpovsky, M.G., Kulikowski, K.J., Taubin, A.: Robust protection against fault-injection attacks on smart cards implementing the advanced encryption standard. In: International Conference on Dependable Systems and Networks (DSN 2004), pp. 93–101. IEEE Comput. Soc. (2004)
213. Karpovsky, M.G., Kulikowski, K.J., Wang, Z.: Robust error detection in communication and computation channels. In: Workshop on Spectral Techniques (2007)
214. Karpovsky, M.G., Kulikowski, K.J., Wang, Z.: On-line self error detection with equal protection against all errors. Int. J. High. Reliab. Electron. Syst. Des. (2008)
215. Karpovsky, M.G., Nagvajara, P.: Optimal codes for the minimax criteria on error detection. IEEE Transac. Info. Theory **35**(6), 1299–1305 (1989)
216. Karpovsky, M.G., Taubin, A.: New class of nonlinear systematic error detecting codes. IEEE Trans. Info. Theory **50**(8), 1818–1820 (2004)
217. Karri, R., Wu, K., Mishra, P., Kim, Y.: Concurrent error detection of fault-based side-channel cryptanalysis of 128-bit symmetric block ciphers. In: 38th Design Automation Conference (DAC 2001), pp. 579–585. ACM Press (2001)
218. Karri, R., Wu, K., Mishra, P., Kim, Y.: Concurrent error detection schemes for fault-based side-channel cryptanalysis of symmetric block ciphers. IEEE Trans. CAD Integr. Circuits Syst. **21**(12), 1509–1517 (2002)
219. Kermani, M.M., Reyhani-Masoleh, A.: Parity-based fault detection architecture of S-box for Advanced Encryption Standard. In: 21st IEEE International Symposium on Defect and Fault-Tolerance in VLSI Systems (DFT 2006), pp. 572–580. IEEE Computer Society (2006)
220. Kermani, M.M., Reyhani-Masoleh, A.: A structure-independent approach for fault detection hardware implementations of the Advanced Encryption Standard. In: Breveglieri et al. [64], pp. 47–53
221. Kermani, M.M., Reyhani-Masoleh, A.: A lightweight concurrent fault detection scheme for the AES S-boxes using normal basis. In: Oswald and Rohatgi [314], pp. 113–129
222. Kermani, M.M., Reyhani-Masoleh, A.: Fault detection structures of the S-boxes and the inverse S-boxes for the Advanced Encryption Standard. J. Electron. Test. **25**(4/5), 225–245 (2009)
223. Khelil, F., Hamdi, M., Guilley, S., Danger, J.L., Selmane, N.: Fault analysis attack on an FPGA AES implementation. In: Aggarwal, A., Badra, M., Massacci, F. (eds.) 2nd International Conference on New Technologies, Mobility and Security (NTMS 2008), pp. 1–5. IEEE Press (2008)
224. Kim, C.H.: Differential fault analysis against AES-192 and AES-256 with minimal faults. In: Breveglieri et al. [67], pp. 3–9
225. Kim, C.H., Bulens, P., Petit, C., Quisquater, J.J.: Fault attacks on public key elements: Application to DLP-based schemes. In: Mjølsnes, S.F., Mauw, S., Katsikas, S.K. (eds.) Public Key Infrastructure (EuroPKI 2008). Lecture Notes in Computer Science, vol. 5057, pp. 182–195. Springer, Berlin (2008)

226. Kim, C.H., Quisquater, J.J.: Fault attacks for CRT based RSA: New attacks, new results, and new countermeasures. In: Sauveron, D., Markantonakis, C., Bilas, A., Quisquater, J.J. (eds.) Information Security Theory and Practices (WISTP 2007), Lecture Notes in Computer Science, vol. 4462, pp. 215–228. Springer, Berlin (2007)
227. Kim, C.H., Quisquater, J.J.: Faults, injection methods, and fault attacks. IEEE Des. Test Comput. **24**(6), 544–545 (2007)
228. Kim, C.H., Quisquater, J.J.: How can we overcome both side channel analysis and fault attacks on RSA-CRT?. In: Breveglieri et al. [64], pp. 21–29
229. Kim, C.H., Quisquater, J.J.: New differential fault analysis on AES key schedule: Two faults are enough. In: Grimaud, G., Standaert, F.X. (eds.) Smart Card Research and Advanced Applications (CARDIS 2008), Lecture Notes in Computer Science, vol. 5189, pp. 48–60. Springer, Berlin (2008)
230. Kim, T.H., Takagi, T., Han, D.G., Kim, H.W., Lim, J.: Side channel attacks and countermeasures on pairing based cryptosystems over binary fields. In: Pointcheval, D., Mu, Y., Chen, K. (eds.) Cryptology and Network Security (CANS 2006), Lecture Notes in Computer Science, vol. 4301, pp. 168–181. Springer, Berlin (2006)
231. Kircanski, A., Youssef, A.M.: Differential fault analysis of HC-128. In: Bernstein, D.J., Lange, T. (eds.) Progress in Cryptology—AFRICACRYPT 2010, Lecture Notes in Computer Science, vol. 6055, pp. 261–278. Springer, Berlin (2010)
232. Kleinjung, T., Aoki, K., Franke, J., Lenstra, A.K., Thomé, E., Bos, J.W., Gaudry, P., Kruppa, A., Montgomery, P.L., Osvik, D.A., te Riele, H.J.J., Timofeev, A., Zimmermann, P.: Factorization of a 768-bit RSA modulus. In: Rabin, T. (ed.) Advances in Cryptology—CRYPTO 2010, Lecture Notes in Computer Science, vol. 6223, pp. 333–350. Springer, Berlin (2010)
233. Koblitz, N.: Elliptic curve cryptosystems. Math. Comput. **48**(177), 203–209 (1987)
234. KoÇ, Ç.K.: High-speed RSA implementation. Technical Report TR 201, RSA Laboratories (1994)
235. Koç, Ç. K., Acar, T.: Montgomery multiplication in $GF(2^k)$. Des. Codes Cryptogr. **14**(1), 57–69 (1998)
236. Koç, Ç.K., Naccache, D., Paar, C. (eds.): Cryptographic Hardware and Embedded Systems – CHES 2001, Lecture Notes in Computer Science, vol. 2162. Springer, Berlin (2001)
237. Koç, Ç.K., Paar, C. (eds.): Cryptographic Hardware and Embedded Systems – CHES '99, Lecture Notes in Computer Science, vol. 1717. Springer, Berlin (1999)
238. Koç, Ç.K., Paar, C. (eds.): Cryptographic Hardware and Embedded Systems – CHES 2000, Lecture Notes in Computer Science, vol. 1965. Springer, Berlin (2000)
239. Kocher, P.C.: Timing attacks on implementations of Diffie-Hellman, RSA, DSS, and other systems. In: Koblitz, N. (ed.) Advances in Cryptology – CRYPTO '96, Lecture Notes in Computer Science, vol. 1109, pp. 104–113. Springer, Berlin (1996)
240. Kocher, P.C., Jaffe, J., Jun, B.: Differential power analysis. In: Wiener M.J. (ed.) Advances in Cryptology – CRYPTO '99, Lecture Notes in Computer Science, vol. 1666, pp. 388–397. Springer, Berlin (1999)
241. Kulikowski, K.: Codes and circuits for secure hardware design. Ph.D. thesis, Boston University (2009)
242. Kulikowski, K.J., Karpovsky, M.G., Taubin, A.: Fault attack resistant cryptographic hardware with uniform error detection. In: Breveglieri et al. [69], pp. 185–195
243. Kulikowski, K.J., Karpovsky, M.G., Taubin, A.: Power attacks on secure hardware based on early propagation of data. In: 12th IEEE International On-Line Testing Symposium (IOLTS 2006), pp. 131–138. IEEE Computer Society (2006)
244. Kulikowski, K.J., Venkataraman, V., Wang, Z., Taubin, A., Karpovsky, M.G.: Asynchronous balanced gates tolerant to interconnect variability. In: IEEE International Symposium on Circuits and Systems (ISCAS 2008), pp. 3190–3193. IEEE Press (2008)

245. Kulikowski, K.J., Wang, Z., Karpovsky, M.G.: Comparative analysis of robust fault attack resistant architectures for public and private cryptosystems. In: Breveglieri et al. [65], pp. 41–50
246. Kwon, S.: Efficient Tate pairing computation for supersingular elliptic curves over binary fields. Cryptology ePrint Archive, Report 2004/303 (2004)
247. Lala, P.K.: Self-Checking and Fault-Tolerant Digital Design. Morgan Kaufmann (2001)
248. Lee, C.Y., Chiou, C.W., Lin, J.M.: Concurrent error detection in a bit-parallel systolic multiplier for dual basis of $GF(2^m)$. J. Electron. Test. Theory Appl. **21**(5), 539–549 (2005)
249. Lee, C.Y., Chiou, C.W., Lin, J.M.: Concurrent error detection in a polynomial basis multiplier over $GF(2^m)$. J. Electron. Test. Theory Appl. **22**(2), 143–150 (2006)
250. Lee, E., Lee, H.S., Park, C.M.: Efficient and generalized pairing computation on abelian varieties. Cryptology ePrint Archive, Report 2008/040 (2008)
251. Lenstra, A.K., Lenstra, H.W., Lovász, L.: Factoring polynomials with rational coefficients. Mathematische Annalen **261**(4), 515–534 (1982)
252. Lenstra, A.K., Lenstra Jr., H.W. (eds.): The Development of the Number Field Sieve, *Lecture Notes in Mathematics*, vol. 1554. Springer, Berlin (1993)
253. Leveugle, R.: Early analysis of fault-based attack effects in secure circuits. IEEE Trans. Comput. **56**(10), 1431–1434 (2007)
254. Li, Y., Sakiyama, K., Kawamura, S., Komano, Y., Ohta, K.: Security evaluation of a DPA-resistant S-box based on the Fourier transform. In: Qing, S., Mitchell, C.J., Wang, G. (eds.) Information and Communications Security (ICICS 2009), Lecture Notes in Computer Science, vol. 5927, pp. 3–16. Springer Berlin (2009)
255. Lichtenbaum, S.: Duality theorems for curves over p-adic fields. Inventiones Mathematicae **7**, 120–136 (1969)
256. Liskov, M., Rivest, R.L., Wagner, D.: Tweakable block ciphers. In: Yung, M. (ed.) Advances in Cryptology – CRYPTO 2002, Lecture Notes in Computer Science, vol. 2442, pp. 31–46. Springer Berlin (2002)
257. López, J., Dahab, R.: Fast multiplication on elliptic curves over $GF(2^m)$ without precomputation. In: Koç and Paar [237], pp. 316–327
258. López-Buedo, S., Garrido, J., Boemo, E.: Thermal testing on reconfigurable computers. IEEE Des. Test Comput. **17**(1), 84–91 (2000)
259. Macé, F., Standaert, F.X., Quisquater, J.J.: Information theoretic evaluation of side-channel resistant logic styles. In: Paillier and Verbauwhede [319], pp. 427–442
260. MacWilliams, F.J., Sloane, N.J.A.: The Theory of Error-Correcting Codes. North-Holland (1977)
261. Maingot, V., Leveugle, R.: Error detection code efficiency for secure chips. In: 13th IEEE International Conference on Electronics, Circuits and Systems (ICECS 2006), pp. 561–564. IEEE Press (2006)
262. Maingot, V., Leveugle, R.: On the use of error correcting and detecting codes in secured circuits. In: 2007 Ph.D. Research in Microelectronics and Electronics Conference (PRIME 2007), pp. 245–248. IEEE Press (2007)
263. Maistri, P., Leveugle, R.: Double-data-rate computation as a countermeasure against fault analysis. IEEE Trans. Comput. **57**(11), 1528–1539 (1982)
264. Maistri, P., Vanhauwaert, P., Leveugle, R.: Evaluation of register-level protection techniques for the Advanced Encryption Standard by multi-level fault injections. In: Bolchini, C., Kim, Y.B., Salsano, A., Touba, N.A. (eds.) 22nd IEEE International Symposium on Defect and Fault-Tolerance in VLSI Systems (DFT 2007), pp. 499–507. IEEE Computer Society (2007)
265. Maistri, P., Vanhauwaert, P., Leveugle, R.: A novel double-data-rate AES architecture resistant against fault injection. In: Breveglieri et al. [64], pp. 54–61
266. Malkin, T., Standaert, F.X., Yung, M.: A comparative cost/security analysis of fault attack countermeasures. In: Breveglieri et al. [69], pp. 159–172

267. Mangard, S., Aigner, M., Dominikus, S.: A highly regular and scalable AES hardware architecture. IEEE Trans. Comput. **52**(4), 483–491 (2003)
268. Mangard, S., Oswald, E., Popp, T.: Power Analysis Attacks: Revealing the Secrets of Smart Cards. Springer, Heidelberg (2007)
269. Mangard, S., Oswald, E., Standaert, F.X.: One for all - all for one: Unifying standard DPA attacks. Cryptology ePrint Archive, Report 2009/449 (2009)
270. Mangard, S., Pramstaller, N., Oswald, E.: Successfully attacking masked AES hardware implementations. In: Rao and Sunar [334], pp. 157–171
271. Mangard, S., Standaert, F.X. (eds.): Cryptographic Hardware and Embedded Systems – CHES 2010, Lecture Notes in Computer Science, vol. 6225. Springer, Berlin (2010)
272. Manohar, R., Martin, A.J.: Quasi-delay-insensitive circuits are Turing-complete. Technical Report CS-TR-95-11, California Institute of Technology (1995)
273. Massey, J.L., Omura, J.K.: Computational method and apparatus for finite field arithmetic. US Patent #4,587,627 (1986)
274. Matsuda, S., Kanayama, N., Hess, F., Okamoto, E.: Optimised versions of the Ate and Twisted Ate pairings. In: S.D. Galbraith (ed.) Cryptography and Coding, Lecture Notes in Computer Science, vol. 4887, pp. 302–312. Springer Berlin (2007)
275. Matsui, M.: On correlation between the order of S-boxes and the strength of DES. In: De Santis, A. (ed.) Advances in Cryptology—EUROCRYPT '94, Lecture Notes in Computer Science, vol. 950, pp. 366–375. Springer Berlin (1995)
276. Maxwell, M.S.: Almost perfect nonlinear functions and related combinatorial structures. Ph.D. thesis, Iowa State University (2005)
277. May, T.C., Woods, M.H.: A new physical mechanism for soft errors in dynamic memories. In: 16th Annual Reliability Physics Symposium, pp. 33–40. IEEE Press (1978)
278. McEvoy, R.P., Tunstall, M., Whelan, C., Murphy, C.C., Marnane, W.P.: All-or-nothing transforms as a countermeasure to differential side-channel analysis. Cryptology ePrint Archive, Report 2009/185 (2009)
279. Medien, Z., Machhout, M., Belgacem Bouallegue, L.K., Baganne, A., Tourki, R.: Design and hardware implementation of QoSS-AES processor for multimedia applications. Trans. Data Priv. **3**(1), 43–64 (2010)
280. Medwed, M., Schmidt, J.M.: A generic fault countermeasure providing data and program flow integrity. In: Breveglieri et al. [65], pp. 68–73
281. Medwed, M., Schmidt, J.M.: A continuous fault countermeasure for AES providing a constant error detection rate. In: Breveglieri et al. [67], pp. 66–71
282. Medwed, M., Standaert, F.X., Großschädl, J., Regazzoni, F.: Fresh re-keying: Security against side-channel and fault attacks for low-cost devices. In: Bernstein, D.J., Lange, T. (eds.) Progress in Cryptology—AFRICACRYPT 2010, Lecture Notes in Computer Science, vol. 6055, pp. 279–296. Springer Berlin (2010)
283. Menezes, A.J., van Oorschot, P.C., Vanstone, S.A.: Handbook of Applied Cryptography. CRC Press, Boca Raton (1997)
284. Messerges, T.S.: Using second-order power analysis to attack DPA resistant software. In: Koç and Paar [238], pp. 238–251
285. Micciancio, D., Goldwasser, S.: Complexity of Lattice Problems: A Cryptographic Perspective, The Kluwer International Series in Engineering and Computer Science, vol. 671. Kluwer Academic Publishers Boston (2002)
286. Miller, V.: Short programs for functions on curves. http://crypto.stanford.edu/miller/miller.pdf
287. Miller, V.: The Weil pairing, and its efficient calculation. J. Cryptol. **17**(4), 235–261 (2004)
288. Miller, V.S.: Use of elliptic curves in cryptography. In: Williams, H.C. (ed.) Advances in Cryptology—CRYPTO '85, Lecture Notes in Computer Science, vol. 218, pp. 417–426. Springer, Berlin (1985)
289. Minkowski, H.: Geometrie der Zahlen. Teubner-Verlag (1896)

290. Misarsky, J.F.: A multiplicative attack using LLL algorithm on RSA signatures with redundancy. In: Kaliski, B.S. Jr. (ed.) Advances in Cryptology—CRYPTO '97, Lecture Notes in Computer Science, vol. 1294, pp. 221–234. Springer, Berlin (1997)
291. Monnet, Y., Renaudin, M., Leveugle, R.: Asynchronous circuits sensitivity to fault injection. In: 10th IEEE International On-Line Testing Symposium (IOLTS 2004), pp. 121–128. IEEE Computer Society (2004)
292. Monnet, Y., Renaudin, M., Leveugle, R., Clavier, C., Moitrel, P.: Case study of a fault attack on asynchronous DES crypto-processors. In: Breveglieri et al. [69], pp. 88–97
293. Montgomery, P.L.: Modular multiplication without trial division. Math. Comput. **44**(170), 519–521 (1985)
294. Montgomery, P.L.: Speeding up the Pollard and elliptic curve methods of factorization. Math. Comput. **48**(177), 243–264 (1987)
295. Moore, S.W., Anderson, R.J., Cunningham, P., Mullins, R., Taylor, G.: Improving smart card security using self-timed circuits. In: 8th International Symposium on Asynchronous Circuits and Systems (ASYNC 2002), pp. 211–218. IEEE Computer Society (2002)
296. Moradi, A., Shalmani, M.T.M., Salmasizadeh, M.: A generalized method of differential fault attack against AES cryptosystem. In: Goubin and Matsui [167], pp. 91–100
297. Morain, F., Olivos, J.: Speeding up the computations on an elliptic curve using addition-subtraction chains. RAIRO Theor. Inform. Appl. **24**, 531–543 (1990)
298. Muir, J.A.: Seifert's RSA fault attack: Simplified analysis and generalizations. In: Ning, P., Qing, S., Li, N. (eds.) Information and Communications Security (ICICS 2006), Lecture Notes in Computer Science, vol. 4307, pp. 420–434. Springer, Berlin (2006)
299. Mukhopadhyay, D.: An improved fault based attack of the Advanced Encryption Standard. In: Preneel, B. (ed.) Progress in Cryptology—AFRICACRYPT 2009, Lecture Notes in Computer Science, vol. 5580, pp. 421–434. Springer, Berlin (2009)
300. Naccache, D., Nguyen, P.Q., Tunstall, M., Whelan, C.: Experimenting with faults, lattices and the DSA. In: Vaudenay, S. (ed.) Public Key Cryptography—PKC 2005, Lecture Notes in Computer Science, vol. 3386, pp. 16–28. Springer, Berlin (2005)
301. Nagel, L.W., Pederson, D.O.: SPICE (Simulation Program with Integrated Circuit Emphasis). Memorandum ERL-M382 EECS Department University of California, Berkeley (1973)
302. Nguyen, P.Q.: Can we trust cryptographic software? Cryptographic flaws in GNU Privacy Guard v1.2.3. In: C. Cachin, J. Camenisch (eds.) Advances in Cryptology—EUROCRYPT 2004, Lecture Notes in Computer Science, vol. 3027, pp. 555–570. Springer, Heidelberg (2004)
303. Nguyen, P.Q.: Public-key cryptanalysis. In: I. Luengo (ed.) Recent Trends in Cryptography, Contemporary Mathematics, vol. 477, pp. 67–120. American Mathematical Society (2009)
304. Nguyen, P.Q., Shparlinski, I.: The insecurity of the digital signature algorithm with partially known nonces. J. Cryptol. **15**(3), 151–176 (2002)
305. Nguyen, P.Q., Shparlinski, I.: The insecurity of the elliptic curve digital signature algorithm with partially known nonces. Des. Codes and Cryptogr. **30**(2), 201–217 (2003)
306. Nguyen, P.Q., Stehlé, D.: LLL on the average. In: F. Hess, S. Pauli, M.E. Pohst (eds.) Algorithmic Number Theory (ANTS-VII), Lecture Notes in Computer Science, vol. 4076, pp. 238–256. Springer, Heidelberg (2006)
307. Nguyen, P.Q., Stern, J.: Merkle-Hellman revisited: A cryptoanalysis of the Qu-Vanstone cryptosystem based on group factorizations. In: B.S. Kaliski Jr. (ed.) Advances in Cryptology—CRYPTO '97, Lecture Notes in Computer Science, vol. 1294, pp. 198–212. Springer, Heidelberg (1997)
308. Nguyen, P.Q., Stern, J.: Cryptanalysis of a fast public key cryptosystem presented at SAC '97. In: S.E. Tavares, H. Meijer (eds.) Selected Areas in Cryptography (SAC '98), Lecture Notes in Computer Science, vol. 1556, pp. 213–218. Springer, Heidelberg (1999)
309. Nicholson, W.K.: Introduction to Abstract Algebra. PWS Publishing Company, Boston (1993)

310. NIST: Announcing request for candidate algorithm nominations for the Advanced Encryption Standard (AES). Federal Register **62**(177), 48,051–48,058 (1997)
311. Novak, R.: SPA-based adaptive chosen-ciphertext attack on RSA implementation. In: D. Naccache, P. Paillier (eds.) Public Key Cryptography (PKC 2002), Lecture Notes in Computer Science, vol. 2274, pp. 252–262. Springer, Heidelberg (2002)
312. O'hEigeartaigh, C.: Pairing computation on hyperelliptic curves of genus 2. Ph.D. thesis, Dublin City University (2006)
313. Oswald, E., Mangard, S., Pramstaller, N., Rijmen, V.: A side-channel analysis resistant description of the AES S-box. In: H. Gilbert, H. Handschuh (eds.) Fast Software Encryption (FSE 2005), Lecture Notes in Computer Science, vol. 3557, pp. 413–423. Springer, Heidelberg (2005)
314. Oswald, E., Rohatgi, P. (eds.): Cryptographic Hardware and Embedded Systems CHES 2008, Lecture Notes in Computer Science, vol. 5154. Springer, Heidelberg (2008)
315. Otto, M.: Fault attacks and countermeasures. Ph.D. thesis, Institut für Informatik, Universität Paderborn (2004)
316. Öztürk, E.: Efficient and tamper-resilient architectures for pairing based cryptography. Ph.D. thesis, Worcester Polytechnic Institute (2008)
317. Öztürk, E., Gaubatz, G., Sunar, B.: Tate pairing with strong fault resiliency. In: Breveglieri, L., et al. vol. 64, pp. 103–111
318. Page, D., Vercauteren, F.: A fault attack on pairing based cryptography. IEEE Trans. Comput. **55**(9), 1075–1080 (2006)
319. Paillier, P., Verbauwhede, I. (eds.): Cryptographic Hardware and Embedded Systems—CHES 2007, Lecture Notes in Computer Science, vol. 4727. Springer, Berlin (2007)
320. Patel, J.H., Fung, L.Y.: Concurrent error detection in ALU0's by recomputing with shifted operands. IEEE Trans. Comput. **31**(7), 589–595 (1982)
321. Patel, J.H., Fung, L.Y.: Concurrent error detection in multiply and divide arrays. IEEE Trans. Comput. **32**(4), 417–422 (1983)
322. Peacham, D., Thomas, B.: A DFA attack against the AES key schedule. White Paper 001, SiVenture (2006)
323. Phan, R.C.W., Yen, S.M.: Amplifying side-channel attacks with techniques from block cipher cryptanalysis. In: J. Domingo-Ferrer, J. Posegga, D. Schreckling (eds.) Smart Card Research and Advanced Applications (CARDIS 2006), Lecture Notes in Computer Science, vol. 3928, pp. 135–150. Springer, Berlin (2006)
324. Piret, G., Quisquater, J.J.: A differential fault attack technique against SPN structure, with application to the AES and KHAZAD. In: Walter, C.D. et al. vol. 411, pp. 77–88
325. Pohlig, S., Hellman, M.: An improved algorithm for computing logarithms over $GF(p)$ and its cryptographic significance. IEEE Trans. Info. Theor. **24**(1), 106–110 (1978)
326. Pollard, J.M.: Monte Carlo methods for index computation (mod p). Mathematics of Computation **32**(143), 918–924 (1978)
327. Popp, T., Kirschbaum, M., Zefferer, T., Mangard, S.: Evaluation of the masked logic style MDPL on a prototype chip. In: Paillier and Verbauwhede [319], pp. 81–94
328. Pramstaller, N., Mangard, S., Dominikus, S., Wolkerstorfer, J.: Efficient AES implementations on ASICs and FPGAs. In: H. Dobbertin, V. Rijmen, A. Sowa (eds.) Advanced Encryption Standard — AES (AES 2004), Lecture Notes in Computer Science, vol. 3373, pp. 98–112. Springer (2005)
329. Proudler, I.K.: Idempotent AN codes. In: IEE Colloquium on Signal Processing Applications of Finite Field Mathematics, pp. 8/1–8/5. IEEE Press (1989)
330. Quisquater, J.J., Couvreur, C.: Fast decipherment algorithm for RSA public-key cryptosystem. Electron. Lett. **18**(21), 905–907 (1982)
331. Quisquater, J.J., Samyde, D.: Electromagnetic analysis (EMA): Measures and countermeasures for smart cards. In: I. Attali, T.P. Jensen (eds.) Smart Card Programming and Security (E-smart 2001), Lecture Notes in Computer Science, vol. 2140, pp. 200–210. Springer, Berlin (2001)

332. Quisquater, J.J., Samyde, D.: Radio frequency attacks. In: H.C. van Tilborg (ed.) Encyclopedia of Cryptography and Security, pp. 503–509. Springer, Berlin (2005)
333. Rackoff, C., Simon, D.R.: Non-interactive zero-knowledge proof of knowledge and chosen ciphertext attack. In: J. Feigenbaum (ed.) Advances in Cryptology—CRYPTO '91, Lecture Notes in Computer Science, vol. 576, pp. 433–444. Springer (1992)
334. Rao, J.R., Sunar, B. (eds.): Cryptographic Hardware and Embedded Systems—CHES 2005, Lecture Notes in Computer Science, vol. 3659. Springer (2005)
335. Rao, T., Garcia, O.: Cyclic and multiresidue codes for arithmetic operations. IEEE Trans. Info. Theor. **17**(1), 85–91 (1971)
336. Redelmeier, R.: CPUburn, CPU testing utilities (2001). http://pages.sbcglobal.net/redelm/
337. Regazzoni, F., Cevrero, A., Standaert, F.X., Badel, S., Kluter, T., Brisk, P., Leblebici, Y., Ienne, P.: A design flow and evaluation framework for DPA-resistant instruction set extensions. In: Clavier, C., Gaj, K. vol. 59, pp. 205–219
338. Regazzoni, F., Eisenbarth, T., Breveglieri, L., Ienne, P., Koren, I.: Can knowledge regarding the presence of countermeasures against fault attacks simplify power attacks on cryptographic devices? In: Bolchini, C., Kim, Y.B., Gizopoulos, D., Tehranipoor, M. (eds.) 23rd IEEE International Symposium on Defect and Fault-Tolerance in VLSI Systems (DFT 2008), pp. 202–210. IEEE Comput. Soc. (2008)
339. Regazzoni, F., Eisenbarth, T., Großschädl, J., Breveglieri, L., Ienne, P., Koren, I., Paar, C.: Power attacks resistance of cryptographic S-boxes with added error detection procedures. In: Bolchini, C., Kim, Y.B., Salsano, A., Touba, N.A. (eds.) 22nd IEEE International Symposium on Defect and Fault-Tolerance in VLSI Systems (DFT 2007), pp. 508–516. IEEE Comput. Soc. (2007)
340. Research Center for Information Security, National Institute of Advanced Industrial Science and Technology (AIST): Cryptographic LSI for SASEBO-R specification, version 1.0 (English). http://www.rcis.aist.go.jp/special/SASEBO/CryptoLSI-en.html
341. Research Center for Information Security, National Institute of Advanced Industrial Science and Technology (AIST): ISO/IEC 18033-3 standard cryptographic LSI specification, version 1.0 (English). http://www.rcis.aist.go.jp/files/special/SASEBO/CryptoLSI-en/CryptoLSI_Spec_Ver1.0_English.pdf
342. Research Center for Information Security, National Institute of Advanced Industrial Science and Technology (AIST): Side-channel attack standard evaluation board (SASEBO). http://www.rcis.aist.go.jp/special/SASEBO/index-en.html
343. Reyhani-Masoleh, A., Hasan, M.A.: Fault detection architectures for field multiplication using polynomial bases. IEEE Trans. Comput. **55**(9), 1089–1103 (2006)
344. Riscure: Diode laser station. Inspector Datasheet (2010). http://www.riscure.com/fileadmin/images/Inspdatasheet/dls_datasheet.pdf
345. Rivain, M.: Differential fault analysis on DES middle rounds. In: Clavier, C., Gaj, K. vol. 95, pp. 457–469
346. Rivain, M.: Securing RSA against fault analysis by double addition chain exponentiation. In: Fischlin, M. (ed.) Topics in Cryptology—CT-RSA 2009. Lecture Notes in Computer Science, vol. 5473, pp. 459–480. Springer, Berlin (2009)
347. Rivain, M., Prouff, E.: Provably secure higher-order masking of AES. In: Mangard, S., Standaert, F.X. (eds.) CHES 2010. Lecture Notes in Computer Science, vol. 6225, pp. 413–427. Springer, Berlin (2010)
348. Rivest, R., Robshaw, M., Sidney, R., Yin, Y.L.: The RC6 block cipher (1998). http://theory.lcs.mit.edu/~rivest/rc6.pdf
349. Rivest, R.L., Shamir, A., Adleman, L.M.: Method for obtaining digital signatures and public-key cryptosystems. Commun. ACM **21**(2), 120–126 (1978)
350. Rosenfeld, K., Karri, R.: Attacks and defenses for JTAG. IEEE Des. Test Comput. **27**(1), 36–47 (2010)

351. Rudra, A., Dubey, P.K., Jutla, C.S., Kumar, V., Rao, J.R., Rohatgi, P.: Efficient implementation of Rijndael encryption with composite field arithmetic. In: Koç., et al. (eds.) CHES 2001. Lecture Notes in Computer Science, vol. 2162, pp. 171–184. Springer, Berlin (2001)
352. Saha, D., Mukhopadhyay, D., RoyChowdhury, D.: A diagonal fault attack on the advanced encryption standard. Cryptology ePrint Archive, Report 2009/581 (2009)
353. Sakiyama, K., Yagi, T., Ohta, K.: Fault analysis attack against an AES prototype chip using RSL. In: Fischlin, M. (ed.) Topics in Cryptology—CT-RSA 2009. Lecture Notes in Computer Science, vol. 5473, pp. 429–443. Springer, Heidelberg (2009)
354. Sarmadi, S.B., Hasan, M.A.: Detecting errors in a polynomial basis multiplier using multiple parity bits for both inputs. In: 25th International Conference on Computer Design (ICCD 2007), pp. 368–375. IEEE Press (2007)
355. Sarmadi, S.B., Hasan, M.A.: On concurrent detection of errors in polynomial basis multiplication. IEEE Trans. Very Larg. Scale Integr. Syst. **15**(4), 413–426 (2007)
356. Sarmadi, S.B., Hasan, M.A.: Run-time error detection in polynomial basis multiplication using linear codes. In: IEEE International Conference on Application-Specific Systems, Architectures and Processors (ASAP 2007), pp. 204–209. IEEE Comput. Soc. (2007)
357. Satoh, A., Morioka, S., Takano, K., Munetoh, S.: A compact Rijndael hardware architecture with S-box optimization. In: Boyd, C. (ed.) Advances in Cryptology—ASIACRYPT 2001. Lecture Notes in Computer Science, vol. 2248, pp. 239–254. Springer, Berlin (2001)
358. Satoh, A., Sugawara, T., Homma, N., Aoki, T.: High-performance concurrent error detection scheme for AES hardware. In: Oswald, E., Rohatgi, P. (eds.) CHES 2008. Lecture Notes in Computer Science, vol. 5154, pp. 100–112. Springer, Berlin (2008)
359. Schmidt, J.M., Herbst, C.: A practical fault attack on square and multiply. In: Breveglieri, L., et al. (eds.), Fault Diagnosis andTolerance in Cryptography—FDTC 2008, pp. 53–58. IEEE Comput. Soc. (2008)
360. Schmidt, J.M., Hutter, M.: Optical and EM fault-attacks on CRT-based RSA: Concrete results. In: Posch, K., Wolkerstorfer, C.J. (eds.) 15th Austrian Workshop on Microelectronics (Austrochip 2007), pp. 61–67. Verlag der Technischen Universität Graz (2007)
361. Schmidt, J.M., Hutter, M., Plos, T.: Optical fault attacks on AES: a threat in violet. In: Breveglieri, L., et al. (eds.) Fault Diagnosis andTolerance in Cryptography—FDTC 2009, pp. 13–22. IEEE Comput. Soc. (2009)
362. Schneier, B.: Description of a new variable-length key, 64-bit block cipher (Blowfish). In: Anderson, R.J. (ed.) Fast Software Encryption (FSE '93). Lecture Notes in Computer Science, vol. 809, pp. 191–204. Springer, Berlin (1994)
363. Schneier, B., Kelsey, J., Whiting, D., Wagner, D., Hall, C., Ferguson, N.: Twofish: a 128-bit block cipher (1998). http://www.schneier.com/paper-twofish-paper.pdf
364. Scott, M.: Implementing cryptographic pairings. ftp://ftp.computing.dcu.ie/pub/resources/crypto/pairings.pdf
365. Scott, M.: Computing the Tate pairing. In: Menezes, A. (ed.) Topics in Cryptology—CT-RSA 2005. Lecture Notes in Computer Science, vol. 3376, pp. 293–304. Springer, Berlin (2005)
366. Scott, M., Costigan, N., Abdulwahab, W.: Implementing cryptographic pairings on smartcards. In: Goubin, L., Matsui, M. (eds.) Cryptographic Hardware and Embedded Systems—CHES 2006. Lecture Notes in Computer Science, vol. 4249, pp. 134–147. Springer, Berin (2006)
367. Seifert, J.P.: On authenticated computing and RSA-based authentication. In: Atluri, V., Meadows, C., Juels, A. (eds.) 12th ACM Conference on Computer and Communications Security (CCS 2005), pp. 122–127. ACM Press (2005)
368. Selmane, N., Bhasin, S., Guilley, S., Graba, T., Danger, J.L.: WDDL is protected against setup time violation attacks. In: Breveglieri, L., et al. (eds.) Fault Diagnosis andTolerance in Cryptography—FDTC 2009, pp. 73–83. IEEE Comput. Soc. (2009)

369. Selmane, N., Guilley, S., Danger, J.L.: Practical setup time violation attacks on AES. In: Seventh European Dependable Computing Conference (EDCC-7), pp. 91–96. IEEE Comput. Soc. (2008)
370. Shamir, A.: A polynomial time algorithm for breaking the basic Merkle-Hellman cryptosystem. In: 23rd Annual Symposium on Foundations of Computer Science, pp. 145–152. IEEE Press (1982)
371. Shamir, A.: Identity-based cryptosystems and signature schemes. In: G.R. Blakley, D. Chaum (eds.) Advances in Cryptology—CRYPTO '84. Lecture Notes in Computer Science, vol. 196, pp. 47–53. Springer, Berlin (1985)
372. Shamir, A.: Method and apparatus for protecting public key schemes from timing and fault attacks. US Patent #5,991,415 (1999), Presented at the rump session of EUROCRYPT '97
373. Shamir, A.: Protecting smart cards from passive power analysis with detached power supplies. In: Kocù, Cù.K., Paar, C. (eds.) Cryptographic Hardware and Embedded Systems—CHES 2000. Lecture Notes in Computer Science, vol. 1965, pp. 71–77, Springer, Berlin (2000)
374. Shanks, D.: Class number, a theory of factorization and genera. In: Lewis, D.J. (ed.) Proceedings of Symposia in Pure Mathematics, vol. 20, pp. 415–440. American Mathematical Society (1971)
375. Shedletsky J.J.: Error correction by alternate-data retry. IEEE Trans.Comput. **27**(2), 106–112 (1978)
376. Shirase, M., Takagi, T., Okamoto, E.: An efficient countermeasure against side channel attacks for pairing computation. In: Chen, L., Mu, Y., Susilo, W. (eds.) Information Security Practice and Experience (ISPEC 2008). Lecture Notes in Computer Science, vol. 4991, pp. 290–303. Springer, Berlin (2008)
377. Skorobogatov, S.P.: Semi-invasive attacks: A new approach to hardware security analysis. Technical Report UCAM-CL-TR-630, Computer Laboratory, University of Cambridge (2005)
378. Skorobogatov, S.P.: Using optical emission analysis for estimating contribution to power analysis. In: Breveglieri, L., et al. (eds.) Fault Diagnosis andTolerance in Cryptography—FDTC 2009, pp. 111–119. IEEE Comput. Soc. (2009)
379. Skorobogatov, S.P., Anderson, R.J.: Optical fault induction attacks. In: Kaliski Jr.., et al. (eds.) Cryptographic Hardware and Embedded Systems—CHES 2002. Lecture Notes in Computer Science, vol. 2523, pp. 2–12, Springer, Berin (2002)
380. Smith, M.J.S.: Application-Specific Integrated Circuits. Addison-Wesley, Boston (1997)
381. Solinas, J.A.: Generalized Mersenne numbers. Technical Report CORR99-39, University of Waterloo (1999)
382. Sollins, K.: The TFTP protocol. RFC 1350 (1992). http://tools.ietf.org/html/rfc1350
383. Stam, M.: On Montgomery-like representations for elliptic curves over $GF(2^k)$. In: Desmedt, Y. (ed.) Public Key Cryptography—PKC 2003. Lecture Notes in Computer Science, vol. 2567, pp. 240–253. Springer, Berlin (2003)
384. Standaert, F.X., Malkin, T., Yung, M.: A unified framework for the analysis of side-channel key recovery attacks. In: Joux, A. (ed.) Advances in Cryptology—EUROCRYPT 2009. Lecture Notes in Computer Science, vol. 5479, pp. 443–461. Springer, Berlin (2009)
385. Standaert, F.X., Veyrat-Charvillon, N., Oswald, E., Gierlichs, B., Medwed, M., Kasper, M., Mangard, S.: The world is not enough: another look on second-order DPA. In: Abe, M. (ed.) Advances in Cryptology—ASIACRYPT 2010. Lecture Notes in Computer Science, vol. 6477, pp. 112–129. Springer, Berlin (2010)
386. Stern, J.: Secret linear congruential generators are not cryptographically secure. In: 28th Annual Symposium on Foundations of Computer Science, pp. 421–426. IEEE Press (1987)
387. Stern, R., Joshi, N., Wu, K., Karri, R.: Register transfer level concurrent error detection in elliptic curve crypto implementations. In: Breveglieri, L., et al. (eds.) Fault Diagnosis andTolerance in Cryptography—FDTC 2007, pp. 112–119. IEEE Comput. Soc. (2007)

388. STMicroelectronics: Clock 90 GPLVT 1.2V 2.2 Standard Cell Library User Manual and Databook (2006)
389. STMicroelectronics: SPEAr Head200, ARM926, 200k Customizable eASIC Gates, Large IP Portfolio SoC (2009). http://www.st.com/stonline/products/literature/bd/14388/spear-09-h042.htm
390. Suzuki, D., Saeki, M., Ichikawa, T.: DPA leakage models for CMOS logic circuits. In: Rao, J.R., Sunar, B. (eds.) Cryptographic Hardware and Embedded Systems—CHES 2005. Lecture Notes in Computer Science, vol. 3659, pp. 366–382, Springer, Berlin (2005)
391. Synopsis Corporation: Design Compiler Ultra. http://www.synopsys.com/Tools/Implementation/RTLSynthesis/Pages/DCUltra.aspx
392. Takahashi, J., Fukunaga, T.: Differential fault analysis on the AES key schedule. Cryptology ePrint Archive, Report 2007/480 (2007). Improved version of DFA mechanism on the AES key schedule. In: Breveglieri, L., et al. (eds.) Fault Diagnosis andTolerance in Cryptography—FDTC 2007. IEEE Comput. Soc. (2007)
393. Takahashi, J., Fukunaga, T.: Differential fault analysis on AES with 192 and 256-bit keys. Cryptology ePrint Archive, Report 2010/023 (2010). Published in SCIS 2010 (in Japanese)
394. Takahashi, J., Fukunaga, T., Yamakoshi, K.: DFA mechanism on the AES key schedule. In: Breveglieri, L., et al. vol. 64, pp. 62–74
395. Takahashi, J., Toshinori, F.: Improved differential fault analysis on CLEFIA. In: Breveglieri, L., et al. vol. 65, pp. 25–34
396. Tate, J.: WC-groups over $\mathfrak{p}$-adic fields. In: Séminaire Bourbaki, Exposé 156, pp. 265–277. Sécrétariat Mathématique, Paris (1995)
397. Teuwen, P.: How to make smartcards resistant to hackers' lightsabers? In: Guajardo, J., Preneel, B., Sadeghi, A.R., Tuyls P. (eds.) Foundations for Forgery-Resilient Cryptographic Hardware, no. 09282 in Dagstuhl Seminar Proceedings. Schloss Dagstuhl - Leibniz-Zentrum für Informatik, Germany (2010)
398. Torrance, R., James, D.: The state-of-the-art in IC reverse engineering. In: Clavier C., Gaj K. vol. 95, pp. 363–381
399. Trichina, E., Korkikyan, R.: Multi fault laser attacks on protected CRT-RSA. In: Breveglieri, L., et al. vol. 67, pp. 75–86
400. Tunstall, M.: Secure cryptographic algorithm implementation on embedded platforms. Ph.D. thesis, Royal Holloway, University of London (2006)
401. Tunstall, M.: Random order *m*-ary exponentiation. In: C. Boyd, J.M.G. Nieto (eds.) Information Security and Privacy (ACISP 2009), Lecture Notes in Computer Science, vol. 5594, pp. 437–451. Springer, Heidelberg (2009)
402. Tunstall, M., Benoît, O.: Efficient use of random delays in embedded software. In: D. Sauveron, C. Markantonakis, A. Bilas, J.J. Quisquater (eds.) Information Security Theory and Practices (WISTP 2007), Lecture Notes in Computer Science, vol. 4462, pp. 27–38. Springer, Heidelberg (2007)
403. Tunstall, M., Mukhopadhyay, D.: Differential fault analysis of the Advanced Encryption Standard using a single fault. Cryptology ePrint Archive, Report 2009/575 (2009)
404. Vargas, F., Cavalcante, D.L., Gatti, E., Prestes, D., Lupi, D.: On the proposition of an EMI-based fault injection approach. In: 11th IEEE International On-Line Testing Symposium (IOLTS 2005), pp. 207–208. IEEE Comput. Soc. (2005)
405. Vater, F., Peter, S., Langendörfer, P.: Combinatorial logic circuitry as means to protect low cost devices against side channel attacks. In: D. Sauveron, C. Markantonakis, A. Bilas, J.J. Quisquater (eds.) Information Security Theory and Practices (WISTP 2007), Lecture Notes in Computer Science, vol. 4462, pp. 244–253. Springer, Heidelberg (2007)
406. Vercauteren, F.: The hidden root problem. In: S.D. Galbraith, K.G. Paterson (eds.) Pairing-Based Cryptography—Pairing 2008, Lecture Notes in Computer Science, vol. 5209, pp. 89–99. Springer, Heidelberg (2008)
407. Vercauteren, F.: Optimal pairings. IEEE Transac. Info. Theor. **56**(1), 455–461 (2010)

408. Vertanen, O.: Java type confusion and fault attacks. In: Breveglieri, L., et al. vol. 69, pp. 237–251
409. Vigilant, D.: RSA with CRT: A new cost-effective solution to thwart fault attacks. In: Oswald, E., Rohatgi, P. vol. 314, pp. 130–145
410. Wagner, D.: Cryptanalysis of a provably secure CRT-RSA algorithm. In:Atluri, V., Pfitzmann, B., McDaniel, P.D. (eds.) 11th ACM Conference on Computer and Communications Security (CCS 2004), pp. 92–97. ACM Press, Washington (2004)
411. Walter, C.D., Koç, Ç.K., Paar, C. (eds.): Cryptographic Hardware and Embedded Systems—CHES 2003, Lecture Notes in Computer Science, vol. 2779. Springer, Heidelberg (2003)
412. Wang, Z., Karpovsky, M.: Robust FSMs for cryptographic devices resilient to strong fault injection attacks. In: 16th IEEE International On-Line Testing Symposium (IOLTS 2010), pp. 240–245. IEEE Comput. Soc. (2010)
413. Wang, Z., Karpovsky, M., Sunar, B.: Multilinear codes for robust error detection. In: 15th IEEE International On-Line Testing Symposium (IOLTS 2009), pp. 164–169. IEEE Comput. Soc. (2009)
414. Wang, Z., Karpovsky, M.G., Joshi, N.: Reliable MLC NAND flash memories based on nonlinear t-error-correcting codes. In: 2010 IEEE/IFIP International Conference on Dependable Systems and Networks (DSN 2010), pp. 41–50. IEEE Comput. Soc. (2010)
415. Wang, Z., Karpovsky, M.G., Kulikowski, K.J.: Replacing linear Hamming codes by robust nonlinear codes results in a reliability improvement of memories. In: 2009 IEEE/IFIP International Conference on Dependable Systems and Networks (DSN 2009), pp. 514–523. IEEE Comput. Soc. (2009)
416. Wang, Z., Karpovsky, M.G., Sunar, B., Joshi, A.: Design of reliable and secure multipliers by multilinear arithmetic codes. In: Qing, S., Mitchell, C.J., Wang, G. (eds.) Information and Communications Security (ICICS 2009), Lecture Notes in Computer Science, vol. 5927, pp. 47–62. Springer, Heidelberg (2009)
417. Watanabe, D., Furuya, S., Yoshida, H., Takaragi, K., Preneel, B.: A new keystream generator MUGI. In: Daemen, J., Rijmen, V. (eds.) Fast Software Encryption (FSE 2002), Lecture Notes in Computer Science, vol. 2365, pp. 179–194. Springer (2002)
418. Weil A. (1940) Sur les fonctions algébriques àà corps de constantes fini. Comptes Rendus de l'Académie des Sciences de Paris 210:592–594
419. Weste, N.H.E., Harris, D.: CMOS VLSI Design: A Circuits and Systems Perspective, 3rd edn. Addison-Wesley, Boston (2004)
420. Whelan, C., Scott, M.: The importance of the final exponentiation in pairings when considering fault attacks. In: T. Takagi, T. Okamoto, E. Okamoto, T. Okamoto (eds.) Pairing-Based Cryptography—Pairing 2007, Lecture Notes in Computer Science, vol. 4575, pp. 225–246. Springer, Heidelberg (2007)
421. Wolkerstorfer, J., Oswald, E., Lamberger, M.: An ASIC implementation of the AES Sboxes. In: B. Preneel (ed.) Topics in Cryptology—CT-RSA 2002, Lecture Notes in Computer Science, vol. 2271, pp. 67–78. Springer, Heidelberg (2002)
422. Wright, P.: Spycatcher: The Candid Autobiography of a Senior Intelligence Officer. Heinemann (1987)
423. Wu, H.: The stream cipher HC-128. In: M. Robshaw, O. Billet (eds.) New Stream Cipher Designs, Lecture Notes in Computer Science, vol. 4986, pp. 39–47. Springer, Berlin (2008)
424. Wu, K., Karri, R., Kuznetsov, G., Gössel, M.: Low cost concurrent error detection for the advanced encryption standard. In: 2004 International Test Conference (ITC 2004), pp. 1242–1248. IEEE Press (2004)
425. Yen, C.H., Wu, B.F.: Simple error detection methods for hardware implementation of advanced encryption standard. IEEE Transac. Comput. **55**(6), 720–731 (2006)
426. Yen, S.M., Chen, J.Z.: A DFA on Rijndael. In: Proc. of the 12th Information Security Conference (ISC 2002). Taichung (2002)

427. Yen, S.M., Joye, M.: Checking before output may not be enough against fault-based cryptanalysis. IEEE Transac. Comput. **49**(9), 967–970 (2000)
428. Yen, S.M., Kim, D.: Cryptanalysis of two protocols for RSA with CRT based on fault injection. In: L. Breveglieri, I. Koren (eds.) 1st Workshop on Fault Diagnosis and Tolerance in Cryptography—FDTC 2004, pp. 381–385. Florence, Italy (2004)
429. Yen, S.M., Kim, S., Lim, S., Moon, S.J.: A countermeasure against one physical cryptanalysis may benefit another attack. In: K. Kim (ed.) Information Security and Cryptology—ICISC 2001, Lecture Notes in Computer Science, vol. 2288, pp. 269–294. Springer, Heidelberg (2002)
430. Yen, S.M., Kim, S., Lim, S., Moon, S.J.: RSA speedup with Chinese remainder theorem immune against hardware fault cryptanalysis. IEEE Transac. Comput. **52**(4), 461–472 (2003)
431. Yen, S.M., Ko, L.C., Moon, S.J., Ha, J.: Relative doubling attack against Montgomery ladder. In: D. Won, S. Kim (eds.) Information Security and Cryptology—ICISC 2005, Lecture Notes in Computer Science, vol. 3935, pp. 117–128. Springer, Heidelberg (2006)
432. Yen, S.M., Moon, S.J., Ha, J.: Hardware fault attack on RSA with CRT revisited. In: P.J. Lee, C.H. Lim (eds.) Information Security and Cryptology—ICISC 2002, Lecture Notes in Computer Science, vol. 2587, pp. 374–388. Springer, Heidelberg (2003)
433. Zhang, X., Parhi, K.K.: Implementation approaches for the advanced encryption standard algorithm. IEEE Circu. Syst. Mag. **2**(4), 24–46 (2002)
434. Zhao, C.A., Zhang, F., Huang, J.: A note on the Ate pairing. Int. J. Info. Secur. **7**(6), 379–382 (2008)
435. Ziegler, J.F., Lanford, W.A.: Effect of cosmic rays on computer memories. Science **206**(4420), 776–788 (1979)

MIX
Papier aus verantwortungsvollen Quellen
Paper from responsible sources
FSC® C105338

If you have any concerns about our products,
you can contact us on
ProductSafety@springernature.com

In case Publisher is established outside the EU,
the EU authorized representative is:
Springer Nature Customer Service Center GmbH
Europaplatz 3, 69115 Heidelberg, Germany

Printed by Libri Plureos GmbH
in Hamburg, Germany